Encyclopedia of
Genetics

Encyclopedia of
Genetics

Volume II

Hermaphrodites – XYY Syndrome
Index

Editor
Jeffrey A. Knight
Mount Holyoke College

Project Editor
Robert McClenaghan

Salem Press, Inc.
Pasadena, California
Hackensack, New Jersey

Executive Editor: Dawn P. Dawson
Managing Editor: Christina J. Moose
Project Editor: Robert McClenaghan
Copy Editor: Douglas Long
Acquisitions Editor: Mark Rehn
Production Editor: Joyce I. Buchea
Photograph Editor: Karrie Hyatt
Design and Layout: James Hutson

Illustrations by: *Electronic Illustrators Group,* 24, 33, 94-95, 102, 119, 159-160, 170, 175, 205, 211, 323, 334, 369, 380, 384-385, 390, 404, 438, 444; *Hans & Cassidy, Inc.,* 3-4, 9, 11, 15, 18, 39, 43, 68, 71, 75, 132, 138, 141, 149, 167, 171, 260, 271, 300, 322, 340, 346, 358-359, 431, 506, 528

Library of Congress Cataloging-in-Publication Data

Encyclopedia of genetics / editor Jeffrey A. Knight; project editor Robert McClenaghan.
   p. cm.
Complete in 2 vols.
Includes bibliographical references and index.
   1. Genetics—Encyclopedias. I. Knight, Jeffrey A., 1948-   . II. McClenaghan, Robert, 1961-   .

QH427.E53 1999
576.5'03—dc21   98-31952
                                              CIP

ISBN 0-89356-978-X (set)
ISBN 0-89356-980-1 (vol. 2)

Third Printing

PRINTED IN THE UNITED STATES OF AMERICA

Contents

Encyclopedia of
Genetics

Hermaphrodites

Field of study: Human genetics

Significance: *Hermaphrodites are people born with both male and female sexual parts. Early identification and thorough medical evaluation of these individuals can help them lead relatively normal lives.*

Key terms

CHROMOSOME: a structure located inside the cell's nucleus that contains the genetic information of the cell

GENOTYPE: an organism's complete set of genes

GONAD: an organ that produces reproductive cells and sex hormones; termed ovaries in females and testes in males

KARYOTYPE: the chromosomal content of an individual's cells (normal female is 46,XX and normal male is 46,XY)

PHENOTYPE: the physical appearance of an individual based on the interaction of genotype and environment

Early Human Sexual Development

Up to the ninth week of gestation, the external genitalia (external sexual organs) are identical in appearance in both male and female human embryos. There is a phallus that will become a penis in males and a clitoris in females and labioscrotal swelling that will become a scrotum in males and labial folds in females. A person's development into a male or female is governed by his or her sex chromosome constitution (the X and Y chromosomes). An individual who has two X chromosomes normally develops into a female, and one who has one X and one Y chromosome normally develops into a male. It is the Y chromosome that determines the development of a male. The Y chromosome causes the primitive gonads (the gonads that have not developed into either an ovary or a testis) to develop into testes and to produce testosterone (the male sex hormone). It is testosterone that acts on the early external genitalia and causes the development of a penis and scrotum. If testosterone is not present, regardless of the chromosome constitution of the embryo, normal female external genitalia will develop.

Hermaphrodites

Hermaphrodites are individuals who have both male and female gonads. At birth, hermaphrodites can have various combinations of external genitalia, ranging from completely female to completely male genitalia. Most hermaphrodites have external genitalia that are ambiguous (genitalia somewhere between normal male and normal female) and often consist of what appears to be an enlarged clitoris or a small penis, hypospadias (urine coming from the base of the penis instead of the tip), and a vaginal opening. The extent to which the genitalia are masculinized depends on how much testosterone was produced by the testicular portion of the gonads during development. The gonadal structures of a hermaphrodite can range from a testes on one side and an ovary on the other side, to testes and ovaries on each side, to an ovotestis (a single gonad with both testicular and ovarian tissue) on one or both sides.

Hermaphroditism has different causes. The chromosomal or genotypic sex of a hermaphrodite can be 46,XX (58 percent have this karyotype), 46,XY (12 percent), or 46,XX/46,XY (14 percent), while the rest have different types of mosaicism such as 46,XX/47,XXY or 45,X/46,XY. Individuals with a 46,XX/46,XY karyotype are known as chimeras. Chimerism usually occurs through the merger of two different cell lines (genotypes), such as when two separate fertilized eggs fuse together to produce one embryo. This can result in a single embryo with some cells being 46,XX and some being 46,XY. Mosaicism means having at least two different cell lines present in the same individual, but the different cell lines are caused by losing or gaining a chromosome from some cells early in development. An example would be an embryo that starts out with all cells having a 47,XXY chromosome constitution and then loses a single Y chromosome from one of its cells, which then produces a line of 46,XX-containing cells. This individual would have a karyotype written as 46,XX/47,XXY. In a chimera or mosaic individual, the proportion of developing gonadal cells with Y chromosomes determines the appearance of the external genitalia. More cells with a Y chromosome mean that more testicular cells are formed and more testosterone is produced.

The cause of hermaphroditism in the majority of affected individuals (approximately 70 percent) is unknown, although it has been postulated that those hermaphrodites with normal male or female karyotypes may have hidden chromosome mosaicism in just the gonadal tissue.

Impact and Applications

Hermaphrodites with ambiguous genitalia are normally recognized at birth. It is essential that these individuals have a thorough medical evaluation, since other causes of ambiguous genitalia besides hermaphroditism can be life threatening if not recognized and treated promptly. Once hermaphroditism is diagnosed in a child, the decision must be made whether to raise the child as a boy or a girl. This decision is made by specialists in genetics, endocrinology, psychology, and urology working with the child's parents. Typically, the karyotype and appearance of the external genitalia of the child are the major factors in deciding the sex of rearing. Previously, most hermaphrodites with male karyotypes who had either an absent or an extremely small penis were reared as females. The marked abnormality or absence of the penis was thought to prevent these individuals from having fulfilling lives as males. This practice has been challenged by adults who are 46,XY but who were raised as females. Some of these individuals believe that their conversion to a female gender was the wrong choice, and they prefer to think of themselves as male. Hermaphrodites with a female karyotype and normal or near-normal female external genitalia are typically reared as females.

The debate over what criteria should be used to decide sex of rearing of a child is ongoing. An increasingly important part of this debate is the concept of gender identity, which describes what makes people male or female in their own minds rather than according to what sex their genitalia are. This is an especially important issue for those individuals with chimerism or mosaicism who have both a male and female karyotype. Currently, the choice of sex of rear-

ing in these patients is based primarily on how masculinized or feminized the external genitalia are.

Those hermaphrodites who have normal female or male genitalia at birth are at risk for developing abnormal masculinization in the phenotypic females or abnormal feminization in the phenotypic males at puberty if both testicular and ovarian tissue remains present. Thus it is usually necessary to remove the gonad that is not specific for the desired sex of the individual. An additional reason to remove the abnormal gonad is that the cells of the gonad(s) that have a 46,XY karyotype are at an increased risk of becoming cancerous.

—*Patricia G. Wheeler*

See Also: Biological Determinism; Gender Identity; Pseudohermaphrodites.

Further Reading: Normal human sexual development is described in *The Developing Human: Clinically Oriented Embryology* (1988), 4th ed., by Keith Moore. "Abnormalities of Gonadal Determination and Differentiation," by Gary D. Berkovitz, in *Seminars in Perinatology* 16 (October, 1992), gives a general overview of abnormal gonadal development. The concept of gender identity is addressed by John Money in "The Concept of Gender Identity Disorder in Childhood and Adolescence After 39 Years," in *Journal of Sex and Marital Therapy* 20 (1994).

High-Yield Crops

Field of study: Genetic engineering and biotechnology

Significance: *The health and well-being of the world's large population is primarily dependent on the ability of the agricultural industry to produce high-yield food and fiber crops. Advances in the production of high-yield crops will have to continue at a rapid rate to keep pace with the needs of an ever-increasing population.*

Key terms

BENEFICIAL MUTATION: a change in genetic character that provides some competitive advantage (such as increased yield) over other plants of the same species

CULTIVAR: a subspecies or variety of plant developed through controlled breeding techniques

GREEN REVOLUTION: the introduction of scientifically bred or selected varieties of grain (such as rice, wheat, and maize) that, with high enough inputs of fertilizer and water, can greatly increase crop yields

MONOCULTURE: the agricultural practice of continually growing the same cultivar on large tracts of land

The Historical Development of High-Yield Crops

No one knows for certain when the first crops were cultivated, but by six thousand years ago, humans had discovered that seeds from certain plants could be collected, planted in land, controlled, and later gathered for food. As human populations continued to grow, it was necessary to select and produce higher-yielding crops. The green revolution of the twentieth century helped to make this possible. Agricultural scientists developed new, higher-yielding varieties of numerous crops, particularly the seed grains that supply most of the calories necessary for maintenance of the world's population. Tremendous increases in the world's food supply resulted from these higher-yielding crop varieties. The new crop varieties also led to an increased reliance on monoculture, the practice of growing only one crop over a vast number of acres. The production of high-yield crops in the modern agricultural unit is highly mechanized and highly reliant on agricultural chemicals such as fertilizers and pesticides, requires relatively few employees, and devotes large amounts of land to the production of only one crop.

Methods of Developing High-Yield Crops

The major high-yield crops, in terms of land devoted to their culture and the total amount of produce, are wheat, corn, soybeans, rice, potatoes, and cotton. These crops originated from a low-yield native plant but have been converted into the highest-yielding crops in the world. There are two major ways to improve yield in agricultural plants. One way is to produce more of the plant or plant part, such as the fruit or leaf material, to be harvested. The second way is to produce larger plants or plant parts to be harvested. For example, to increase yield in corn, the grower must either produce more ears of corn per plant or produce larger ears on each plant. Numerous agricultural practices are required to produce higher yields, but one of the most important is the selection and breeding of genetically superior cultivars.

Throughout most of history, any improvement in yield was primarily based on the propagation of genetically favorable mutants. When a grower observed a plant with a potentially desirable gene mutation that produced a change that improved some yield characteristic such as more or bigger fruit, the grower would collect seeds or take cuttings (if the plant could be propagated vegetatively) and grow additional plants that produced the higher yields. This selection process is still one of the major means of improving yield in agricultural crops; advances in the understanding of genetics in the early part of the twentieth century have made it possible to breed some of the desirable characteristics resulting from mutation into plants that lack the characteristic. If the plants are closely related, traditional breeding techniques are used to crossbreed the plant with the desirable trait with the plant that does not possess the characteristic.

Until the advent of recombinant deoxyribonucleic acid (DNA) technology, the use of traditional breeding techniques between two very closely related species was the only means of transferring heritable characteristics such as increased yield from one plant to another. The advent of new DNA technology, however, has made it possible to transfer genetic characteristics from any plant. The simplest method for accomplishing this transfer generally involves the insertion of a gene or genes that might increase yield into a piece of circular DNA called a plasmid. The plasmid is then inserted into a bacteria, and the bacteria is then used as a vector to transfer the gene to another plant where it will also be expressed.

Impact and Applications

As population levels increase, pressure on the world's food supply increase as well; as a consequence, researchers are continually seeking ways to increase food production. In order to accomplish this goal, advances in the production of high-yield crops will have to continue at a rapid rate to keep pace with the needs of an ever-increasing population. New technologies will have to be developed, and much of this new technology will center on advances in genetic engineering. Such advances will lead to the development of new crop varieties that will provide higher yields and reduce the dependency on chemical pesticides by exhibiting greater resistance to a variety of pests.

—*D. R. Gossett*

See Also: Biofertilizers; Biopesticides; Biotechnology; Classical Transmission Genetics.

Further Reading: *The Standard Cyclopedia of Horticulture* (1942), edited by L. H. Bailey, has an excellent horticultural prospectus and provides the general reader with a firm understanding of techniques used to develop high-yield crops. *Plants, Genes, and Agriculture* (1994), by Maarten J. Chrispeels and David E. Sadava, is an outstanding treatise on the use of biotechnology in crop production and contains sections related to the use of biotechnology to transfer desirable traits from one plant to another. *Horticulture Science* (1986), by Jules Janick, contains sections on horticultural biology, environment, technology, and industry and covers the fundamentals associated with the production of high-yield crops. *Crop Production: Principles and Practices* (1980), by D. S. Metcalfe and D. M. Elkins, is a text for the introductory agriculture student and serves as one of the most valuable sources available on the practical aspects of the production of high-yield crops.

Homeotic Genes

Field of study: Developmental genetics
Significance: *The discovery of the homeotic gene has provided the key to understanding the mysterious process of shape determination in multicellular organisms. Knowledge of homeotic genes not only is helping scientists understand the variety and evolution of body shapes but also is providing new insights into genetic diseases and cancer.*

Key terms

DEOXYRIBONUCLEIC ACID (DNA): the genetic material, composed of sequences of organic molecules called bases (adenine, guanine, thymine, and cytosine); the order of the bases in a DNA strand determines its informational content

GENE: a discrete section of DNA with a control region (an "on-off" switch) followed by a much longer DNA sequence coding for a protein molecule; genes are arranged in tandem to form the chromosomes

PROMOTER/OPERATOR: the control switch in genes where transcription factors bind to activate or repress the conversion of DNA information into proteins

TRANSCRIPTION FACTOR: a protein with specialized structures that binds specifically to the promoters/operators in genes and controls the gene's activity

The Discovery of Homeotic Genes

One of the most powerful tools in genetic research is the application of certain mutagenic agents (such as X rays) that cause base changes in the deoxyribonucleic acid (DNA) of genes to create mutant organisms. These mutants display altered appearances, or phenotypes, giving the geneticist clues about how the normal organism functions. Few geneticists have used this powerful research tool as well as 1995 Nobel Prize in Physiology or Medicine recipient Christiane Nüsslein-Volhard (who shared the award with Edward B. Lewis and Eric Wieschaus). She and her colleagues, analyzing thousands of mutant *Drosophila melanogaster* fruit flies, discovered many of the genes that functioned early in embryogenesis; however, Lewis, in 1978, and Thomas Kaufman, in 1980, identified a homeotic gene that determined body shape during a later stage of development called organogenesis.

Among the many mutant *Drosophila* flies studied by these and other investigators, two were particularly distinctive in phenotype. One mutant had two sets of fully normal wings; the second set of wings, just behind the first set, displaced the normal halteres (flight balancers). The other mutant had a pair of legs protruding from its head in place of its antennae. These mutants were termed "homeotic" because major body parts were displaced to other regions. Using such mutants, Lewis was able to identify a clustered set of genes responsible for the extra wings and map or locate them on the third chromosome of *Drosophila*. He called this gene cluster with three genes the bithorax complex (*BX-C*). The second mutation was called antennapedia, and its complex with five genes was called *ANT-C*. If all the *BX-C* genes were removed, the fly larvae had normal head structures, partially normal middle or thoracic structures (where wings and halteres are located), but very abnormal abdominal structures that appeared to be nothing more than the last thoracic structure repeated several times. From these genetic studies, it was concluded that the *BX-C* genes controlled the development of parts of the thorax and all of the abdomen and that the *ANT-C* genes controlled the rest of the thorax and most of the head.

The *BX-C* and *ANT-C* genes were called homeotic selector genes: "homeotic" because their mutated versions caused the appearance of fully formed body parts at wrong locations and "selector" because they acted as major switch points to select or activate whole groups of other genes for one developmental pathway or another (for example, formation of legs, antennae, or wings from small groups of larval cells in special compartments called "imaginal disks"). Although geneticists knew that these homeotic selector genes were arranged tandemly in two clusters on the third *Drosophila* chromosome, they did not know the molecular details of these genes or understand how these few genes

functioned to cause such massive disruptions in the *Drosophila* body parts.

The Molecular Properties of Homeotic Genes

With so many mutant embryos and adult flies available, and with precise knowledge about the locations of the homeotic genes on the third chromosome, the stage was set for an intensive molecular analysis of the genes in each cluster type. In 1983, William Bender's laboratory used new, powerful molecular methods to isolate and thoroughly characterize the molecular details of *Drosophila* homeotic genes. He showed that the three bithorax genes constituted only 10 percent of the whole *BX-C*

Edward B. Lewis, who shared a 1995 Nobel Prize for helping to identify homeotic genes. (California Institute of Technology)

cluster. What was the function of the other 90 percent if it did not contain genes? Then William McGinnis's and J. Weiner's laboratories made another startling discovery: All the homeotic genes they examined by the molecular sequencing methods to determine the order of bases in the DNA contained nearly the same sequence in the terminal 180 bases. This conserved 180-base sequence was termed the "homeobox." What was the function of this odd but commonly found DNA sequence? What kind of protein did this homeobox-containing gene make?

Soon it was discovered that homeotic genes and homeoboxes were not confined to *Drosophila*: All animals had them, both vertebrates, such as mice and humans, and invertebrates, such as worms and even sea sponges. The homeobox sequence was not only conserved within homeotic and other developmental genes, but it was also conserved throughout the entire animal kingdom. All animals seemed to possess versions of an ancestral homeobox gene that had duplicated and diverged over evolutionary time.

New discoveries about homeobox genes flowed out of laboratories all over the world in the late 1980's and early 1990's; it was discovered that the order of the homeobox genes in the gene clusters from all animals was roughly the same as the order of the eight genes found in the original *BX-C* and *ANT-C* homeotic clusters of *Drosophila*. In more complex animals such as mice and humans, the two *Drosophila*-type clusters were duplicated on four chromosomes instead of just one. Mice have thirty-two homeotic genes, plus a few extra not found in *Drosophila*. Frank Ruddle hypothesized that the more anatomically complex the animal, the more homeotic genes it will have in its chromosomes. Experimental evidence from several laboratories supported Ruddle's hypothesis.

The questions posed earlier about the functions of extra DNA in the homeotic clusters and the role of the homeobox in gene function were finally answered. It seems that all homeotic genes code for transcription factors, or proteins that control the activity or expression of other genes. The homeobox portion codes for a section of protein that binds to base sequences in the promoters of other genes, thus stimulating those genes to express their proteins. The earlier idea of homeotic genes as selector genes makes sense; the protein products of homeotic genes bind to the promoter control regions of many other genes and activate them to make complex structures such as legs and wings. The homeotic genes themselves are under the control of other genes making transcription factors that bind to the extra DNA in the homeotic clusters. The bound transcription factors control the differential expression of homeotic genes in many different cellular environments throughout the developing embryo, all along its anterior to posterior axis. Embryonic development and organogenesis proceed by way of a complex series of cascaded gene activities, which culminate in the activation of the homeotic genes to specify the final identities of body parts and shapes.

Impact and Applications

In a 1997 episode of the television series *The X-Files*, a mad scientist transforms his brother into a monster with two heads. Federal Bureau of Investigation (FBI) agent Dana Scully patiently explains to her partner Fox Mulder that the scientist altered his brother's homeobox genes, causing the mutant phenotype. Science fiction indeed—but with the successful cloning of Dolly the sheep in 1997, the prospect of manipulating homeobox genes in embryos is no longer far-fetched. The first concern of scientists is to elucidate more molecular details about the actual process by which discrete genes transform an undifferentiated egg cell into a body with perfectly formed, bilateral limbs. Sometimes mutations in homeobox genes cause malformed limbs, extra digits on the hands or feet, or fingers fused together, conditions known as synpolydactyly; often limb and hand deformities are accompanied by genital abnormalities. Several reports in 1997 provided experimental evidence for mutated homeobox genes in certain leukemias and cancerous tumors. Beginning in 1996, the number of reports describing correlations between mutated homeobox genes and specific cancers and other developmental abnormalities increased dramatically. Although no specific gene-based therapies have been proposed for treating such diseases, the merger between the accumulated molecular knowledge of homeotic genes and the practical gene manipulation technologies spawned by the Dolly sheep cloning will undoubtedly lead to new treatments for limb deformities and certain cancers.

—*Chet S. Fornari*

See Also: Developmental Genetics; Protein Structure; RNA Structure and Function; Sheep Cloning.

Further Reading: Rudolf Raff's *The Shape of Life* (1996) is a detailed but readable account of how genes and evolution influence animal form. A comprehensive but clear discussion of homeotic genes with great illustrations can be found in Benjamin Lewin's *Genes* (1997). Eddy DeRobertis, "Homeobox Genes and the Vertebrate Body Plan," *Scientific American* 269 (July, 1990), is a classic article on homeobox gene studies.

Human Genetics

Field of study: Human genetics

Significance: *The study of human genetics is concerned with how human characteristics are inherited and how they are, in turn, passed along to offspring. Human geneticists are also interested in how genetic material controls or relates to the expression of human traits. Understanding the mechanisms of inheritance for characteristics ranging from blood types to genetic disorders may be crucial to biomedical decision making and genetic counseling.*

Key terms

GENOTYPE: the particular combination of genes, or the actual genetic makeup, for a particular trait found within an individual

PHENOTYPE: the visible or measurable manifestation of a trait

GENE: a specific area of a chromosome that contains genetic information that codes for a particular protein responsible for a specific trait

ALLELE: an alternative variety of a gene for a particular characteristic

CHROMOSOME: a threadlike structure consist-

ing of DNA that is found in the nucleus of a cell

LOCUS: the specific location of a gene on a particular chromosome

HOMOLOGOUS PAIR: two chromosomes that contain the same set of genetic loci; the genetic information contained on members of homologous pairs may be identical or consist of alleles

Mendelian Genetics

Basic rules that govern the inheritance of characteristics were first delineated by Austrian botanist Gregor Mendel. While a monk in an Augustinian monastery, Mendel conducted hybridization experiments with garden peas. Mendel's initial experiments involved crossing varieties of peas that displayed two alternate manifestations, or phenotypes, for each of seven characteristics. For example, stem length in the garden pea took the form of either a tall or a dwarf phenotype, while seed coat texture manifested as either a smooth or a wrinkled phenotype. Mendel's subsequent experiments involved crosses of plants that differed in two characteristics, such as plants that displayed tall stem length and green seeds with plants that had dwarf stems and yellow seeds.

These experiments led Mendel to state a theory of inheritance known as "particulate inheritance," in which characteristics are determined by a pair of tiny, discrete, and invisible particles. Since the early 1900's, the particles Mendel envisioned have been referred to as "genes." Unlike his predecessors, Mendel believed that these particles maintained their structural integrity during inheritance. In other words, the genetic determiners that were contributed from one parent did not break down or blend with those contributed from the other parent, as some scientists prior to Mendel had believed.

Mendel outlined some of the most basic rules concerning behavior and interactions of these genetic determiners; these rules, in turn, govern the inheritance of traits. These basic principles of heredity include the principle of dominance, which holds that the dominant form of a genetic determiner for a given trait tends to mask or hide the presence of an alter-

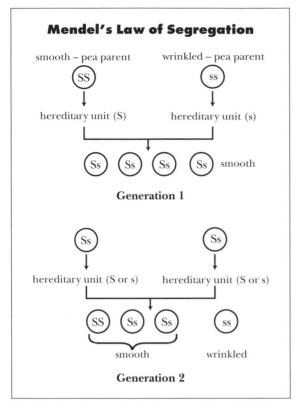

Mendel's Law of Segregation

smooth – pea parent

SS

hereditary unit (S)

wrinkled – pea parent

ss

hereditary unit (s)

Ss Ss Ss Ss smooth

Generation 1

Ss

hereditary unit (S or s)

Ss

hereditary unit (S or s)

SS Ss Ss   ss

smooth   wrinkled

Generation 2

Mendel's law of segregation is demonstrated by an initial cross between true-breeding plants with smooth peas and plants with wrinkled peas. The smooth trait is dominant, and the wrinkled trait is recessive. The second generation consists of smooth-pea plants and wrinkled-pea plants produced in a ratio of 3:1.

native, or recessive, form when the two occur together in the same genotype. This principle is based on Mendel's inference that there are two alternative varieties of determiners, now referred to as "alleles," that correspond to the two alternative phenotypes for a given trait. Mendel's experiments showed, for example, that the allele for tall stem length was dominant over the allele for the dwarf variety. Thus, a hybrid plant that had a genotype consisting of one allele for the tall stem length and another for the dwarf stem length expressed only the tall allele in its phenotype. Such a genotype is known as a "heterozygous" genotype, while genotypes that consist of identical genetic information from both parents are known as "homozygous" genotypes.

Mendel's experiments yielded two additional principles governing the inheritance of

characteristics. The law of segregation refers to the normal tendency for the parental gene pairs to segregate or separate before being passed along to an offspring. This mechanism ensures that an individual offspring inherits one and only one member of a given gene pair from each parent. Finally, Mendel's law of independent assortment relates to his observation that different characteristics were inherited independently of each other. In his experiments with the garden pea, for example, the inheritance of stem length did not influence the inheritance of seed coat texture.

Genes, Chromosomes, and Traits

An understanding of how Mendel's principles relate to human heredity began to emerge during the early 1900's when his work was independently rediscovered by three botanists. Around the same time, scientists such as Wilhelm Johannsen began applying the terms "genes" and "alleles" and recognizing that these important sources of genetic information had a special relationship to chromosomes, threadlike structures located in the nucleus of the cell. The research of August Weismann and Theodor Boveri in the late 1800's and Walter Sutton in the early 1900's demonstrated that some of the characteristics and behavior of chromosomes follow Mendel's predictions with respect to hereditary particles. For example, chromosomes, like Mendel's genetic determiners, occur in pairs. During the formation of eggs and sperm, collectively known as "sex cells" or "gametes," chromosomal pairs segregate in a manner consistent with Mendel's description of the law of segregation. This process, which takes place during a specialized form of cell division known as "meiosis," ensures that each parent contributes one and only one member of a gene pair to an individual offspring.

Other important research in the early 1900's helped clarify important relationships among genes, chromosomes, and the expression of characteristics. Working with the fruit fly *Drosophila melanogaster*, geneticist Thomas Hunt Morgan and his collaborators at Columbia University in New York City demonstrated that specific regions of chromosomes are associated with particular traits. The term "genetic locus" is used to describe the area on a chromosome that encodes for a specific characteristic. Morgan and his coworkers also described other important aspects of the relationships between genes and chromosomes, including "linkage" and "crossing-over." Linkage refers to the tendency for some characteristics to be inherited together because their genetic loci are in close proximity. Crossing-over occurs during meiosis when individual chromosomal pairs exchange segments. Although traits that are linked closely are more likely to be exchanged together and inherited together, crossing-over helps to produce new combinations of genes in the gametes by reshuffling genetic information.

An understanding of how genes mediate the expression of characteristics and how these factors relate to inherited disorders in human populations was furthered by the work of English physician Archibald Garrod in 1909. Garrod identified a genetic disorder, alkaptonuria, that he classified as an "inborn error of metabolism." His and subsequent work helped to demonstrate that such disorders, including albinism and phenylketonuria, are caused by a pair of recessive genes inherited in a Mendelian fashion. The inheritance of the recessive condition often results in the inability to produce a particular enzyme (a substance that catalyzes, or accelerates, a chemical reaction). The absence of the enzyme causes a breakdown, or metabolic block, in a sequence of events that is essential for carrying out one or more biological processes. As a result of this metabolic block, toxic or useless substances may accumulate in the individual's tissues. In the case of alkaptonuria, the enzyme homogentisic acid oxidase is lacking, causing the buildup of excessive quantities of alkaptones. Although most of these chemicals are excreted in the urine, causing the blackening of urine upon exposure to air, some may accumulate in such tissues as cartilage, producing arthritic symptoms.

Important discoveries relating to the relationships among genes, enzymes, chromosomes, and characteristics continued in the 1940's and 1950's. During the former decade,

George Beadle and Edward Tatum, working with *Neurospora* (pink bread mold), demonstrated the importance of genes in the synthesis of specific enzymes and amino acids. The essential link between deoxyribonucleic acid (DNA) and the expression of characteristics was established through the work of bacteriologist Oswald Avery in 1944. Avery and his collaborators established that DNA was the substance that effected the transformation of a harmless form of pneumococcus bacteria into a lethal variety. The physical structure of the DNA molecule was delineated by biologist James Watson and physicist Francis Crick in 1953. In the following decade, research by François Jacob and Jacques Monod with *Escherichia coli* helped establish how DNA controls the synthesis of proteins.

Taken collectively, the work of these and other researchers established essential links between the genetic message encoded in the DNA molecule and the expression of traits as regulated by the production of enzymes and proteins. The basic building blocks of DNA are structures known as "nucleotides," each of which consists of a phosphate, a sugar, and one of four chemical bases (adenine, guanine, cytosine, and thymine). Of greatest importance to protein synthesis is the ability of a codon, a sequence of three chemical bases of DNA, to code for specific amino acids. Different combinations of amino acids, in turn, are assembled into long chemical chains, or polypeptide chains, that form the chemical basis for proteins. To accomplish protein synthesis, the genetic message contained on the DNA molecule must be copied and transported out of the nucleus of the cell to the ribosomes. This essential step is performed by messenger ribonucleic acid (mRNA). Once positioned along the ribosomes, the genetic information is read codon-by-codon. This process attracts transfer RNA (tRNA), a molecule that brings complementary three-base sequences, known as anticodons, and the corresponding amino acids to the ribosomes, thereby effecting the assembly of a polypeptide chain.

Human Chromosomes, Genes, and Biological Characteristics

Chromosomes lie at the essence of the relationships among DNA, proteins, and characteristics. Chromosomes occur in pairs and are located in the nucleus of cells. The number of chromosomes is specific to the species in question. In humans, for example, the total number of chromosomes, also known as the "diploid" number, is twenty-three pairs, for a total of forty-six chromosomes. While normal body cells contain the diploid number, human gametes, consisting of sperm and ova, contain the haploid number of twenty-three. This number represents one member of each chromosomal pair. Thus, with the fertilization of a human ovum by a sperm cell, the full complement of twenty-three pairs of chromosomes is again formed.

In humans, the first twenty-two pairs of chromosomes are known as "autosomes" and encode genetic information for a majority of human traits. These pairs are also known as "homologous" pairs, since members of a given pair contain the same genetic loci for the same characteristics. Depending on an individual's genotype, members of a homologous pair may carry either identical genetic information or alleles for a given trait. The last chromosomal pair, consisting of the X and Y chromosomes, are known as the "sex chromosomes." In addition to carrying genetic information for a number of traits, the sex chromosomes determine the sex of the individual. Females normally have two X chromosomes, forming a homologous pair, while males normally have a genotype consisting of one X and one Y chromosome.

Human biological characteristics include traits that are inherited in the manner predicted by Mendel as well as those that have a much more complicated pattern of inheritance. Inheritance of characteristics that occurs in a Mendelian fashion is often referred to as "monogenic," or simple, inheritance, since an individual's phenotype is governed by a single gene pair.

Among the Mendelian traits of humans are a number of blood group systems and genetic disorders. Of these, the ABO blood group system has certainly been one of the most extensively studied. This system was first described by physician Karl Landsteiner in 1900. Landsteiner discovered two proteins that may occur on the surfaces of human red blood cells. The presence or absence of these proteins, known as antigens *A* and *B*, controls an individual's blood type or phenotype for the ABO system. For example, individuals who have antigen *A* but not antigen *B* have blood type A, while those who have antigen *B* but not antigen *A* belong to blood type B. On the other hand, individuals who lack both antigens belong to blood type O, while those in possession of both antigens *A* and *B* belong to blood type AB. Landsteiner also demonstrated that the presence of the antigens was sometimes associated with naturally occurring antibodies. The latter substances are proteins that are capable of producing agglutination of the opposing, or foreign, blood group antigens. For example, individuals with antigen *A* also possess antibodies against the *B* antigen. Individuals with antigen *B*, on the other hand, have anti-*A* antibodies. Individuals of blood type AB have neither type of antibodies, while those with blood type O have both types of antibodies.

While Landsteiner's discovery of the *A* and *B* antigens was crucial to the process of typing and matching blood prior to administering blood transfusions, the genetic mechanisms underlying this system were discovered by Emil von Dungern and Ludwig Hirszfeld several years later. Essentially, the inheritance of ABO

A Punnett Square Showing Alleles for Blood Type

		Father's Sperm Cells	
		B	O
Mother's Egg Cells	A	AB (AB blood)	AO (A blood)
	O	BO (B blood)	OO (O blood)

A heterozygous AO mother and a heterozygous BO father can produce children with any of the four blood types.

blood type is controlled by three alleles, designated as *A*, *B*, and *O*, that occur in pairs in an individual's genotype. The six possible genotypes of the ABO system, representing various combinations of these alleles, are *AA, AO, BB, BO, AB*, and *OO*. The blood type, or phenotype, that corresponds to each genotype is a function of the rules governing the interaction between the alleles. For example, the *A* and *B* alleles are dominant to the *O* allele. Thus, individuals who have the heterozygous *AO* genotype belong to blood type A and, consequently, produce the *A* antigen on their red blood cells. Similarly, individuals with the heterozygous *BO* genotype would belong to the B blood group and possess the *B* antigen on their red blood cells. On the other hand, the *A* and *B* alleles are codominant to each other. This means that when both alleles occur together in the genotype, both are expressed phenotypically. Thus, individuals who inherit one *A* and one *B* allele from their parents produce both the *A* and *B* antigens and belong to blood group AB. Finally, the O blood type is inherited in a recessive manner such that the only genotype that corresponds to the O blood type is the homozygous recessive genotype, or *OO*.

Since the discovery of the ABO system, the presence of a number of other blood group antigen systems has been documented. Many of these systems have no immediate medical application, but they can aid researchers who study variation in, and the significance of, frequencies with which genes occur in human populations. However, one such system, known as the Rh system (for the rhesus monkey, in which was discovered), must often be considered in relation to pregnancy and childbirth. While the mechanisms governing the inheritance of this system are not completely understood, the crucial issue relates to the presence or absence of the Rh antigen in the blood systems of the mother and infant. Generally speaking, the Rh antigen is inherited as a dominant, while the Rh negative condition is inherited as a recessive. Problems are most common when an Rh negative mother gives birth to an Rh positive infant. Problems usually arise following the birth of an initial Rh positive infant, since it is during the birth of this infant that the

Rh negative mother forms antibodies against the Rh antigen. During the birth of subsequent Rh positive infants, the mother's antibodies may cause agglutination, a clumping reaction that causes the destruction of the infant's red blood cells. If the process goes unchecked, the infant can become severely oxygen deprived and die. Fortunately, preventive measures can now be taken following the birth of a woman's first Rh positive infant to prevent the formation of antibodies in the mother's bloodstream.

Sex-Linked Traits and Sickle-Cell Anemia

While most Mendelian human traits are located on the autosomes, some are governed by genes on the sex chromosomes. Such characteristics are caused chiefly by the presence of genes on the X chromosome and are referred to as "X-linked" traits. Again, the presence of such genetic loci is indicated by the abnormal consequences that arise when certain genotypic combinations occur. For example, red-green color blindness and a blood-clotting disorder known as hemophilia are inherited as X-linked recessives. This means that females, whose pair of X chromosomes constitute an homologous pair, can only manifest these conditions if they inherit the recessive alleles for these traits from each parent. Males, however, have only a single X and thus will express an X-linked recessive trait if they inherit the responsible allele from their mother.

Sickle-cell anemia is another disease that is inherited in a Mendelian fashion and is of great medical importance. Knowledge of this disorder helps demonstrate the associations among specific DNA bases, amino acids, and their phenotypic consequences. The consequences in this instance consist of an abnormal form of adult hemoglobin that is synthesized when individuals inherit a recessive allele (*HbS*) from each of their parents. Normal hemoglobin provides the primary means to transport molecules of oxygen to the various tissues of the body. The defective hemoglobin in sickle-cell disease results in inadequate quantities of oxygen, reflected most vividly in the shapes of red blood cells. In the absence of adequate oxygen, these cells collapse from their normal disc shape and form a crescent or sickle shape. The biochemical basis for this phenomenon is situated at position 6 on the beta polypeptide chain of hemoglobin, where individuals who have sickle-cell disease contain the amino acid valine instead of glutamic acid. This error is caused by the substitution of the wrong DNA base for the base that is specified for that position in normal adult hemoglobin.

Aside from monogenic traits, many of the characteristics that humans express are under the control of more than one gene pair. This phenomenon, known as "polygenic inheritance," includes such traits as skin color, hair color, and stature. Rather than manifesting as two discrete alternatives, phenotypes for these characteristics typically display a continuous pattern of variation. Moreover, polygenic traits are generally influenced to some degree by factors external to the individual's genotype. For example, stature is influenced not only by the contents of genetic makeup but also by such factors as adequate nutrition. The term "heritability" is used to describe the proportion of a given trait that is controlled by genetic makeup as opposed to environmental factors.

Impact and Applications

The study of the inheritance of human traits has several major areas of application. First, an understanding of human genetics can be vital to the prevention and treatment of genetic diseases. Second, the study of biochemical consequences of genetic disorders can be used to gain a better understanding of the mechanisms responsible for the expression of biological characteristics. Finally, by focusing on the changes in the frequencies and occurrences of alleles over time, humans can arrive at a better understanding of the genetic structure of human populations and how such populations are evolving.

—*Mary K. Sandford*

See Also: Hereditary Diseases; Human Genome Project; Inborn Errors of Metabolism; Mendel, Gregor, and Mendelism; Population Genetics.

Further Reading: *Human Variation: Races, Types, and Ethnic Groups* (1983), by Stephen Molnar, and *People and Races* (1990), by Alice M. Brues, provide very clear explanations of the principles of Mendelian and polygenic genetics, focusing on the inheritance of and variation in human characteristics. *Genetics and Evolution* (1995), by Jill Bailey, contains a very basic introduction to the field of modern genetics, complete with colorful illustrations, that links discoveries in genetics with those in evolutionary biology. A very readable account of the pioneers in the study of evolution and genetics is found in *Blueprints: Solving the Mystery of Evolution* (1990), by Maitland A. Edey and Donald C. Johanson. More advanced treatises on the historical development of genetics with concise explanations of key concepts are found in *The Search for the Gene* (1992), by Bruce Wallace, and *The Foundations of Human Genetics* (1989), by Krishna R. Dronamraju. *The Human Genome Project: Deciphering the Blueprint of Heredity* (1994), edited by Necia Grant Cooper, contains a well-illustrated overview of fundamental concepts in molecular genetics as well as the goals and techniques of the Human Genome Project. A more succinct discussion of the objectives, methods, and implications of the efforts to sequence the human genome is found in *Mapping and Sequencing the Human Genome* (1988), prepared by the National Research Council.

Human Genome Project

Field of study: Molecular genetics
Significance: *The Human Genome Project is a research effort focusing on the analysis of human DNA structure and the determination of the precise location of an estimated 100,000 human genes.*

Key terms

DEOXYRIBONUCLEIC ACID (DNA): a double-stranded substance composed of units called nucleotides that comprise the genetic material of humans and most organisms

DEPARTMENT OF ENERGY (DOE): a federal program fostering genetic research directed at improving assessment of the effects of radiation and energy-related chemicals on human health

GENE MAPS: physical maps that isolate and characterize individual genes and other DNA regions of interest and provide the substrate for DNA sequencing

NATIONAL INSTITUTES OF HEALTH (NIH): a federal program fostering research in molecular biology as an integral part of its mission to improve human health

Background and Objectives of the Human Genome Project

The Human Genome Project is an international, centrally coordinated basic and clinical research effort focusing on the analysis of the structure of human deoxyribonucleic acid (DNA). Although genome mapping began decades earlier, the Human Genome Project database formally began operation at The Johns Hopkins University in Baltimore, Maryland, in October, 1990, with its most challenging goal being to determine the precise locations of an estimated 100,000 human genes by the year 2005. Another important part of this mission is to make current genome data quickly and conveniently accessible to members of the scientific and medical community, legislators, and the general public. Other original project objectives are to determine the complete sequence of an estimated 3 billion DNA subunits (bases) and to study DNA from model organisms such as the bacterium *Escherichia coli* for comparative information on genome function.

Following the development of the atomic bomb, the U.S. Congress charged the predecessor agencies of the Department of Energy (DOE)—the Atomic Energy Commission and the Energy Research and Development Administration—with studying and analyzing genome structure, replication, damage, and repair and the consequences of genetic mutations, particularly those caused by radiation. Recognition that the best method by which to study these effects was to analyze the entire human genome and obtain a reference sequence for studying DNA directly became evident, leading to a merger between the DOE and the National Institutes of Health (NIH). In 1990, coordination between these two departments led to the establishment of several specific five-year scientific goals and their rationale, which are reviewed and updated regularly,

in the following areas: mapping and sequencing the human genome; mapping and sequencing the genome of model organisms; data collection and distribution; ethical, legal, and social considerations; research training; technology development; and technology transfer. In 1993, it became clear that advances in technology would enable the Human Genome Project to complete its fifth year ahead of schedule, and the original five-year goals were updated. Development of a genome research infrastructure, which includes data, material resources, and technology, has greatly improved the ability of investigators to do biological research rapidly, efficiently, and cost-effectively.

In January, 1997, the three largest genome research centers (Lawrence Berkeley National Laboratory, Lawrence Livermore National Laboratory, and Los Alamos National Laboratory) were merged into the Joint Genome Institute to integrate their work, shift toward larger-scale sequencing, and combine expertise and resources regarding biological function, dubbed "functional genomics," in addition to the "structural genomics" of generating maps. As of November, 1997, almost 5,800 genes had been mapped to specific chromosomes and about 1,400 genes were stored in the genome data base whose locations have not yet been determined unequivocally. At any given time, the Human Genome Project funds approximately two hundred separate principal investigators at various universities and laboratories throughout the United States.

The speed with which human genes are being identified, particularly those responsible for genetic diseases, continues to increase rapidly. When the genome project was first initiated, the longest DNA sequence described was the 250,000-base-pair (bp) cytomegalovirus sequence. In 1997, the longest contiguous human sequence was 685 kilobase pairs (kb) from the human T-cell beta receptor locus, a chromosomal region involved in immune responses. Several countries have established collaborative human genome research projects with the United States, including Australia, Brazil, Canada, China, Denmark, France, Germany, Israel, Italy, Japan, Korea, Mexico, Netherlands, Russia, Sweden, and the United Kingdom.

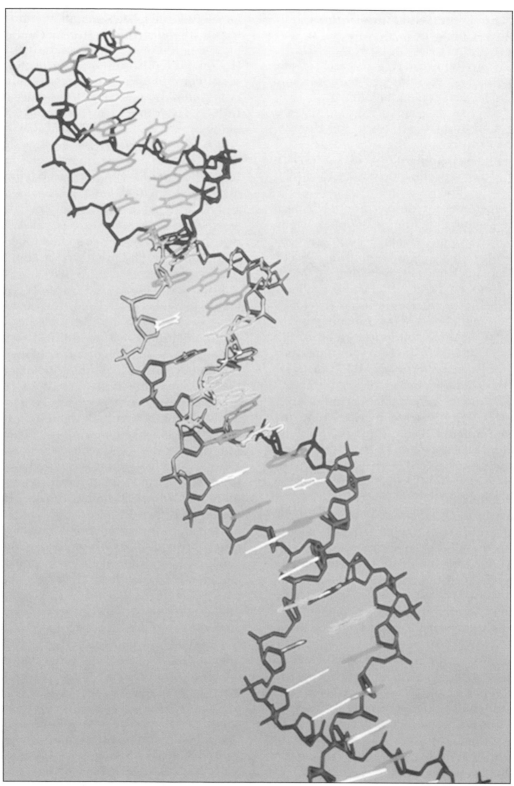

The Human Genome Project seeks to identify the complete sequence of human DNA. (Langridge/McCoy/ Rainbow)

Ethical, Legal, and Social Implications

Organizers of the Human Genome Project have recognized from the beginning that the increase in knowledge about personal genetic information will raise complex ethical and policy issues for individuals and society. A research and education grant program supported by 3 to 5 percent of funds from the project's overall budget is concentrated on identifying and addressing ethical, legal, and social issues arising from genetic research, responsible clinical integration of new genetic technologies, privacy and the fair use of genetic information, and professional and public education. Ongoing committees have been established to explore and propose guidelines for professional and public policies related to genome research and its applications.

While information from genetic testing can certainly benefit a patient by enabling the possibility of advanced interventions, it may also bring unwelcome consequences such as increased anxiety, altered family relationships, and stigmatization and discrimination on the basis of genotype with respect to employability and insurability. In 1993, the Task Force on Genetic Information and Insurance was established to assess the potential impact of human genetic advances on the U.S. health-care system and to make recommendations for managing that impact. In 1994, the Institute of Medicine published a study of the clinical integration of new genetic tests that offered a number of recommendations for quality control of DNA diagnostics and for genetic testing in the clinical setting. In 1997, the Equal Employment Opportunity Commission ruled that genetic discrimination in employment decisions was illegal.

Education programs have greatly increased understanding of the nature and appropriate use of genetic information by health-care professionals, policymakers, and the public. These projects include references to assist federal and state judges in understanding genetic evidence, educational materials and curriculum modules for schools, teacher training workshops, and radio and television programs on science and ethical issues of the genome project.

Impact and Applications

Data generated by the Human Genome Project will benefit humankind by revolutionizing biomedical exploratory research and enhancing medical treatment of more than four thousand genetic diseases. Detailed information regarding the structure, organization, and function of DNA within chromosomes will enable future generations of researchers to develop technologies that will enable a better understanding of the genetic contribution to diseases such as cancer. Genome maps of less complex organisms often provide the basis for comparative studies that are important to understanding human biological systems. The ability to sequence DNA directly and quickly will continue to enhance mutation research by allowing researchers to directly study the relationships between disease development and exposure to various agents and provide information that can be used to diagnose the onset of disease and develop therapeutic strategies.

New technologies based on DNA diagnostics have the potential to dramatically improve clinical practice by shifting emphasis from a treatment-based approach to a prevention-based approach by identifying individuals predisposed to particular diseases. Therapeutic regimens based on more powerful classes of medications will be devised, as will immunotherapy techniques. Increased knowledge will allow people to avoid environmental conditions that may enhance disease risk and may lead to the ability to replace defective genes through gene therapy. Technologies, databases, and biological resources developed in genome research will have an enormous impact on a variety of related industrial fields such as agriculture, energy production, waste control, and environmental cleanup.

The funding required to complete the Human Genome Project was estimated at $200 million per year for a total of fifteen years. The private sector is actively involved in the Human Genome Project at all levels, from participation in the advisory committees to the receipt of grants and contracts. The potential for commercial development generated by the sale of biotechnology products was projected to exceed $20 billion by the year 2000. The early

1990's experienced a shift from ongoing arguments about whether the Human Genome Project was a worthy enough effort to continue its funding to debates regarding the most efficient methods to reap its benefits.

—*Daniel G. Graetzer*

See Also: Bioethics; Developmental Genetics; DNA Structure and Function; Gene Therapy; Genetic Testing.

Further Reading: *To Know Ourselves: The United States Department of Energy and the Human Genome Project* (1996), edited by Douglas Vaughan, gives an excellent brief review of chromosome mapping and the Human Genome Project for the nonscientist. Tom Wilkie, *Perilous Knowledge: The Human Genome Project and Its Implications* (1993), investigates the many arguments surrounding the potential moral and ethical consequences of advanced molecular biology. In *The Book of Man: The Human Genome Project and the Quest to Discover Our Genetic Heritage* (1995), Walter Fred Bodmer and Robin McKie review human gene mapping and genetic engineering. *The Human Genome Project: Deciphering the Blueprint of Heredity* (1994), edited by Necia Grant Cooper, contains chapters by recognized experts describing classical and molecular genetics and their implications into the twenty-first century. *The Gene Wars: Science, Politics, and the Human Genome* (1994), by Robert M. Cook-Deegan, reviews the politics of human gene mapping, research, and government policy.

Human Growth Hormone

Field of study: Human genetics

Significance: *Human growth hormone (HGH) determines a person's height; abnormalities in the amount of HGH in a person's body may cause conditions such as dwarfism, giantism, and acromegaly. Genetic research has led to the means to manufacture enough HGH to correct such problems and expand the understanding of HGH action and endocrinology.*

Key terms

ENDOCRINE GLAND: a gland that secretes hormones into the circulatory system

PITUITARY GLAND: an endocrine gland located at the base of the brain; also called the hypophysis

HYPOPHYSECTOMY: surgical removal of the pituitary gland

PROTEIN: a chainlike substance composed of conjoined amino acids

Growth Hormones and Disease Symptoms

The pituitary (hypophysis) is an acorn-sized gland located at the base of the brain that makes important hormones and disseminates stored hypothalamic hormones. The hypothalamus controls the activity of the pituitary

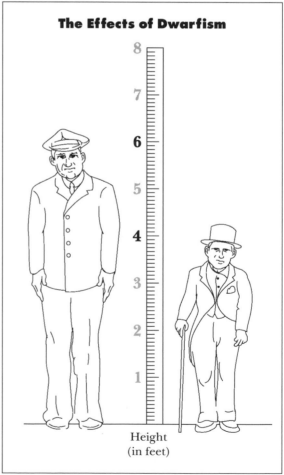

The Effects of Dwarfism

Height
(in feet)

Human growth hormone is used to treat dwarfism, which is caused by an endocrine disfunction. Adult males less than sixty inches tall and adult females less than fifty-eight inches tall are generally classified as dwarfs.

gland by sending signals along a network of blood vessels and nerves that connects them. The main portion of the pituitary gland, the adenohypophysis, makes six trophic hormones that control many body processes by causing other endocrine glands to produce hormones. The neurohypophysis, the remainder of the pituitary, stores two hypothalamic hormones for dissemination.

Dwarfism is caused by the inability to produce growth hormone. When humans lack only human growth hormone (HGH), resultant dwarfs have normal to superior intelligence. However, if the pituitary gland is surgically removed (hypophysectomy), the absence of other pituitary hormones causes additional mental and gender impairment. The symptoms of dwarfism are inability to grow at a normal rate or attain adult size. Many dwarfs are two to three feet tall. In contrast, some giants have reached heights of over eight feet. The advent of gigantism often begins with babies born with pituitary tumors that cause the production of too much HGH, resulting in continued excess growth. People who begin oversecreting HGH as adults (also caused by tumors) do not grow taller. However, the bones in their feet, hands, skull, and brow ridges overgrow, causing disfigurement and pain, a condition known as acromegaly.

Dwarfism that is uncomplicated by the absence of other pituitary hormones is treated with growth hormone injections. Humans undergoing such therapy can be treated with growth hormones from humans or primates. Growth hormone from all species is a protein made of approximately two hundred amino acids strung into a chain of complex shape. However, differences in amino acids and chain arrangement in different species cause shape differences; therefore, growth hormone used for treatment must be extracted from a related species. Treatment for acromegaly and gigantism involves the removal of the tumor. In cases where it is necessary to remove the entire pituitary gland, other hormones must be given in addition to HGH. Their replacement is relatively simple. Such hormones usually come from animals in a fashion similar to use of pig, sheep, or cow insulin for diabetes. Until re-

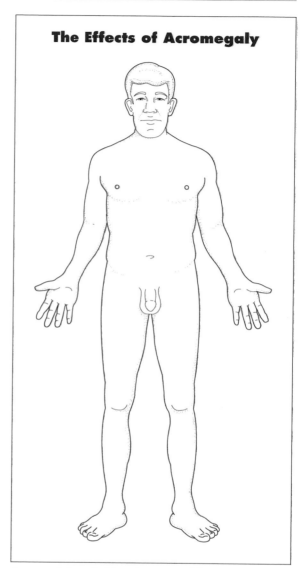

The Effects of Acromegaly

Acromegaly is marked by enlargement of the hands, feet, chest, jaw, and other facial features.

cently, the sole source of HGH was pituitaries donated to science. This provided the ability to treat fewer than one thousand individuals per year. Molecular genetics has solved that problem by devising the means to manufacture large amounts of HGH.

Growth Hormone Operation and Genetics

In the mid-1940's, growth hormone was isolated and used to explain why pituitary extracts increase growth. One process associated with HGH action involves cartilage cells at the ends of long bones (such as those in arms and legs). HGH injection causes these epiphysial plate cells (EPCs) to rapidly reproduce and stack up. The EPCs then die and leave a layer of protein, which becomes bone. From this it has been concluded that growth hormone acts to cause all body bones to grow until adult size is reached. It is unclear why animals and humans from one family exhibit adult size variation. The differences are thought to be genetic and related to production and cooperation of HGH, other hormones, and growth factors.

Genetic research has produced HGH isolated from bacteria through the use of genetic engineering technology called recombinant deoxyribonucleic acid (DNA) research. A piece of DNA that causes HGH production is pasted into the DNA of a bacterium, which is then grown on an industrial scale. After bacterial growth ends, a huge number of cells are harvested and HGH is isolated. The complicated procedure enables isolation of enough HGH to treat anyone who needs it.

Impact and Applications

One use of genetically engineered HGH is the treatment of acromegaly, dwarfism, and gigantism. The availability of large quantities of HGH has also led to other biomedical advances in growth and endocrinology. For example, growth hormone does not affect EPCs in tissue culture. Ensuing research, first with animal growth hormone and later with HGH, uncovered the EPC stimulant somatomedin. Somatomedin stimulates growth in other tissues as well and belongs to a protein group called "insulinlike growth factors." Many researchers have concluded that the small size of women compared to men is caused by estrogen-diminished somatomedin action on EPCs. Estrogen, however, stimulates female reproductive system growth by interacting with other insulin-like growth factors.

Another interesting experiment involving HGH and genetic engineering is the production of rat-sized mice. This venture, accomplished by putting the HGH gene into a mouse chromosome, has important implications for understanding such mysteries as the basis for species specificity of growth hormones and

maximum size control for all organisms. Hence, experiments with HGH and advancements in genetic engineering technology have had, and continue to have, valuable uses in the study of growth and other aspects of life science.

—Sanford S. Singer

See Also: Biotechnology; Cloning; Genetic Engineering; Molecular Genetics.

Further Reading: *The Endocrine System* (1990), by Marjorie Little, describes the pituitary gland and growth hormones for general readers. *Regulation of Growth Hormone and Somatic Growth* (1992), by Luis de la Cruz, contains much technical data. *The Merck Manual* (1991), 16th ed., by Robert Berkow and Andrew Fletcher, provides medical information. *Principles of Biochemistry* (1993), by Albert Lehninger et al., describes endocrinology and HGH molecular biology.

Huntington's Chorea

Field of study: Human genetics

Significance: *Huntington's chorea occurs worldwide and in all ethnic groups, especially Caucasians. It is a known human genetic aberration of chromosome 4 and is always fatal, usually ten to fifteen years after onset. Symptoms do not usually develop until after thirty years of age.*

Key terms

CEREBRAL CORTEX: the surface layer of the cerebrum that controls skeletal muscles and coordination of voluntary muscular movements

CEREBRUM: the upper part of the brain that controls conscious mental processes

CHOREA: a nervous affliction marked by uncontrollable muscular twitching referred to as "choreic movements"

CHROMOSOMES: microscopic bodies in the nucleus of the cell that contain all genes and hereditary traits

DEMENTIA: an irrecoverable, deteriorative mental state; it is the common end result of many diseases

Definition and Diagnosis

Huntington's chorea is a hereditary disease in which degeneration in the cerebral cortex and basal ganglia of the brain causes chronic, progressive chorea and mental deterioration ending in dementia. Huntington's chorea usually strikes people between the ages of twenty-five and fifty-five (the average age is thirty-five); however, 2 percent of all cases are in children, and 5 percent develop as late as age sixty. The Huntington's chorea gene aberration is identified on the short arm of chromosome 4. (There are twenty-three pairs of chromosomes in humans.) Huntington's chorea is manifested by loss of musculoskeletal control and choreic movements. It is named after George Huntington, the physician who first identified the disease in 1872.

Huntington's chorea can be diagnosed by a computed tomography (CT) scan of the brain, a careful clinical history, confirmation of a genetic link, and observation of symptoms such as dementia. CT scans and magnetic resonance imaging (MRI) show definite brain atrophy. Genetic testing would allow affected individuals the choice regarding future childbearing. Gross pathologic changes are evident on postmortem exams. Overall, the brain appears shrunken, with severe atrophy of the anterior portions of the frontal and temporal lobes and dilation of the ventricles. Comfort measures are the only known means of treatment for this dreaded hereditary disease. Medications may aid in controlling the psychiatric components of the disease such as depression and paranoid thought processes.

Causes and Incidence

Even though Huntington's chorea is a genetic disorder, the direct cause is unknown. Either gender can transmit or inherit it. Each child with a biological parent with this disease has a 50 percent chance of inheriting it. Because of its hereditary transmission, Huntington's chorea is more concentrated in areas where carrier families have lived for generations. Genetic testing for the disease was developed in the 1990's. Children whose parents have Huntington's chorea can determine if they are also carriers of the defective chromosome 4. The results allow these individuals to then make an informed decision about child bearing, fertility, and birth control.

The onset of Huntington's chorea is gradual. The person develops progressive loss of musculoskeletal control and manifests severe choreic movements similar to an uncontrollable dance. These movements are rapid and can be violent and purposeless. Initially, they occur on one side of the body and are observed more in the face and arms than in the legs. This progresses to mild fidgeting and grimacing, tongue smacking, slurred and indistinct speech, and spasmodic contractions of the neck muscles that draw the afflicted person's head to one side. The individual has no control over these overt body movements.

Treatment and Psychosocial Implications

There is no known cure for Huntington's chorea. Any care rendered should be supportive and protective, allowing the individual to live his or her remaining life in a safe environment. Because of the unsteady, dancelike gait characteristic of Huntington's chorea, the person afflicted would be appropriately suited to an environment free of hazards such as sharp dresser corners, glass objects, unsteady chairs, glass showers, and other objects of potential self-harm. A walker may help the person maintain balance. Psychological counseling and emotional support to the victim's family may help decrease frustration. Observation for suicide attempts is often a necessary part of care.

A person with Huntington's chorea develops progressive dementia. The dementia may not progress at the same rate as the chorea, causing the unfortunate patient to be aware of the worsening malady. Dementia may be mild at first but eventually disrupts the individual's personality. Such personality changes include moodiness, apathy, obstinacy, carelessness, inappropriate behaviors, and loss of memory, cognition, and concentration. The individual progresses at times to severe paranoia and psychiatric disturbances. The emotional disorders of Huntington's disease may come before the spasmodic choreic movements are ever observed. Mania, hallucinations, delusions, schizophrenia-like thinking, impulsiveness, hostility, and extreme agitation can develop. Very common beginning signs and symptoms also include withdrawn behaviors, a decrease in

The legendary folk singer Woody Guthrie inherited Huntington's chorea from his mother. (Frank Driggs/Archive Photos)

socialization and conversation, and inattention to personal hygiene. Depression is rampant: Approximately 10 percent of people with this disease commit suicide.

Changes in cognition appear later and are characterized by impairment of recent memory and judgment, loss of capacity to plan and organize, and intellectual decline. Dementia also leads to urinary and fecal incontinence and a marked inability to handle activities of daily living. Treatment of the severe depression and psychotic thinking can be achieved with antidepressants and antipsychotic agents. Use of these drugs, combined with supervision of the person's daily activities, can allow the person to stay at home in the early stages of disease. As the disease advances, the individual often must be placed in a locked nursing home or psychiatric facility.

—*Lisa Levin Sobczak*

See Also: Developmental Genetics; Genetic Screening; Genetic Testing; Genetic Testing: Ethical and Economic Issues; Hereditary Diseases.

Further Reading: An overview of medical diagnosis and treatments is provided in *Current Medical Diagnosis and Treatment in 1997* (1997), edited by Lawrence M. Tierney, Jr. et al. The *Cecil Textbook of Medicine* (1988), edited by James B. Wyngaarden and Lloyd H. Smith, Jr., and *Harrison's Principles of Internal Medicine* (1994), edited by Kurt J. Isselbacher et al., are classic sources used to educate physicians. Each of these sources provides a brief summary of Huntington's chorea, along with many other ailments, as does *Merritt's Textbook of Neurology* (1995), edited by Lewis P. Rowland.

Hybridization and Introgression

Field of study: Genetics
Significance: *Hybridization is the mating of organisms that differ from each other in one or more genetic traits. It may occur either naturally or by human-derived methods. Introgression is the assimilation of genes of one species into the gene pool of another species by means of successful hybridization. Hybridization ensures the development of new plant and animal species and the continued diversification of life on Earth.*

Key terms

CROSSBREEDING: mating of differing individuals from the same species (interstrain), different species (interspecies), or different families (interfamily) to produce hybrids

GENE: a piece of hereditary material that encodes a genetic trait

HYBRID SWARM: a large, complex, naturally occurring group of progeny produced in hybridization

HYBRID VIGOR: the ability of many hybrids to grow faster and become larger and stronger than their parents; also called "heterosis"

POLYPLOIDY: doubling of the number of chromosomes in an organism for use in asexual reproduction; it is usually seen in plants

TRANSGENICS: a human-made form of hybridization in which genetic engineers insert related or foreign genetic material into embryonic cells that are then grown into hybrids

Hybridization

Hybridization, the mating of parent organisms that differ in one or more genetic traits, yields offspring that share or blend parental traits. Such hybrids can be produced with individuals of the same species (interstrain), different species (interspecies), or different families (interfamily). These new organisms, which may be animals or plants, are often developed by humans for specific purposes, although they also occur in nature without human intervention.

An example of an interspecies hybrid is the mule, the offspring of a male donkey and a female horse. Mules are produced to provide an organism that blends the appearance of the horse and donkey while being stronger and having more stamina than either parent. Most of the corn and wheat that are grown as crops throughout the world are also hybrids. Both of these crop plants are interstrain hybrids. Such plant hybrids are produced because they grow much faster, become larger, and produce higher crop yields than the parents from which they are derived. A great many other plants and some animals are hybridized for similar reasons: to produce a better food plant or animal, to yield a disease-resistant organism, or to create more ornamental organisms.

Hybrid Vigor and Sterility

Often hybrids are larger or grow more vigorously than their parents. This hybrid vigor, or heterosis, is very useful in the hybrid organism produced, whether it is a mule, a corn or wheat plant, or some other crossbred organism. However, heterosis is not a phenomenon that yields lasting adaptive genetic superiority to the hybrid strain. This is because most hybrids are sterile and die without producing offspring. Also, when hybrids made in the first (F_1) generation do reproduce, they yield a second generation (F_2) of numerous progeny that are variable or inferior compared to the organisms crossbred to produce F_1. Because of this, hybrid corn must be reproduced by seed companies each year by crossbreeding the original parent strains if farmers are to be satisfied with their crop yields.

The reason for such problems with F_2 generations arises from the fact that all crossbreeding efforts result in the transferral not only of desired genetic traits but also of many deleterious or harmful genes. These genes—a result of previous natural mutation—usually do not appear without inbreeding, which allows two such genes to be brought together in one organism. The resultant hybrid variability usually causes loss of hybrid vigor in the F_2 generation.

Natural Hybrids

Hybridization also occurs in nature, where successful natural hybrids engender a great percentage of the process of evolutionary change. Some naturally formed hybrids may be able to reproduce even if they are completely sterile by means of polyploidy and asexual reproduction. The term "polyploidy" indicates that the total number of chromosomes in the hybrid can double in number and be used in asexual reproduction. Polyploidy is very common in flowering plants but exceedingly rare in animals.

The sexual reproduction of two natural hybrids may result in the production of a huge number of variant organisms within several generations. The resultant, very complex group of progeny is called a "hybrid swarm." In most instances, nearly all individuals in a hybrid swarm will be poorly designed for survival and die off because of natural selection. However, some members of the hybrid swarm will inherit a favorable combination of genetic attributes and begin a new species.

Various factors prevent most hybrids from forming a hybrid swarm. First, it is very unusual for more than a few natural hybrids to arise at any time, and these hybrid individuals are outnumbered by members of the parent species. Hence, most hybrids mate with the parental species (backcrosses), and the hybrid characteristic is soon lost. In addition, hybrids are usually partially fertile at best, so the small number of hybrids produced will be swallowed up by competition for food and living space. However, when a successful series of backcrosses occurs and genes flow from one species to another by such hybridization, a new species arises, a process referred to as "introgression." Introgression is common in plants, and some cases have been reported in simple organisms; however, it is unclear whether it occurs in higher animals.

Impact and Applications

The survival of natural hybrids or their offspring, which may yield new species, depends upon the availability of habitats in which they can flourish. Stable or static environments are not conducive to the creation of new species. However, in the past several millennia a mixture of both natural phenomena and human activities has led to situations in which many new species have developed by hybridization, particularly by plant introgression. The ecological upsets produced by humans may therefore be perceived as being useful to the survival of some hybrid species. Changes caused by many natural phenomena can also be perceived as acting in a similar fashion. In the case of hybrids designed by humans for agriculture, species continuity is achieved by repeated backcrosses between parental strains in plant or animal farms maintained by agricultural industries for such purposes.

Evolution is largely dependent upon genetic diversity, which entails having many new genes available for natural selection. While mutation accounts for some of this added diversity, most mutants hold genes with little or no survival potential. Hence, the chief source for relatively rapid evolution, especially in plants, is thought to be hybridization, giving this phenomenon a tremendous impact on evolutionary processes.

Transgenic experimentation has led to other means to produce hybrids. Molecular biologists involved in genetic engineering are able to insert related or foreign genetic material into embryonic cells. These cells are then grown in vitro or in a host mother to produce new organisms that would otherwise be impossible. Examples include a rat-sized mouse, tobacco plants that grow in the dark, and pesticide-resistant plants. It is believed that continuation of human-derived transgenics and more usual types of hybridization, as well as hybridization by natural selection, will engender an ever richer and more satisfactory world via the production of new organisms.

—*Sanford S. Singer*

See Also: Genetic Engineering: Agricultural Applications; Inbreeding and Assortative Mating; Natural Selection; Speciation; Transgenic Organisms.

Further Reading: *Human Heredity: Principles and Issues* (1997), by Michael R. Cummings, covers many human heredity topics clearly. *Plant Speciation* (1971), by Verne Grant, covers speciation, hybridization, and introgression. *The Business of Breeding: Hybrid Corn in Illinois* (1990), by Deborah Fitzgerald, describes the development of agricultural hybrids. *Introgressive Hybridization of the Common Carp and the Goldfish in the Western Basin of Lake Erie* (1977), by Ronald L. Crunkilton, describes a search for animal introgression. *Essentials of Genetics* (1997), by William S. Klug and Michael R.

Mules derive their exceptional strength from their hybrid nature. (Ben Klaffke)

Cummings, covers interesting hereditary concepts from the molecular biologist's point of view.

Hybridomas and Monoclonal Antibodies

Field of study: Immunogenetics
Significance: *In 1975, George Kohler and Cesar Milstein reported that fusion of spleen cells from an immunized mouse with a cultured plasmacytoma cell line resulted in the formation of hybrid cells called "hybridomas" that secreted the antibody molecules that the spleen cells had been stimulated to produce. Clones of hybrid cells producing antibodies with a desired specificity are called "monoclonal antibodies" and can be used as a reliable and continuous source of that antibody. These well-defined and specific antibody reagents have a wide range of biological uses, including basic research, industrial applications, and medical diagnostics and therapeutics.*

Key terms

ANTIGEN: a foreign molecule or microorganism that stimulates an immune response in an animal

ANTIBODY: a protein produced by plasma cells (matured B cells) that binds specifically to an antigen

ANTISERA: a complex mixture of heterogeneous antibodies that react with various parts of an antigen; each type of antibody protein in the mixture is made by a different type (clone) of plasma cells

PLASMACYTOMA: a plasma cell tumor that can be grown continuously in a culture

REAGENT: a substance used because of its chemical or biological activity

A New Way to Make Antibodies

Because of their specificity, antisera have long been used as biological reagents to detect or isolate molecules of interest. They have been useful for biological research, industrial separation applications, clinical assays, and immunotherapy. One disadvantage of conventional antisera is that they are heterogeneous collections of antibodies against a variety of antigenic determinants present on the antigen that has elicited the antibody response. In an animal from which antisera is collected, the mixture of antibodies changes with time so that the types and relative amounts of particular antibodies are different in samples taken at different times. This variation makes standardization of reagents difficult and means that the amount of characterized and standardized antisera is limited to that available from a particular sample.

The publication of a report by Georges Köhler and Caesar Milstein in the journal *Nature* in 1975 describing production of the first monoclonal antibodies provided a method to produce continuous supplies of antibodies against specific antigenic determinants. Milstein's laboratory had been conducting basic research on the synthesis of immunoglobulin chains in plasma cells, mature B cells that produce large amounts of a single type of immunoglobulin. As a model system, they were using rat and mouse plasma cell tumors (plasmacytomas). Prior to 1975, Kohler and Milstein had completed a series of experiments in which they had fused rat and mouse plasmacytomas and determined that the light and heavy chains from the two species associate randomly to form the various possible combinations. In these experiments they used mutant plasmacytoma lines that would not grow in selective culture media, while the hybrid cells complemented each others' deficiencies and multiplied in culture.

After immunizing mice with sheep red blood cells (SRBC), Kohler and Milstein removed the spleen cells from the immunized mice and fused them with a mouse plasmacytoma cell line. Again, the selective media did not allow unfused plasmacytomas to grow, and unfused spleen cells lasted for only a short time in culture so that only hybrids between plasmacytoma cells and spleen cells grew as hybrids. These hybrid plasmacytomas have come to be called "hybridomas."

Shortly after the two types of cells are fused by incubation with a fusing agent such as polyethylene glycol, they are plated out into a series of hundreds of small wells so that only a limited number of hybrids grow out together in the same well. Depending on the frequency of hybrids and the number of wells used, it is possible to distribute the cells so that each hybrid cell grows up in a separate cell culture well.

On the basis of the number of spleen cells that would normally be making antibodies against SRBC after mice have been immunized with them, the investigators expected that one well in about 100,000 or more might have a clone of hybrid cells making antibody that reacted against this antigen. The supernatants (liquid overlying settled material) from hundreds of wells were tested, and the large majority were found to react with the immunizing antigen. Further work with other antigens confirmed that a significant fraction of hybrid cells formed with spleen cells of immunized mice produce antibodies reacting with the antigen recently injected into the mouse. The production of homogeneous antibodies from clones of hybrid cells thus became a practical way to obtain reliable supplies of well-defined immunological reagents.

The antibodies can be collected from the media in which the cells are grown, or the hybridomas can be injected into mice so that larger concentrations of monoclonal antibodies can be collected from fluid that collects in the abdominal cavity of the animals.

Specific Antibodies Against Antigen Mixtures

One advantage of separating an animal's antibody response into individual antibody components by hybridization and separation of cells derived from each fusion event is that antibodies that react with individual antigenic components can be isolated even when the mouse is immunized with a complex mixture of antigens. For example, human tumor cells injected into a mouse stimulate the production of many different types of antibodies. A few of these antibodies may react specifically with tumor cells or specific types of human cells, but, in a conventional antisera, these antibodies would be mixed with other antibodies that react with any human cell and would not be easily separated from them. If the tumor cells are injected and hybridomas are made and screened to detect antibodies that react with tumor cells and not with most normal cells, it

is possible to isolate antibodies that are useful for detection and characterization of specific types of tumor cells. Similar procedures can also be used to make antibodies against a single protein after the mouse has been immunized with this protein included in a complex mixture of other biological molecules such as a cell extract.

Following the first report of monoclonal antibodies, biologists began to realize the implications of being able to produce a continuous supply of antibodies with selected and well-defined reactivity patterns. There was discussion of "magic bullets" that would react specifically with and carry specific cytotoxic agents to tumor cells without adverse effects on normal cells. Biologists working in various experimental systems realized how specific and reliable sources of antibody reagent might contribute to their investigations, and entrepreneurs started several biotechnology companies to develop and apply monoclonal antibody methods. This initial enthusiasm was quickly moderated as some of the technical difficulties involved in production and use of these antibodies became apparent; with time, however, many of the projected advantages of these reagents have become a reality.

Monoclonal Reagents

A survey of catalogs of companies selling products used in biological research confirms that many of the conventional antisera commonly used as research reagents have been replaced with monoclonal antibodies. These products are advantageous to the suppliers, being produced in constant supply with standardized protocols from hybrid cells, and the users, who receive well-characterized reagents with known specificities free of other antibodies that could produce extraneous and unexpected reactions when used in some assay conditions. Antibodies are available against a wide range of biomolecules reflecting current trends in research; examples include antibodies against cytoskeletal proteins, protein kinases, and oncogene proteins, gene products involved in the transition of normal cells to cancer cells.

Immunologists were among the first to take

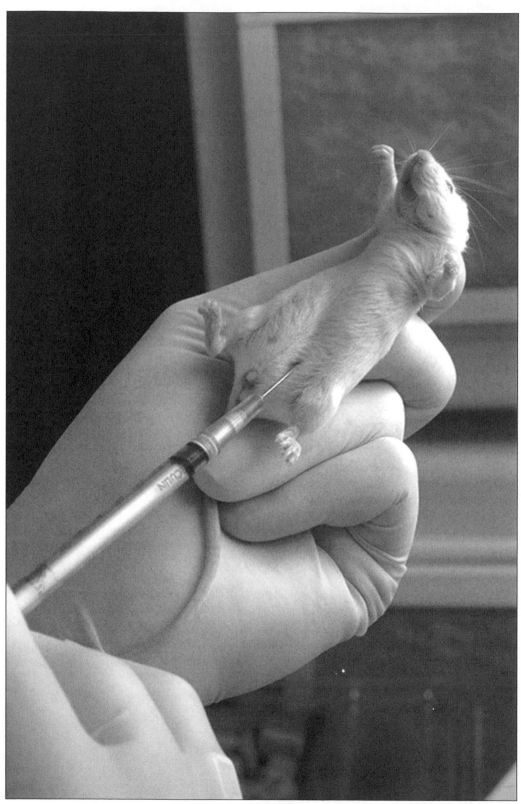

A mouse is injected with antigens to produce monoclonal antibodies. (Dan McCoy/Rainbow)

advantage of monoclonal antibody technology. They were able to use them to "trap" the spleen cells making antibodies against small, well-defined molecules called "haptens" and to then characterize the antibodies produced by the hybridomas. This enabled them to define classes of antibodies made against specific antigenic determinants and to derive information about the structure of the antibody-binding sites and how they are related to the determinants they bind. Other investigators produced antibodies that reacted specifically against subsets of lymphocytes playing specific roles in the immune responses of animals and humans. These reagents were then used to study the roles that these subsets of immune cells play in responses to various types of antigens.

Antibodies that react with specific types of immune cells have also been used to modulate the immune response. For example, antibodies that react with lymphocytes that would normally react with a transplanted tissue or organ can be used to deplete these cells from the circulation and thus reduce their response against the transplanted tissue.

Monoclonal Antibodies as Diagnostic Reagents

Monoclonal antibodies have been used as both in vitro and in vivo diagnostic reagents. By the 1980's, many clinical diagnostic tests such as assays for hormone or drug levels relied upon antisera as detecting reagents. Antibodies reacting with specific types of bacteria and viruses have also been used to classify infections so that the most effective treatment can be determined. In the case of production of antibodies for typing microorganisms, it has frequently been easier to make type-specific monoclonal antibodies than it had been to produce antisera that could be used to identify the same microorganisms.

Companies supplying these diagnostic reagents have gradually switched over to the use of monoclonal antibody products, thus facilitating the standardization of the reactions and the protocols used for the clinical tests. The reproducibility of the assays and the reagents has made it possible to introduce some of these tests that depend upon measurement of concentrations of substances in urine as kits that can be used by consumers in their own homes. Kits have been made available for testing glucose levels of diabetics, for pregnancy, and for the presence of certain drugs.

Although the much-hoped-for "magic bullet" that would eradicate cancer has not been found, there are several antibodies in use for tumor detection and for experimental forms of cancer therapy. Monoclonal antibodies that react selectively with cancer cells but not normal cells can be used to deliver cytotoxic molecules to the cancer cells. Monoclonal reagents are also used to deliver isotopes that can be used to detect the presence of small concentrations of cancer cells that would not normally be found until the tumors grew to a larger size.

Human Monoclonal Antibodies

The majority of monoclonal antibodies made against human antigens were mouse antibodies derived from the spleens of immunized mice. When administered to humans in clinical settings, the disadvantage of the animal origin of the antibodies soon became apparent. The human immune system recognized the mouse antibodies as foreign proteins and produced an immune response against them, limiting their usefulness. Even when the initial response to an antibody's administration was positive, the immune reaction against the foreign protein quickly limited its effectiveness. In an attempt to avoid this problem, human monoclonal antibodies have been developed using several methods. The first is the hybridization of human lymphocytes stimulated to produce antibodies against the antigen of interest with mouse plasmacytomas or later with human plasmacytoma cell lines. This method has been used successfully, although it is limited by the ability to obtain human B cells or plasma cells stimulated against specific antigens because it is not possible to give an individual a series of immunizations and then remove stimulated cells from the spleen. Limited success has resulted from the fusion of circulating lymphocytes from immunized individuals or fusion of lymphocytes that have been stimulated by the antigen in cell cultures. Investigators have reported some success in making antitumor monoclonal antibodies by fusing lymph node cells from cancer patients with plasmacytoma cell lines and screening for antibodies that react with the tumor cells.

There has also been some success at "humanizing" mouse antibodies using molecular genetic techniques. In this process, the portion of the genes that make the variable regions of the mouse antibody protein that reacts with a particular antigen is spliced in to replace the variable region of a human antibody molecule being produced by a cultured human cell or human hybridoma. What is produced is a human antibody protein that has the binding specificity of the original mouse monoclonal antibody. When such antibodies are used for human therapy, the reaction against the injected protein is reduced compared to the administration of the whole mouse antibody molecules.

Another application of antibody engineering is the production of bispecific antibodies.

This has been accomplished by fusing two hybridomas making antibodies against two different antigens. The result is an antibody that contains two types of binding sites and thus binds and cross-links two antigens, bringing them into close proximity to each other.

Recombinant Antibodies

Advances in molecular genetic techniques and in the characterization of the genes for the variable and constant regions of antibody molecules have made it possible to produce new forms of monoclonal antibodies. The generation of these recombinant antibodies is not dependent upon the immunizing of animals but on the utilization of combinations of antibody genes generated using the in vitro techniques of genetic engineering. Geneticists discovered that genes inserted into the genes for fibers expressed on the surface of bacterial viruses called "bacteriophages" are expressed and detectable as new protein sequences on the surface of the bacteriophage. Investigators working with antibody genes found that they could produce populations of bacteriophage expressing combinations of antibody-variable genes. Molecular genetic methods have made it possible to generate populations of bacteriophage expressing different combinations of antibody-variable genes with frequencies approaching the number present in an individual mouse or human immune system. The population of bacteriophage can be screened for binding to an antigen of interest, and the bacteriophage expressing combinations of variable regions binding to the antigen can be multiplied and then used to generate recombinant antibody molecules in culture.

Researchers have also experimented with introducing antibody genes into plants, resulting in plants that produce quantities of the specific antibodies. Hybridomas or bacteriophages expressing specific antibodies of interest may be a potential source of the antibody gene sequences introduced into these plant antibody factories.

—*Roger H. Kennett*

See Also: Antibodies; Cancer; Genetic Engineering; Genetic Engineering: Medical Applications; Immunogenetics.

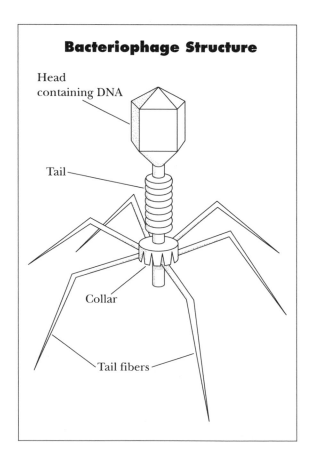

Bacteriophage Structure

Head containing DNA

Tail

Collar

Tail fibers

Further Reading: H. R. Hoogenboom, "Designing and Optimizing Library Selection Strategies for Generating High-Affinity Antibodies," *Trends in Biotechnology* 15 (1997), contains detailed information about laboratory techniques used to engineer monoclonal antibodies. T. Stigbrand et al., "Twenty Years with Monoclonal Antibodies: State of the Art," *Acta Oncologica* 35 (1996), provides an overview of the development of monoclonal antibodies. W. W. Gibbs, "Plantibodies: Human Antibodies Produced by Field Crops Enter Clinical Trials," *Scientific American* 277 (November, 1997), details experiments in introducing antibody genes into plants. J. G. van de Winkel et al., "Immunotherapeutic Potential of Bispecific Antibodies," *Immunology Today* 18 (December, 1997), takes a look at the potential uses of bispecific antibodies.

Immune Deficiency Disorders

Field of study: Immunogenetics

Significance: *The human immune system, a complex of interacting cells and secreted proteins, has evolved to provide protection from disease-causing organisms. Individuals who have an inherited or acquired defect in any part of their immune system are thus susceptible to a variety of maladies, termed opportunistic diseases, that healthy individuals would be able to recover from or resist.*

Key terms

PATHOGENS: organisms that cause disease

ANTIGENS: molecules recognized as foreign by the immune system, including those found on pathogens

ANTIBODIES: soluble proteins that specifically bind to pathogens

B LYMPHOCYTES: cells of the immune system that secrete antibodies

T LYMPHOCYTES: the central cell of the adaptive immune system; this class of lymphocytes stimulates a variety of immune responses

OPPORTUNISTIC DISEASES: maladies associated with the immunosuppressed, causing little or no disease in individuals with healthy immune systems

The Human Immune System and Primary Immunodeficiencies

Deficiencies in the immune system are classified as primary or secondary. Primary (inherited) immunodeficiencies are the result of genetic defects that prevent the immune system from functioning normally. Since primary immunodeficiencies are present at birth, these disorders usually exhibit themselves in children and young adults. On the other hand, secondary (acquired) immunodeficiencies are the result of an illness in previously healthy individuals. Thus, secondary immunodeficiencies can exhibit themselves at any age.

The immune response is a complex system of interacting cells and secreted factors that has evolved to protect the body from disease-causing, pathogenic organisms. Humans can fight these organisms using a variety of nonspecific mechanisms that provide innate immunity. These mechanisms include physical barriers such as the skin that prevent microorganisms from entering the body, and phagocytic cells such as macrophages and neutrophils, which have the ability to rapidly recognize, ingest, and destroy certain types of bacteria. Innate immune responses are the first line of defense against pathogens.

The most effective class of immunity, however, is known as adaptive immunity. These mechanisms, although slower than innate immunity, are more specific to pathogen-associated molecules known as antigens. It is this aspect of the immune response that leads to the phenomenon of immunological memory. Immunological memory is the characteristic of the adaptive immune response that allows stronger and faster immunity to develop after the first exposure to a pathogen, preventing reinfection. Additionally, immunological memory is the basis of vaccination, in which an initial immune response to pathogen-associated antigens is induced by inoculation with a killed or weakened pathogen that is by itself incapable of causing disease. Vaccination prevents infection with the pathogen, just as if the individual had previously contracted the disease. As many primary immunodeficiencies are the result of defects in adaptive immunity, the effectiveness and safety of vaccinations may be altered in the immunosuppressed.

Defects in B lymphocytes account for approximately 50 percent of all primary immunodeficiencies. B lymphocytes act primarily by secreting soluble proteins called antibodies, which can bind directly to pathogen-specific antigens and lead to the destruction of the pathogen and recovery from the disease. Antibody-mediated immunity is most successful against extracellular pathogens such as many bacteria. Therefore, defects in the functionality of B lymphocytes will make the individual susceptible to a variety of bacterial diseases.

T lymphocyte defects account for approximately 30 percent of primary immunodeficiencies. T lymphocytes are the central cell of the adaptive immune response and function in several ways, depending on the subpopulation involved. Cytotoxic T lymphocytes can directly kill living cells, including virally infected cells. Helper T lymphocytes act by regulating other effectors of immunity, including the production of antibodies by B lymphocytes, the induction of cellular inflammatory responses known as "delayed hypersensitivity," and the development of cytotoxic T lymphocytes. T lymphocytes are extremely important in the body's defense against intracellular pathogens such as viruses and certain types of bacteria. Because of the interaction of the two classes of lymphocytes during the production of antibodies, many T-lymphocyte defects cause deficiencies in both B-lymphocyte- and T-lymphocyte-associated immunity. This class of disease is known as severe combined immunodeficiency (SCID). Individuals with SCID or other T-lymphocyte defects are susceptible to infection with a wide variety of pathogens.

Phagocytic cell defects cause about 18 percent of primary immunodeficiencies. Phagocytic cells participate in adaptive immune responses in a variety of ways. Macrophages and other phagocytes are attracted to specific sites within the body by the release of chemical signals in the process known as "chemotaxis" and are especially good at ingesting and destroying bacteria and other organisms coated by specific antibodies. The most common phagocytic cell defects involve either the loss of the ability to destroy pathogenic microorganisms or the loss of the ability to respond to chemotactic signals.

Defects in the complement system are relatively rare, accounting for only 2 percent of primary immunodeficiencies. Complement is a group of serum proteins produced by the liver that can interact with antibodies that have bound to their antigens. The main activity of complement is to protect against bacteria. Thus, one of the major consequences of complement deficiencies is a greater incidence of bacterial infections in these individuals. Also, for reasons that are not understood, an ill-defined syndrome with symptoms resembling certain autoimmune diseases (in which people's immune responses attack the tissues of their own bodies) is sometimes seen in complement-deficient individuals. The overall incidence of symptomatic primary immunodeficiency in the United States is estimated to be one in ten thousand. Approximately four hundred new cases are diagnosed each year. Thus, primary immunodeficiencies make up a small percentage of all immune deficiency disorders.

Secondary Immunodeficiencies

Secondary immunodeficiencies are much more common and can be the result of a variety of injuries to a once-healthy immune system. These include trauma, cancer, immunosuppressive drug therapy, and infectious agents such as human immunodeficiency virus (HIV).

Trauma is often an overlooked source of immune deficiencies. Because of the connection and interaction between the central nervous system and the immune system, any prolonged stress will tend to inhibit the body's ability to fight disease. For example, even though going out in the rain without a raincoat does not by itself cause diseases such as the common cold or influenza, this often-heard parental advice is still sound. The loss of body heat caused by walking unprotected in a cold rain will indeed make the body more susceptible to disease by impairing the immune system.

Cancers, especially leukemia and other cancers of lymphocytes, undermine the immune response by overproducing cells capable of interacting with one or only a few antigens. The large number of cancerous lymphocytes tend to crowd out normal cells, inhibiting one of the most important aspects of the adaptive immune system: the ability to respond to the wide variety of pathogen-associated antigens.

Immunosuppressive drugs are used to prevent tissue rejection in transplant recipients and to inhibit the immune response against self-antigens in patients with autoimmune disease. Individuals receiving such therapy are more susceptible to infectious disease and thus must take special precautions to avoid unnecessary exposure to pathogenic organisms.

Finally, many infectious diseases, including mononucleosis, acute bacterial disease, parasitic infections, and viral infections, will also inhibit the functionality of the immune system. In the 1980's and 1990's, the acquired immu-

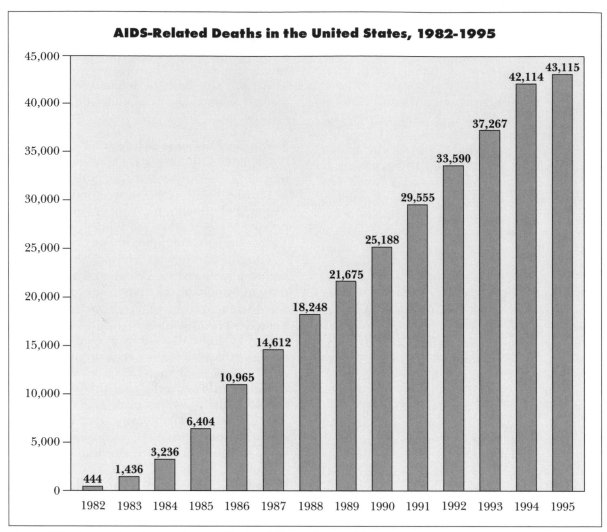

AIDS-Related Deaths in the United States, 1982-1995

(*Source:* U.S. Centers for Disease Control, National Center for Health Statistics)

nodeficiency syndrome (AIDS) pandemic, caused by HIV, brought new focus to the importance of a functional immune response to the maintenance of good health.

Problems of the Immunodeficient

AIDS was recognized as a disease in the early 1980's. Individuals with infections and cancers associated with the immunosuppressed began showing up with increasing frequency in the major cities of the United States. The frequencies of these usually rare diseases, termed opportunistic diseases, was the major factor in the eventual identification of AIDS as a disease and HIV as the virus that causes it. HIV is a virus that

infects and eventually destroys the helper T lymphocytes in the infected individual. The fact that it is the lack of immunity and not the HIV virus itself that kills patients with AIDS focused attention on the importance of maintaining a strong immune response whenever possible.

Some vaccination procedures also lead to health problems in immunosuppressed individuals. Vaccines consisting of live, weakened viruses, such as the Sabin vaccine used for the prevention of poliomyelitis, can cause disease in immunosuppressed individuals. As vaccinations are usually given during childhood, susceptible individuals are usually those with primary immunodeficiencies. Even though the number of children infected with polio by vaccination in the United States is extremely low, the fact that the risk of naturally acquired poliomyelitis in this country is practically nonexistent has led some to propose using the older, less effective Salk vaccine instead. The Salk vaccine is made from a killed poliovirus, which is incapable of causing disease in anyone, including the immunosuppressed.

—*James A. Wise*

See Also: Antibodies; Immunogenetics; Organ Transplants and HLA Genes.

Further Reading: "Life, Death, and the Immune System," *Scientific American* 269 (September, 1993), is a special issue that provides an excellent overview of the immune system. Charles Janeway and Paul Travers provide an excellent review of immunodeficiency diseases in *Immunobiology: The Immune System in Health and Disease* (1994). *Introduction to Medical Immunology* (1990), edited by Gabriel Virella et al., describes the symptoms and treatment of immunodeficiencies.

Immunogenetics

Field of study: Immunogenetics
Significance: *Immunogenetics is the study of the genetics of transplantation and tissue rejection, immunologic response, antibody structure, and immunosuppression (including the genetic basis for diseases of the immune system). It also encompasses the study of other systems of the vertebrate body through the use of the proteins of the immune system such as monoclonal antibodies.*

Key terms

ANTIBODY: a protein produced by immune system cells in response a foreign substance and capable of binding specifically to that foreign substance

ANTIGEN: a foreign substance that stimulates the production of antibodies when introduced into a vertebrate animal

COMPLEMENT: a group of plasma enzymes that can be activated immunologically or nonimmunologically and that participates in a variety of immunological functions

IMMUNOGLOBULIN: an antibody secreted by mature immune system cells

LYMPHOCYTES: cells found in the lymph nodes, thymus, spleen, bone marrow, and blood; includes B-cell and T-cell subpopulations

MONOCLONAL ANTIBODIES: immunoglobulins derived from a single clone of plasma cells; a chemically and structurally pure population

Overview of the Immune System

The function of the immune system is to recognize and eliminate foreign organisms that attack the body. To do this, it must be rapid, specific, and aimed exclusively against foreign molecules (or host-cell molecules that are inappropriate). The system must be prepared for contact with a variety of organisms and molecular structures. In mammals, an extremely efficient, genetically complex immune system has evolved to tackle this problem. The mammalian immune system consists of peripheral blood lymphocytes (white blood cells) and the lymphatic organs and tissues. The primary lymphatic organs are the thymus and the bone marrow. Secondary organs include the spleen and lymph nodes, along with decentralized tissues such as the tonsils and Peyer's patches (intestines).

The immune system is an example of a dynamic genetic process. In the new cells that are constantly being produced (millions per day), changes arise in the genes of the immune system. This variability results from deoxyribonucleic acid (DNA) splicing and point mutations. It benefits the organism by providing the cells of the immune system with the widest possible range of specificities. As soon as the organism is invaded by a foreign agent (usually a bacteria or virus), the immune system is ready and rapidly becomes activated. Those immune system cells that carry the specific immune receptors for interacting with the foreign agent are stimulated to multiply. In a matter of days, clones of immune system cells with the appropriate specificity have been produced, and the organism fends off the invader with the tools provided by that clone of cells. A portion of these clones are held back and persist, often for years at a time, thus accounting for the ability of the organism to respond even faster the second time the same foreign invader makes its presence known.

There are basically two types of immune response. The first, called the "humoral" or antibody response, involves the concerted action of two classes of lymphocytes known respectively as B cells and T cells. (B cells mature in the bone marrow and then migrate to the secondary lymphoid organs; T cells, on the other hand, mature in the thymus.) The active agents of this response are antibodies, or immunoglobulins, proteins secreted by the B cells into the bloodstream. This combined B-T immune response is characteristic of most vertebrates. The second type of immune response is called the cell-mediated response and involves only the T cells. The active agent in this re-

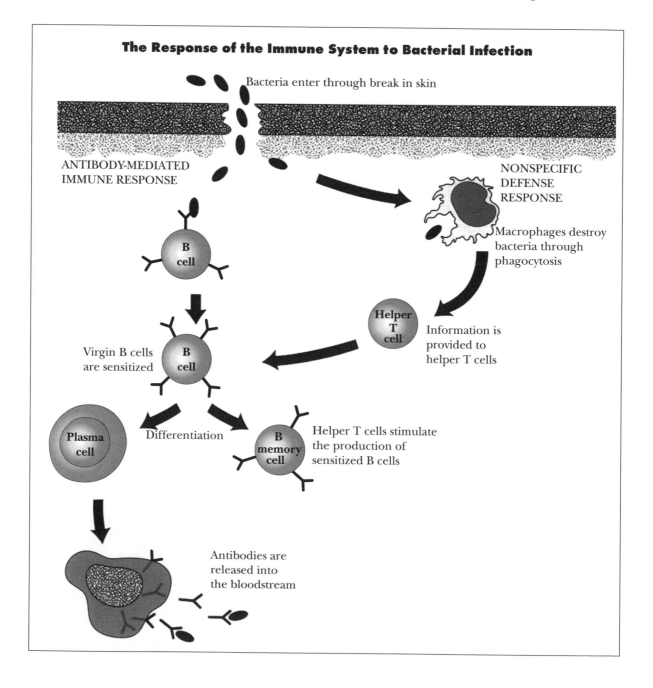

The Response of the Immune System to Bacterial Infection

Bacteria enter through break in skin

ANTIBODY-MEDIATED IMMUNE RESPONSE

NONSPECIFIC DEFENSE RESPONSE

Macrophages destroy bacteria through phagocytosis

B cell

Helper T cell

Information is provided to helper T cells

Virgin B cells are sensitized

B cell

Plasma cell

Differentiation

B memory cell

Helper T cells stimulate the production of sensitized B cells

Antibodies are released into the bloodstream

sponse is usually the circulating T cell itself, which attacks the foreign agent directly. This response is found only in mammals and a few other select vertebrates.

Immunoglobulins and Their Diversity

The immunoglobulins are a family of structurally related proteins that mediate the humoral responses. They are found in most vertebrates. In higher mammals, five major immunoglobulin (Ig) classes (named for their chemical structures) are recognized—IgA, IgD, IgE, IgG, and IgM—and in some species, subclasses of these Igs have also been described. In all cases, the proteins are manufactured by the plasma cells that are derived from the B lymphocytes. Most of the Ig molecules produced by these cells are found in the serum and secretions of the body, but small quantities also occur in a cell-associated form attached to the surface membranes of lymphocytes, macrophages, mast cells, and basophils (all forms of white blood cells). Although antibody production occurs from B lymphocytes, the involvement of other cell types is usually essential for their response.

The ability of antibodies (immunoglobulins) to combine specifically with antigens on the surface of a foreign organism depends on their three-dimensional structure and the specificity of their binding regions. The structures of the peptides that form the largest class of antibodies, IgG, are composed of four polypeptides: two identical light (L) chains and two identical heavy (H) chains. Differences among IgG antibody molecules are the result of variable amino acid sequences in the terminal sections of both the L and H chains. These "variable" domains consist of approximately 110 amino acids each. The amino acid sequences in the rest of the chains do not change significantly from one IgG molecule to another, and this portion of each chain is referred to as the "constant" domain.

The L-chain variable domain includes a variable (V) region and a short joining (J) region, which joins the constant and variable regions. The H chains have a similar arrangement, with a third region, diversity (D), added between the V and J. The D region consists of a short,

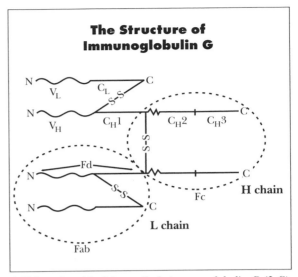

The Structure of Immunoglobulin G

*A Y-shaped model of the antibody immunoglobulin G (IgG). V indicates a region of variability that would permit recognition by a wide variety of antigens. (*Source: *John J. Cebra's "The 1972 Nobel Prize for Physiology or Medicine," Science, 1972)*

hypervariable sequence of amino acids. Each L chain is linked to one of the H chains by a single disulfide (S-S) bond, and the H chains are joined to each other by two disulfide linkages. The variable domains on the two pairs of chains form the active end of the divalent molecule (the sites that combine with a specific antigen). The specificity of the binding reaction (the binding of the appropriate antibody to the antigen) is a result of the complementary nature of the antigen and antibody shapes. The variable amino acid sequences of different IgG molecules cause their variable domains to have different chemical properties and to assume slightly different folding patterns, each of which recognizes only one antigen shape.

To understand this immune response, one must ask how B cells create so many different antibodies. In other words, how can the specific molecules needed to recognize all possible invading organisms be encoded in the human genome? A human only has approximately 100,000 genes (of which only a fraction are devoted to the immune system) yet makes at least one billion specific antibody molecules and can have up to 100 million different antibody proteins available at any one time. This enormous diversity of proteins is generated from a limited amount of genetic information with the help of two mechanisms: somatic recombination and somatic mutation.

Somatic Recombination and Class Switching

In 1976, Susumu Tonegawa showed that antibody genes are inherited as gene fragments. These fragments are joined together to form a complete gene only in individual B cells as they develop. In addition, the enzymes that help combine these gene segments together add random DNA bases to the ends of the pieces as they are joined. Thus, new genes are constantly being formed. Further diversity results during the assembly of the protein chains into a complete receptor.

The rearrangement of genes for Ig molecules involves an excision (cutting) mechanism. This requires precise recognition of the pieces in order to ensure formation of the correct code. Noncoding DNA between genes for different regions of the molecule is excised, and the rest is subsequently joined. Unlike recombination during meiosis, Ig rearrangements involve nonhomologous DNA joining. In the human, chromosomes 2, 22, and 14 contain those gene clusters associated with the kappa (κ) light chain, the lambda (λ) light chain, and the heavy chains, respectively.

Recombination between the DNA segments coding for an Ig molecule is mediated by a system of enzymes called "recombinases." The enzymes are controlled by specific DNA recognition sequences located in adjacent noncoding DNA segments at the end of each V region and at the opposite end of each J segment. The D segments are flanked on both sides by recognition sequences. Recognition sequences are noncoding but highly conserved (the same in everyone) DNA segments. They are separated by precisely defined intervals produced by spacers, the sequences of which are not conserved.

When an H chain is formed, nonhomologous pairing of the D- and J-segment recognition sequences occurs. These D and J segments are then joined (D-J joining) by means of recombination. The spacer and all of the intervening DNA form a loop. This is excised, and the D and J segments are joined. By pairing and recombination of the recognition sequences at the end of a D-J segment and that at the opposite end of a V gene, a V segment is joined to complete the segment.

"Class switching" is a process whereby a B cell that previously expressed one class of Ig is able to change and express a different class of Ig. It involves an additional DNA splicing operation and occurs frequently with heavy chain genes. Thus a B cell, which initially expresses a *mu* constant region, results in IgM antibodies. Subsequently, the same variable region of the genome may become located next to a gamma constant region gene, causing the cell to produce an IgG antibody with the same variable region. In cases of class switching, the second DNA splicing involves removal of the intervening genes. Thus, in the case of IgM to IgG switching, the region containing the IgD constant region, which exists between them, must be removed by the second splicing reaction.

As a rule, the DNA splicing reactions leading to the joining process occur in only one of the alleles for each of these chains. The mechanism for this phenomenon, known as "allelic exclusion," is not well understood. Another type of exclusion process results in the production of only κ or λ light chains in a given B cell. This process is called "isotypic exclusion" (an isotype, in this case κ or λ, is an antigenic determinant common to all members of the same species).

Cell Surface Proteins

The numerous cell-surface and soluble molecules of the immune system that mediate functions such as recognition, rejection, binding, or adhesion show many structural similarities. As a group, they constitute a gene superfamily, probably derived from a common ancestral gene. The molecules that belong to this supergene family are the T-cell receptors (TCR) and the molecules of the major histocompatibility complex (MHC; generally known as the human leukocyte antigens, or HLA, in humans). They all consist of variable, constant, or primordial immunoglobulin-like domains. Although their genes are located on different chromosomes, the gene products form functional complexes with each other. The basic structure of accessory molecules such as CD-2, CD-3, CD-4, CD-8, and Thy-1 are relatively simple by comparison (CD stands for "cluster designation," a term applied to surface proteins of the immune system).

The MHC region is one of highly polymorphic genes spanning about 3,500 kilobases (kb) on the short arm of chromosome 6 in humans and chromosome 17 in mice. Collectively, these genes are called the "immune response" (IR) genes. They are expressed on the surface of various cell types. MHC genes control the immune response to antigen proteins, including graft and transplant rejection, by specific binding to T cells. The gene loci of the MHC region are grouped into three classes (designated I-III). Class I in humans includes HLA-A, HLA-B, and HLA-C. In mice it includes D, L, and K regions of the H-2 system. Class II includes HLA-DP, HLA-DQ, and HLA-DR in humans (I-A and I-E in mice). Class III genes do not technically belong to the MHC (they produce proteins of the complement system) but are often grouped with the MHC because of their physical proximity. Both class I and class II proteins consist of two different polypeptide chains. In class I molecules, an MHC-coded α chain is associated with a non-MHC-coded β chain. Class II MHC molecules consist of two polypeptide chains, α and β, each with two domains (α1, α2, β1, β2). In both, the peptide-binding regions are highly polymorphic.

The cell surface of a T cell contains receptor molecules (TCR) that recognize foreign antigens and cell-surface molecules of the MHC complex. The TCR consists of a complex of several integral plasma membrane proteins. Unlike B cells, T cells recognize only fragments of foreign antigen proteins. In addition, they bind physically to the MHC complex of an antigen-presenting cell. During maturation of the T cells in the thymus, T-cell gene segments are rearranged in a defined order by somatic recombination, similar to the formation of immunoglobulins.

The genes for each chain of the TCR in humans consists of V, D, J, and C segments, similar to those seen in the Igs. In a given T cell, only one of the two α chain loci and one of the two β chain loci become rearranged and expressed. Different mechanisms help to produce diversity of the TCR, just as they do in the Ig genes. The β chains are located on chromosome 7 in humans and chromosome 6 in mice. The α and δ chain loci are located on chromosome 14 in both; γ chain genes are found on chromosome 7 in humans and chromosome 13 in mice.

Two different classes of T cells recognize different types of MHC gene products. T cells with the ability to destroy other cells by cytolysis (cytotoxic T cells) recognize class I MHC molecules by means of CD-8 (CD-8 is a membrane-bound glycoprotein in the Ig superfamily). The second type of T cell (helper T cell) specifically binds MHC class II molecules.

Proteins of the Complement System

The complement system, which is composed of at least twenty plasma proteins (C1-C9, factors B and D, and a series of regulatory proteins), is one of the major activation systems present in the blood of all vertebrate animals. The activation of this cascade system is certainly one of the most important functions of antibodies. Elimination of foreign microorganisms proceeds much less efficiently in individuals with deficiencies of certain key complement components, and such individuals are particularly vulnerable during early life. Numerous peptide fragments are generated during activation and are known to have potent biological activity. Some act as factors that encourage the migration of phagocytic cells, some possess a special receptor for the phagocytes and thereby promote immune adherence, some bring about changes in capillary permeability, and some damage the lipid layer of cell membranes.

The so-called classical pathway of complement activation is still the subject of intense investigation. An antibody-independent mechanism of complement activation (the alternate pathway) was originally described in the 1950's and then rediscovered in the 1970's. Both pathways yield enzymes that cleave the predominant complement protein known as C3. The differences between the pathways are to be found in both the nature of the early-acting components required to generate the respective enzymes and in the distinctive mechanisms of activation.

Most complement components circulate as inactive molecules (often as proenzymes). Activation of any one complement component is frequently achieved by the proteolytic attack of the preceding component, revealing an enzymatically active site that in turn acts on another component. This process led early workers to describe the classical pathway as the complement "cascade." Activated components usually have very short biological half-lives and will decay to an inactive form if the substrate molecule is not encountered. There are several regulator and control proteins that play a critical role in protecting the tissues of the host against the potentially damaging effects of uncontrolled activation.

The classical and alternative pathways converge on the relatively abundant C3 component, and cleavage of this component is probably the most important single event of the activation process. The largest cleavage product (C3b) is also an integral part of the alternate pathway enzyme, and so activation of the classical pathway inevitably leads to recruitment of the alternative pathway and the generation of further enzyme. Thus a marked amplification effect occurs at this stage as a result of possible feedback.

The proteins of the complement system may be divided into four groups according to where they occur and how they relate to the complement pathways: proteins specific to the classical pathway (plus C3), proteins specific to the alternate pathway, proteins of the attack complex, and regulatory proteins. As previously mentioned, these proteins are often referred to as class III MHC proteins.

Autoimmune and Immune Deficiency Diseases

Immune system protection against infectious organisms and their effects after successful invasion is a basic requirement for survival. Functional impairment of the immune system leads to severe illness, collectively called the immune deficiency diseases. The primary cause may be either hereditary (primary immune deficiency) or acquired (secondary immune deficiency).

Many genetically determined defects of the immune system are known. A defect in the common precursor of B and T cells leads to a group of diseases called severe combined immunodeficiency (SCID). In this group of diseases, both B and T cells are affected. This has correspondingly severe clinical consequences. At least one X-chromosomal and one autosomal recessive form are known. A disorder of B-cell maturation leads to the absence of free Igs. Other disorders concern B-cell differentiation (variable immunodeficiency) or alteration of an isotype (such as IgA deficiency). An embryonic developmental disorder of the third and fourth brachial arches leads to an absence or hypoplasia (lack of growth) of the thymus. In this case, T lymphocytes cannot mature. Acquired immunodeficiency is a frequent accompanying sign of diseases and immunosuppressive medications or the effect of a harmful dose of whole-body irradiation. Acquired immunodeficiency syndrome (AIDS), the disease caused by human immunodeficiency virus (HIV) HIV-1 and HIV-2, has become very significant. It affects CD4-expressing T cells (T helper cells).

A plethora of diseases were discovered during the twentieth century in which circulating antibodies can be demonstrated that bind to tissues or cells of the individual from which they are derived. Such conditions are referred to as autoimmune diseases. At one end of the spectrum are those diseases (such as Hashimoto's thyroiditis) in which the autoantibodies specifically attack one organ of the body. At the other end of the spectrum are non-organ-specific or systemic diseases such as systemic lupus erythematosus (SLE), in which the antibodies have widespread reactivity within the body. There is a tendency for more than one autoimmune disease to occur in the same individual. Ample evidence suggests that autoimmune diseases are genetically programmed and that they are genetically complex.

Impact and Applications

The understanding of immunogenetic diversity and mechanisms has led to the use of immune-system proteins to study a variety of other cellular processes and to better understand and deal with tissue typing for grafts and transplants. These applications have come about primarily through the use of monoclonal antibodies. Monoclonal antibodies are the product of cloned hybridoma cells, which are produced through fusion of normal B lymphocytes and cancerous myeloma cells. Once cultured and selected for a specific antibody, these cells produce uniform antibodies that are directed against a single antigenic determinant. The applications of this technology are enormous. Simply put, if a protein molecule can be identified by a monoclonal antibody, its expression and function can be studied. Enough protein can be isolated for an amino acid sequence to be determined, a DNA probe can be defined and synthesized, and the gene encoding the protein can be isolated.

Other applications include possible immunization with a specific antibody rather than with an antigen that causes patients to develop their own antibodies, as most vaccinations do. Monoclonal antibodies can also be used for detecting specific cancers (diagnostic kits using monoclonal antibodies for colon cancer are already in use). Antitumor monoclonal antibodies have been successfully employed in immunotherapy for tumors, and poisons directed against tumors can potentially be linked to specific monoclonal antibodies. The applications of immunogenetic knowledge are widespread and are reaching the forefront of many new strategies in medical care.

—*Kerry L. Cheesman*

See Also: Antibodies; Autoimmune Disorders; Hybridomas and Monoclonal Antibodies; Immune Deficiency Disorders; Organ Transplants and HLA genes.

Further Reading: L. Fugger et al., "The Role of Human Major Histocompatibility Complex (HLA) Genes in Disease," and D. J. Barrett et al., "Antibody Deficiency Diseases," both in *The Metabolic and Molecular Bases of Inherited Disease* (1995), edited by C. R. Scriver et al., give detailed information on specific aspects of the immune system. Philip Leder, "The Genetics of Antibody Diversity," *Scientific American* 246 (May, 1982), and S. Tonegawa, "Somatic Generation of Antibody Diversity," *Nature* 302 (1983), provide an overview of the diversity of proteins involved in the immune system. Susumu Tonegawa, "The Molecules of the Immune System," *Scientific American* 253 (October, 1985), describes the immune system on the molecular level.

In Vitro Fertilization and Embryo Transfer

Field of study: Human genetics
Significance: *The term "in vitro" designates a living process removed from an organism and isolated "in glass" for laboratory study. In vitro fertilization (IVF) provides clinical access to both the paternal and maternal genetic information before it is combined to form a new living creature. It also allows early access to the results of the combination of this information.*

Key terms

CHROMOSOME: one of a pair of microscopic filaments in the nucleus of a cell carrying a specific portion of the total genetic information for the cell and the organism
DIPLOID: a paired set of chromosomes bearing two complete copies of nuclear genetic information for a cell
GAMETE: a germ cell; an egg (ovum or oocyte) or a sperm (spermatozoan)
HAPLOID: a single set of chromosomes; mature gametes are haploid

Natural Fertilization

Fertilization, the union of a male gamete (sperm) with a female gamete (ovum), is fundamentally a genetic process. The creation of a new being, an embryo, requires two complementary sets of genetic instructions, paternal and maternal. A paternal set of genetic information is in the nucleus of each sperm cell; a maternal set is in the nucleus of each ovum. Fertilization brings together these two sets and thereby provides the diploid chromosomal complement for a new living organism.

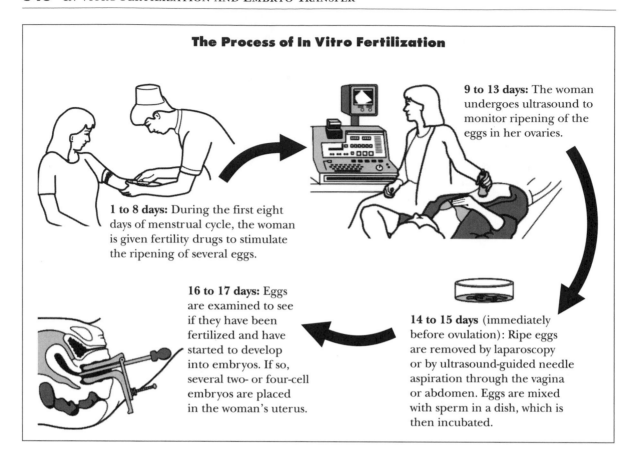

The Process of In Vitro Fertilization

1 to 8 days: During the first eight days of menstrual cycle, the woman is given fertility drugs to stimulate the ripening of several eggs.

9 to 13 days: The woman undergoes ultrasound to monitor ripening of the eggs in her ovaries.

16 to 17 days: Eggs are examined to see if they have been fertilized and have started to develop into embryos. If so, several two- or four-cell embryos are placed in the woman's uterus.

14 to 15 days (immediately before ovulation): Ripe eggs are removed by laparoscopy or by ultrasound-guided needle aspiration through the vagina or abdomen. Eggs are mixed with sperm in a dish, which is then incubated.

Fertilization is the final step in a two-part process that ensures genetic diversity, the key to evolutionary development of life. The first step of this process, meiosis, takes place in male and female gonads. Meiosis causes immature forms of each gamete to lose one of two complete sets of chromosomes. Mature gametes are haploid. In humans, the natural place for fertilization is in a Fallopian tube of a woman, the channel through which an ovum travels to the uterus. A normal adult woman ovulates each month, releasing a single haploid ovum from one of two female gonads (the ovaries). Ovulation is under direct control of hormones called gonadotropins.

Sperm from the male gonads (the testicles) are normally placed in a woman's vagina. In contrast to the one ovum produced each month, normal men release hundreds of millions of sperm into the vagina during each sexual intercourse. From the vagina, these sperm travel through the uterus and into each Fallopian tube in order to meet an ovum. During this trip, the sperm undergo changes called capacitation. To fuse with the ovum, a sperm must penetrate several surrounding barriers. After fusion of sperm and egg, the nuclear membranes of the two cells break down so that the paternal and maternal chromosomes pair up. The egg divides into two new diploid cells, the first cells of a genetically unique new being.

In Vitro Fertilization and Embryo Transfer

The fertilization process that occurs in nature can also take place artificially in laboratory culture dishes. Gametes are collected, brought together, and fertilized in a laboratory. The embryo is then transferred to a uterus for development and birth. This procedure can be done for many species, including humans. The first human conceived by in vitro fertilization (IVF), Louise Brown, was born on July 25, 1978, in England.

IVF is usually performed in human medicine to overcome infertility caused by problems such as blocked Fallopian tubes or low sperm count. IVF is also done in veterinary medicine and for scientific research. IVF could be performed to make genetic diagnoses or even to alter the set of genetic information in an embryo. IVF is usually performed with mature, haploid sperm and ova. Mature sperm for IVF are easily obtained by masturbation. Mature ova are more difficult to obtain. The female is given gonadotropin hormones to stimulate her to superovulate (that is, to produce ten or more mature eggs rather than just one). Ova are later collected by inserting a small suction needle into her pelvic cavity. The ova are inseminated with laboratory-capacitated sperm. Two to four embryos are transferred into the uterus through a catheter. Excess embryos can be saved by a freezing procedure called "cryopreservation." These may be thawed to be transferred during another cycle should the first attempt at IVF fail or should a second pregnancy be desired.

Impact and Applications

Technology such as the polymerase chain reaction permits assessment of genetic information in the nucleus of a single cell, whether diploid or haploid. IVF gives physicians access to sperm, ova, and very early embryos. One or two cells (called blastomeres) can be removed from an eight-cell embryo without damaging the ability of the remaining cells to develop normally following embryo transfer. Thus IVF permits genetic diagnosis at the earliest stages of human development and even allows the possibility of influencing the genetics of an embryo.

Preimplantation genetic diagnosis (PGD) is used clinically to help people with significant genetic risks to avoid giving birth to an abnormal child that might die in infancy or early childhood. PGD detects genetic problems by laboratory studies on blastomeres. If the blastomere is free of the genetic problem in question, its embryo is transferred to the uterus for implantation and pregnancy. If the blastomere

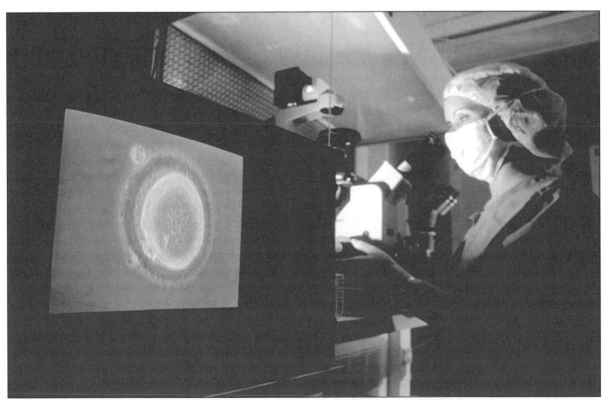

A technician examines a ripe egg during an in vitro fertilization process. (Hank Morgan/Rainbow)

has the genetic problem, its embryo is destroyed prior to transfer and pregnancy never takes place. PGD is successful in avoiding pregnancies with embryos that will develop cystic fibrosis, Huntington's chorea, Lesch-Nyhan disease, Tay-Sachs disease, and other genetic abnormalities. Prior to the development of PGD, avoidance of severe genetic diseases was possible only by prenatal diagnosis after pregnancy was established. When prenatal diagnosis detects the presence of an undesired genetic trait, termination of the pregnancy through elective abortion must be considered. Abortion is an unacceptable choice for many people because of ethical and moral concerns.

Access to gametes prior to fertilization and to embryos prior to implantation also opens the possibility of genetic engineering. IVF, combined with chemical genetic knowledge derived from the Human Genome Project, takes genetic engineering to a sophisticated level. Regardless of how it is done, alterations in the genetics of humans presents major ethical issues. Alterations in the genetics of other species, especially those of commercial interest, is usually of lesser concern and sometimes may be an attractive opportunity.

Access to early embryos opens the possibility of genetic cloning. Cloning is the process of creating multiple individuals with identical genetic characteristics. One way to do this resembles the way nature creates identical twins: by separating an early embryo into blastomeres, allowing each to develop into a separate organism. Using embryo transfer, different females could be impregnated with these identical offspring. Using cryopreservation, these pregnancies could occur years apart. It is even possible to remove the diploid nucleus from an isolated blastomere and replace it with a diploid nucleus taken from an adult. The blastomere with the transplanted nucleus is able, using special procedures in the laboratory, to develop into an embryo that can be transferred to a uterus. The offspring will be genetically identical to the adult source of the transplanted nucleus. Most people recognize cloning technology as inappropriate in human medicine, but it has acceptable applications in veterinary medicine.

—*Armand M. Karow*

See Also: Cloning; Genetic Engineering; Human Genome Project; Meiosis; Prenatal Diagnosis.

Further Reading: Three books have different perspectives on the ethical issues of IVF as influenced by time and professional focus. Clifford Grobstein, a world-renowned embryologist, presents a view of IVF before the advent of PGD in *From Chance to Purpose: An Appraisal of External Human Fertilization* (1981). Andrea L. Bonnicksen, a social scientist, comments on *In Vitro Fertilization: Building Policy from Laboratories to Legislature* (1989). Machelle M. Seibel, a physician, and Susan L. Crockin, a lawyer, edit relevant essays concerning *Family Building Through Egg and Sperm Donation* (1996). Each book gives a description of clinical and laboratory methods in varying degrees of detail.

Inborn Errors of Metabolism

Field of study: Human genetics
Significance: *Inborn errors of metabolism are hereditary genetic defects found in varying frequencies in human populations. Diagnosis and cure of these genetic diseases is a continuing focus of medical research.*

Key terms

ENZYMES: proteins that act as biological catalysts

METABOLISM: the collection of biochemical reactions occurring in an organism

METABOLIC PATHWAY: enzyme-mediated reactions that are connected in a series

PROTEINS: large molecules that serve important functions in organisms; examples include enzymes, transporters, structural elements, and hormones

Early Observations

In 1902, Sir Archibald Garrod, a British physician, presented a classic paper in which he summarized his observations and analyses of a rather benign condition known as alkaptonuria. The condition is easily diagnosed because the major symptom is dark urine caused by the excretion of homogentisic acid. Since homogentisic acid is not normally found in urine and is a by-product of certain amino acids with particular ring structures, Garrod reasoned that individuals with alkaptonuria had a defect in the utilization of these amino acids. Garrod also noted that the condition is often found in two or more siblings and postulated that the occurrence of this condition may be explained by the mechanism of inheritance.

In 1908, in "Inborn Errors of Metabolism," Garrod extended his observations on alkaptonuria to other diseases such as albinism and cystinuria. In each case, he argued that the abnormal or disease condition was caused by a defect in metabolism that resulted in a block of an important metabolic pathway. He speculated that when such a pathway is blocked, there would be an accumulation of products that are not seen in normal individuals, or important substances would be missing or abnormal. Some of these abnormal metabolic events might be harmless, such as in alkaptonuria, but others could lead to serious disease. He traced the inheritance of these conditions and discovered that they could be passed on from one generation to the next. He was the first to use the term "inborn errors of metabolism" to describe these conditions. Other investigators have studied over three thousand additional diseases that can be included in this category. A few of these conditions occur at relatively high frequency in humans. In the U.S. Caucasian population, cystic fibrosis occurs in about 1 in 2,000 births. Some conditions, such as phenylketonuria, are seen at moderate frequency, about 1 in 10,000. Many of the inborn errors are rare, with frequencies less than 1 in 100,000. A generally accepted definition of an inborn error of metabolism is any condition with actual or potential health consequences that can be inherited in the fashion described by Gregor Mendel in the nineteenth century.

Malfunctioning Proteins and Enzymes

The biochemical causes of the inborn errors of metabolism were discovered many years after Garrod presented his ideas. In 1952, von Gierke's disease was found to be caused by the defective enzyme glucose-6 phosphatase. After this discovery, many inborn errors of metabolism were traced to defects in other enzymes. Enzymes are proteins that catalyze biochemical reactions. They are responsible for increasing the rates of reactions that occur in all cells. These reactions are important steps in metabolic pathways that are responsible for processes such as utilization of nutrients, generation of energy, cell division, and biosynthesis of substances that are needed by organisms. There are many metabolic pathways that can be affected if one of the enzymes in the pathway is missing or malfunctions. In addition to enzymes, defective proteins with other functions may also be considered as candidates for inborn errors of metabolism. For example, there are many types of defective hemoglobin, the protein responsible for oxygen transport. These defective hemoglobins are the causes of diseases such as sickle-cell anemia and thalassemia.

Genetic Basis of Inborn Errors

The cause of these defects in enzymes and proteins has been traced to mutations in the genes that code for them. Alterations in the structure or nucleotide composition of deoxyribonucleic acid (DNA) can have various consequences for the structure of the protein coded for by the DNA. Some of the genetic alterations affecting metabolism simply represent normal variation within the population and are asymptomatic. An example of such a genetic alteration is the ability of some individuals to experience a bitter taste after exposure to chemical derivatives of thiourea. Some asymptomatic variations may lead to complications after environmental conditions are changed. There are a few "inborn errors" that can be induced by certain drugs. Another class of alterations may be minor, with the resulting protein having some degree of function. Individuals with such alterations may live long lives but will occasionally experience a range of problems associated with their conditions. Depending on the exact nature of the mutation, some of the alterations in the resulting protein structure can lead to a completely nonfunctional protein or enzyme. Consequences of this type of mutation can be quite severe and may result in death.

Many of the inborn errors of metabolism are inherited as autosomal recessive traits. Individuals are born with two copies of the gene. If one copy is defective and the second copy is normal, enough functioning protein or enzyme can be made to prevent the individual from exhibiting any symptoms of the disease. Such individuals will be classified as carriers for the defect since they can pass on the defective gene to their offspring. About 1 in 20 Caucasians in the U.S. is a carrier for the cystic fibrosis gene, and about 1 in 30 individuals of Eastern Jewish descent carries the gene for the lethal Tay-Sachs disease. When an individual inherits two defective copies of the gene, the manifestations of the disease can be much more severe.

Some inborn errors of metabolism such as Huntington's chorea are manifested as dominant genetic traits. Only one copy of the defective gene is necessary for manifestations of the abnormal condition. There are some inborn errors of metabolism that are sex linked. Diseases that involve mutations carried on the X chromosome may be severe in males because they have only one X chromosome but less severe or nonexistent in females because females carry two X chromosomes.

Diagnosis and Treatment

Significant progress has been made in the diagnosis of inborn errors of metabolism. Prior to 1980, much of the diagnosis for metabolic defects relied on symptoms detected during clinical examination. Biochemical tests are used to detect various substances that accumulate or are missing when an enzymatic defect is present. The commonly used screening for phenylketonuria (PKU) relies on detection of phenylketones in the blood of newborns. For cases in which the genetic defect is known, DNA can often be used for the purpose of genetic testing. Genetic counselors will help parents determine their chances of having a child with a severe defect when parents are identified as carriers. Small samples of cells can be used as a source of DNA, and such cells may even be obtained from amniotic fluid by amniocentesis. This allows diagnosis to be made prenatally. Some parents choose abortion when their fetus is diagnosed with a lethal or debilitating defect.

Although strides have been made in diagnosis, the problem of treatment still remains. For some inborn errors of metabolism such as phenylketonuria, dietary modification will often prevent the serious symptoms of the disease condition. Individuals with phenylketonuria must limit their intake of the amino acid phenylalanine during the critical stages of brain development, generally the first eight years of life. Treatment of other inborn errors may involve avoidance of certain environmental conditions. For example, individuals suffering from albinism, a lack of pigment production, must avoid the sun. For other inborn errors of metabolism, there are no simple cures on the horizon. Since the early 1990's, some medical pioneers have been involved in clinical trials of gene therapy, an attempt to replace a defective gene by insertion of a normal, functioning version. Although theoretically promising, gene therapy has not met with significant success. In addition, there are many ethical issues raised when gene therapy trials are proposed before potential hazards have been completely eliminated. Nevertheless, scientists are looking more and more toward genetic cures to genetic problems such as those manifested as inborn errors of metabolism.

—*Barbara Brennessel*

See Also: Albinism; Cystic Fibrosis; Gene Therapy; Hereditary Diseases; Prenatal Diagnosis.

Further Reading: A comprehensive overview of the development of the concept of inborn errors of metabolism and many examples of such conditions are provided in *The Biochemical Genetics of Man* (1972), edited by D. J. H. Brock and Oliver Mayo. George Feuer and Felix A. de la Inglesia present detailed explanations of the metabolic pathways and symptoms of resulting abnormalities in a collection of inborn errors of metabolism in *Molecular Biochemistry of Human Disease*, vol. 1, (1985). The diagnosis of inborn errors of metabolism, development of molecular methods for diagnosis of these genetic defects, and prospects for treatment of these conditions by gene therapy are highlighted within the context of the Human Genome Project in Thomas F. Lee's *The Human Genome Project: Cracking the Genetic Code of Life* (1991).

Inbreeding and Assortative Mating

Field of study: Population genetics

Significance: *Most population genetic models assume that individuals mate at random. The most common violation of this assumption is inbreeding, in which individuals are more likely to mate with relatives. Inbreeding depression, a reduction in the health and vigor of inbred offspring, is a common and widespread phenomenon. Many traits of organisms, including pollination systems in plants and dispersal in animals, can be understood as mechanisms that reduce the frequency of inbreeding and the cost of inbreeding depression. Inbreeding depression is a potential concern for endangered species and zoo populations, where there are few individuals and inbreeding is hard to avoid.*

Key terms

RANDOM MATING: a mating system in which each male gamete (sperm) is equally likely to combine with any female gamete (egg)

INBREEDING: mating between genetically related individuals

ALLELE: any of a number of possible genetic variants of a particular gene

HOMOZYGOTE: a diploid genotype that consists of two identical alleles

HETEROZYGOTE: a diploid genotype that consists of two different alleles

Random Mating and the Hardy-Weinberg Law

Soon after the rediscovery of Gregor Mendel's rules of inheritance in 1900, British mathematician Godfrey Hardy and German physician Wilhelm Weinberg published a simple mathematical treatment of the effect of sexual reproduction on the distribution of genetic variation. Both men published their ideas in 1908 and showed that there was a simple relationship between allele frequencies and genotypic frequencies in populations. An allele

is simply a genetic variant of a particular gene; for example, blood type in humans is controlled by a single gene with three alleles (*A*, *B*, and *O*). Every individual inherits one allele for each gene from both their mother and father and has a two-allele genotype. In the simplest case with only two alleles (for example, *A* and *a*), there are three different genotypes (*AA*, *Aa*, *aa*). The Hardy-Weinberg predictions specify the frequencies of genotypes (combinations of two alleles) in the population: how many will have two copies of the same allele (homozygotes such as *AA* and *aa*) or copies of two different alleles (heterozygotes such as *Aa*).

One important assumption that underlies the Hardy-Weinberg predictions is that gametes (sperm and egg cells) unite at random to form individuals or that individuals pair randomly to produce offspring. An example of the first case is marine organisms such as oysters that release sperm and eggs into the water; zygotes (fertilized eggs) are formed when a single sperm finds a single egg. Exactly which sperm cell and which egg cell combine is expected to be unrelated to the specific allele each gamete is carrying, so the union is said to be random. In cases in which males and females form pairs and produce offspring, it is assumed that individuals find mates without reference to the particular gene under examination. In humans, people do not choose potential mates at random, but they do mate at random with respect to most genetic variation. For instance, since few people know (or care) about the blood type of potential partners, people mate at random with respect to blood-type alleles.

Inbreeding and assortative mating are violations of this basic Hardy-Weinberg assumption. For inbreeding, individuals are more likely to mate with relatives than with a randomly drawn individual (for outbreeding, the reverse is true). Assortative mating occurs when individuals make specific mate choices based on the genotype or appearance of others. Each has somewhat different genetic consequences. When either occurs, the Hardy-Weinberg predictions are not met, and the relative proportions of homozygotes and heterozygotes are different from what is expected.

The Genetic Effects of Inbreeding

When relatives mate to produce offspring, the offspring may inherit an identical allele from each parent. This is because related parents share many of the same alleles, inherited from their common ancestors. The closer the genetic relationship, the more alleles two individuals will share. Inbreeding increases the number of homozygotes for a particular gene in a population because the offspring are more likely to inherit identical alleles from both parents. Inbreeding also increases the number of different genes in an individual that are homozygous. In either case, the degree of inbreeding can be measured by the level of homozygosity (the percentage or proportion of homozygotes relative to all individuals).

Inbreeding is exploited by researchers who want genetically uniform (completely homozygous) individuals for experiments: Fruit flies or mice can be made completely homozygous by repeated brother-sister matings. The increase in the frequency of homozygotes can be calculated for different degrees of inbreeding. Self-fertilization is the most extreme case of inbreeding, followed by sibling mating, and so forth. Sewall Wright pioneered computational methods to estimate the degree of inbreeding in many different circumstances. For self-fertilization, the degree of homozygosity increases by 50 percent each generation. For repeated generation of brother-sister matings, the homozygosity increases by about 20 percent each generation.

Inbreeding Depression and Assortative Mating

Inbreeding commonly produces inbreeding depression. This is characterized by poor health, lower growth rates, reduced fertility, and increased incidence of genetic diseases. Although there are several theoretical reasons why inbreeding might occur, the major effects are produced by uncommon and deleterious recessive alleles. These alleles produce negative consequences for the individual when homozygous, but when they occur in a heterozygote, their negative effects are masked by the presence of the other allele. Because inbreeding increases the relative proportion of ho-

mozygotes in the population, many of these alleles are expressed, yielding reduced health and vigor. In some cases, the effects can be quite severe. For example, when researchers wish to create homozygous lines of the fruit fly *Drosophila melanogaster* by repeated brother-sister matings, 90 percent or more of the lines fail because of widespread genetic problems.

In assortative mating, the probability of particular pairings is affected by the genotype or appearance of the individuals. In positive assortative matings, individuals are more likely to mate with others of the same genotype or appearance, while in negative assortative mating, individuals are more likely to mate with others that are dissimilar. In both cases, the primary effect is to alter the expected genotypic frequencies in the population from those expected under the Hardy-Weinberg law. Positive assortative mating has much the same effect as inbreeding and increases the relative frequency of homozygotes. Negative assortative mating, as expected, has the opposite effect and increases the relative proportion of heterozygotes. Positive assortative mating has been demonstrated for a variety of traits in humans, including height and hair color.

Impact and Applications

The widespread, detrimental consequences of inbreeding are believed to shape many aspects of the natural history of organisms. Many plant species have mechanisms developed through natural selections to increase outbreeding and avoid inbreeding. The pollen (male gamete) may be released before the ovules (female gametes) are receptive, or there may be a genetically determined self-incompatibility to prevent self-fertilization. In most animals, self-fertilization is not possible, and there are often behavioral traits that further reduce the probability of inbreeding. In birds, males often breed near where they were born, while females disperse to new areas. In mammals, the reverse is generally true, and males disperse more widely. Humans appear to be an exception among the mammals, with a majority of cultures showing greater movement by females. These sex-biased dispersal patterns are best understood as mechanisms to prevent inbreeding.

Animals in zoos often suffer from inbreeding problems caused by limited mating choices. (Robert McClenaghan)

In humans, individuals are unlikely to marry others with whom they were raised. This prevents the potentially detrimental consequences of inbreeding in matings with close relatives. This has also been demonstrated in some birds. Domestic animals and plants may become inbred if careful breeding programs are not followed. Many breeds of dogs exhibit a variety of genetic-based problems (for example, hip problems, skull and jaw deformities, and nervous temperament) that are likely caused by inbreeding. Conservation biologists who manage endangered or threatened populations must often consider inbreeding depression. In very small populations such as species maintained in captivity (zoos) or in isolated natural populations, inbreeding may be hard to avoid. Inbreeding has been blamed for a variety of health defects in cheetahs and Florida panthers.

—*Paul R. Cabe*

See Also: Hardy-Weinberg Law; Monohybrid Inheritance; Population Genetics.

Further Reading: Most population genetics texts provide detailed mathematical analyses of the effects of inbreeding. *A Primer of Population Genetics* (1988), by Daniel Hartl, and *Genetics of Populations* (1985), by Philip Hedrick, are good examples. For discussions of inbreeding avoidance and kin recognition, see *An Introduction to Behavioral Ecology* (1993), by J. Krebs and N. Davies. Good discussions of inbreeding in birds and mammals, the effects of inbreeding depression in plants and animals, and issues related to the conservation of natural heritage can be found in *Conservation Biology: The Science of Scarcity and Diversity* (1986), edited by Michael Soulé. More information on inbreeding in cheetahs and panthers can be found in *Conservation Genetics: Case Histories from Nature* (1996), edited by John Avise and James Hamrick.

Incomplete Dominance

Field of study: Classical transmission genetics

Significance: *In most allele pairs, one allele is dominant and the other recessive; however, other relationships can occur. In incomplete dominance, one allele can only partly dominate or mask the other. Some very important human genes, such as the genes for pigmentation and height, show incomplete dominance of alleles.*

Key terms

ALLELE: one of the alternative forms of a gene

ENZYME: a molecule that can increase the rate of a chemical reaction in a cell

HETEROZYGOUS: having two different alleles of a gene, often symbolized Aa or a^+a

HOMOZYGOUS: having two of the same alleles of a gene, often symbolized AA, aa, or a^+a^+

PHENOTYPE: the outward appearance of an organism

Incomplete Dominance and Complete Dominance

Diploid organisms have two copies of each gene and thus two alleles of each gene. They can have either a homozygous genotype (two of the same alleles of a gene, such as AA, aa, or a^+a^+) or a heterozygous genotype (two different alleles of a gene, such as Aa or a^+a). The phenotype (appearance) of an organism that is homozygous for a particular gene is usually easy to determine. If a pea plant has two tall alleles of the height gene, the plant is tall; if a plant has two dwarf alleles of the height gene, it is small. The phenotype may be less obvious when an organism is heterozygous. If a plant has one tall allele and one dwarf allele, what will the plant look like? In most circumstances, one of the alleles (the dominant) is able to mask or cover the other (the recessive). The phenotype is determined by the dominant allele, so this heterozygous pea will be tall. When Gregor Mendel reported his results in 1866, he reported one dominant and one recessive allele for each gene he had studied. Later researchers, starting with Carl Correns in the early 1900's, discovered alleles that did not follow this pattern.

When a red snapdragon or four-o'clock plant is crossed with a white snapdragon or four-o'clock, the offspring are neither red nor white. Instead, the progeny of this cross are pink. Similarly, when a chinchilla (gray) rabbit is crossed with an albino rabbit, the progeny are neither chinchilla nor albino but an intermediate shade called light chinchilla. This phenomenon is known as incomplete dominance, partial dominance, or semidominance.

If one compares the flower color gene of peas with the flower color gene of snapdragons, the differences and similarities can be seen. The two alleles in peas can be designated W for the purple allele and w for the white allele. Peas that are WW are purple, and peas that are ww are white. Heterozygous peas are Ww and appear purple. In other words, as long as one dominant allele is present, enough purple pigment is made to make the plant's flower color phenotype purple. In snapdragons, R is the red allele and r is the white allele. Homozygous RR plants have red flowers and rr plants have white flowers. The heterozygous Rr plants have the same kind of red pigment as the RR plants but not enough to make the color red. Instead, the

Carl Correns was the first geneticist to investigate the phenomenon of incomplete dominance. (National Library of Medicine)

less-pigmented red flower is designated as pink. Because neither allele shows complete dominance, other symbols are sometimes used. The red allele might be called c^R or C_1, while the white allele might be called c^W or C_2.

The Enzymatic Mechanism of Incomplete Dominance

To understand why this happens, one must understand the way enzymes work. Enzymes are proteins found in cells that are able to increase the rate of chemical reactions in cells without the enzymes themselves being altered. Thus an enzyme can be used over and over again to speed up a particular reaction. Each different chemical reaction in a cell needs its own enzyme. Each enzyme is coded by a gene. Looking again at flower color in peas, the W allele codes for the enzyme that allows the production of purple pigment. Whenever a W allele is present, the purple-pigment-producing enzyme is also present. The w allele has been changed (mutated) in some way so that it no longer codes for a functional enzyme. Thus ww plants have no functional enzyme and cannot produce any purple pigment. Since many biochemicals such as fibrous polysaccharides and proteins found in plants are opaque white, the color of a ww flower is white by default. In a Ww plant, there is only one copy of the allele for functional enzyme. Since enzymes can be used over and over again, one copy of the "good" allele produces sufficient purple enzyme to make enough pigment for the flower to appear purple. In snapdragons, however, that is not quite the case. The R allele, like the W allele, codes for functional enzyme, while the r allele does not. The difference is in the enzyme coded by the R allele. This red-pigment-producing enzyme is not as efficient in assisting the chemical reaction that produces pigment as the purple pigment enzyme was. Thus more of the red enzyme is needed to make the normal amount of pigment. The RR homozygote is red, and the rr homozygote is white, but the Rr heterozygote produces less pigment than the RR because the enzyme coded is not very efficient. When this enzyme is present at lower amounts, it cannot make the normal amount of red pigment. The flower with the lowered amount of red pigment appears pink. The amount of enzyme in this case determines the rate of the reaction. More enzyme produces more product (red pigment), and less enzyme produces less product. The enzyme is often said to be "rate limiting."

Phenotypic Ratios

The appearance is not the only thing different in incomplete dominance. The ratios (and progeny frequencies) also differ slightly. Normally, crossing two heterozygous organisms will produce the following results: $Ww \times Ww \rightarrow \frac{1}{4}WW + \frac{1}{2}Ww + \frac{1}{4}ww$. Since both WW and Ww look the same, the $\frac{1}{4}WW$ and the $\frac{1}{2}Ww$ can be added together to give $\frac{3}{4}$ purple. In other words, when two heterozygotes are crossed, the most common result is to have $\frac{3}{4}$ of the progeny look like the dominant and $\frac{1}{4}$ look like the recessive—the standard 3:1 ratio. With incomplete dominance, each genotype has its own phenotype, so when two heterozygotes are crossed (for example, $Rr \times Rr$), $\frac{1}{4}$ of the progeny will be RR and look like the dominant (in this case red), $\frac{1}{4}$ will be rr and look like the recessive (in this case white), but $\frac{1}{2}$ will be Rr and have an intermediate appearance (in this case pink)—a 1:2:1 ratio. Note that in incomplete dominance, each genotype has its own unique phenotype. There are other differences in ratios. With normal dominance, $WW \times Ww \rightarrow \frac{1}{2}WW + \frac{1}{2}Ww$ as far as genotype, but the phenotype of all progeny is the same (purple). For incompletely dominant genes, $RR \times Rr \rightarrow \frac{1}{2}RR + \frac{1}{2}Rr$ as far as genotype, while $\frac{1}{2}$ of the progeny show the red phenotype and $\frac{1}{2}$ show the pink phenotype. In the cross $WW \times ww$, all progeny show the purple phenotype like the dominant parent. However, in the cross $RR \times rr$, all progeny show the pink phenotype, which is intermediate between the dominant phenotype and the recessive phenotype. The appearance of the heterozygote as intermediate between the dominant and recessive homozygotes is a hallmark of incomplete dominance.

Codominance

One type of inheritance that can be confused with incomplete dominance is codominance. In codominance, both alleles involved are acting like dominant alleles. Good examples are the *A* and *B* alleles of the human ABO blood system. ABO refers to chemicals, in this case short chains of sugars called antigens, that can be found on the surface of cells. Blood classified as A has *A* antigens on the surface, B blood has *B* antigens, and AB blood has both *A* and *B* antigens. (O blood has neither *A* nor *B* antigens on the surface. Since it has zero *A* or *B* antigens, it is called O for zero.)

Genetically, individuals that are homozygous for the A allele, $I^A I^A$, have *A* antigens on their cells and are classified as type A. Those homozygous for the *B* allele, $I^B I^B$, have *B* antigens and are classified as type B. Heterozygotes for these alleles, $I^A I^B$, have both *A* and *B* antigens and are classified as type AB. This is called codominance because both alleles are able to produce enzymes that function. The *A* allele codes for the enzyme that creates the *A* antigen, while the *B* allele codes for the enzyme that creates the *B* antigen. When both enzymes are present, as in the heterozygous $I^A I^B$ individual, both antigens will be formed. At the enzyme level, codominance (in which enzymes coded by both alleles are functional, while phenotype is a result of both enzymes) is quite different from incomplete dominance (in which one allele codes for the functional enzyme and the other allele does not code for the functional enzyme, while phenotype is based on the amount of active enzyme present). The progeny ratios (and frequencies), however, are the same for both of these since in codominance, like in incomplete dominance, each genotype has its own phenotype, and the phenotype of the heterozygote is intermediate between the two homozygotes.

Whether an allele is called completely dominant, incompletely dominant, or codominant often depends on how the observer looks at the phenotype. Consider two alleles of the hemoglobin gene: H^A (which codes for normal hemoglobin) and H^S (which codes for sickle-cell hemoglobin). To the casual observer, both $H^A H^A$ homozygotes and $H^A H^S$ heterozygotes have normal-appearing blood. Only the $H^S H^S$ homozygote shows the sickling of blood cells that is characteristic of the disease. Thus H^A is dominant to H^S. Another observer, however,

may note that under conditions of oxygen deprivation, the blood of heterozygotes does sickle. This looks like incomplete dominance. The phenotype is intermediate between never sickling, as seen in the normal homozygote, and frequently sickling, as seen in the $H^S H^S$ homozygote. A third way of observing, however, would be to look at the hemoglobin itself. In normal homozygotes, all hemoglobin is normal. In $H^S H^S$ homozygotes, all hemoglobin is abnormal. In the heterozygote, both normal and abnormal hemoglobin is present, thus the alleles seem to be codominant.

Incomplete Dominance and Polygenes

In humans and many other organisms, single characteristics are often under the genetic control of several genes. Many times these genes function in an additive manner so that a characteristic such as height is not determined by a single height gene with just two possible alternatives as in tall and dwarf peas. There can be any number of these genes that determine the expression of a single characteristic, and very often the alleles of these genes show incomplete dominance.

Suppose one gene with an incompletely dominant allele determined height. Three genotypes of height could exist: *HH*, which codes for the maximum height possible (100 percent above the minimum height), *Hh*, which codes for 50 percent above the minimum height, and *hh*, which codes for the minimum height. If two height genes existed, there would be five possible heights: *AABB* (maximum height); *AaBB* or *AABb* (75 percent above minimum); *AAbb*, *AaBb*, or *aaBB* (50 percent above minimum); *Aabb* or *aaBb* (25 percent above minimum); and *aabb* (minimum). If there were five genes involved in height, there would be *aabbccddee* individuals with minimum height; *Aabbccddee*, *aaBbccddee*, and other individuals having genotypes with only one of the incompletely dominant alleles at 10 percent above the minimum; *AAbbccddee*, *aaBbccDdee*, and other individuals with two incompletely dominant alleles at 20 percent above the minimum; all the way up to *AABBCCDDEE* individuals that show the maximum (100 percent above the minimum) height. The greater the number of genes with incompletely dominant alleles that affect a phenotype, the more the distribution of phenotypes begins to look like a continuous distribution. Human skin, hair, and eye pigmentation phenotypes are also determined by the additive effects of several genes with incompletely dominant alleles.

Incomplete Dominance and Sex Linkage

In many organisms, sex is determined by the presence of a particular combination of sex chromosomes. Human females, for example, have two of the same kind of sex chromosomes, called X chromosomes, so that all normal human females have the XX genotype. Human males have two different sex chromosomes, thus all normal human males have the XY genotype. The same situation is also seen in the fruit fly *Drosophila melanogaster*. When genes with incompletely dominant alleles are located on the X chromosome, only the female with her two X chromosomes can show incomplete dominance. The apricot (w^a) and white (w) alleles of the eye color gene in *D. melanogaster* are on the X chromosome, and w^a is incompletely dominant to w. Male flies can have either of two genotypes, w^aY or wY, and appear apricot or white, respectively. Females have three possible genotypes: w^aw^a, w^aw, and ww. The first is apricot and the third is white, but the second genotype, w^aw, is an intermediate shade often called light apricot.

In birds and other organisms in which the male has two of the same kind of sex chromosomes and the female has the two different sex chromosomes, only the male can show incomplete dominance. A type of codominance can also be seen in genes that are sex linked. In domestic cats, an orange gene exists on the X chromosome. The alleles are orange (X^O) and not orange (X^+). Male cats can be either black (or any color other than orange, depending on other genes that influence coat color) when they are X^+Y, or they can be orange (or light orange) when they are X^OY. Females show those same colors when they are homozygous (X^+X^+ or X^OX^O) but show a tortoiseshell (or calico) pattern of both orange and not-orange hairs when they are X^+X^O.

—*Richard W. Cheney, Jr.*

See Also: Classical Transmission Genetics; Complete Dominance; Multiple Alleles; One Gene-One Enzyme Hypothesis; Quantitative Inheritance.

Further Reading: V. Grant, *Genetics of Flowering Plants* (1975), thoroughly reviews heredity in plants and covers incomplete dominance. D. J. Nolte, "The Eye-Pigmentary System of *Drosophila*," *Heredity* 13 (1959), covers *Drosophila* eye pigments quite well. For mammalian coat colors, see A. G. Searle, *Comparative Genetics of Coat Color in Mammals* (1968). A. Yoshida, "Biochemical Genetics of the Human Blood Group ABO System," *American Journal of Genetics* 34 (1982), covers the genetics of the ABO system.

Infertility

Field of study: Human genetics

Significance: *Infertility is a disease of the reproductive system that impairs the conception of children. About one in six couples in the United States is infertile. The risk that a couple's infertility may be caused by genetic problems such as abnormal sex chromosomes is approximately one in ten.*

Key Terms

SEX CHROMOSOMES: the chromosomes that control the sexual attributes of males and females; females have two X chromosomes, while males have one X and one Y chromosome

SEX CHROMOSOME ABNORMALITIES: genetic defects that can cause infertility or abnormal development; approximately 1 in 350 children is born with sex chromosome variations

SYNDROME: a set of features or symptoms often occurring together and believed to stem from the same cause

A Reproductive Disease

Infertility is a disease of the reproductive system that impairs a couple's ability to have children. Sometimes infertility has a genetic cause. The conception of children is a complex process that depends upon many factors, including the production of healthy sperm by the man and healthy eggs by the woman, unblocked Fallopian tubes that allow the sperm to reach the egg, the sperm's ability to fertilize the egg when they meet, the ability of the fertilized egg (embryo) to become implanted in the woman's uterus, and sufficient embryo quality. If the pregnancy is to continue to full term, the embryo must be healthy, and the woman's hormonal environment must be adequate for its development. Infertility can result when one of these factors is impaired. Physicians define infertility as the inability to conceive a child after one year of trying.

Genetic Causes of Infertility

The most common male infertility factors include conditions in which few or no sperm cells are produced. Sometimes sperm cells are malformed or die before they can reach the egg. A genetic disease such as a sex chromo-some abnormality can also cause infertility in men. A genetic disorder may be caused by an incorrect number of chromosomes (having more or fewer than the normal forty-six chromosomes). Having a wrong arrangement of the chromosomes may also cause infertility. This situation occurs when part of the genetic material is lost or damaged. One such genetic disease is Klinefelter's syndrome, which is caused by an extra X chromosome in males. The loss of a tiny piece of the male sex chromosome (the Y chromosome) may cause the most severe form of male infertility: the complete inability to produce sperm. This form of infertility can arise from a deletion in one or more genes in the Y chromosome. Fertility problems can pass from father to son, especially in cases in which physicians use a single sperm from an infertile man to inseminate a woman's egg.

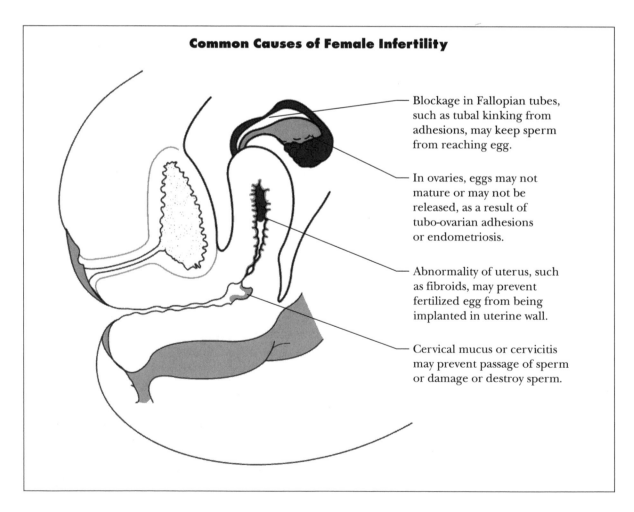

Common Causes of Female Infertility

Blockage in Fallopian tubes, such as tubal kinking from adhesions, may keep sperm from reaching egg.

In ovaries, eggs may not mature or may not be released, as a result of tubo-ovarian adhesions or endometriosis.

Abnormality of uterus, such as fibroids, may prevent fertilized egg from being implanted in uterine wall.

Cervical mucus or cervicitis may prevent passage of sperm or damage or destroy sperm.

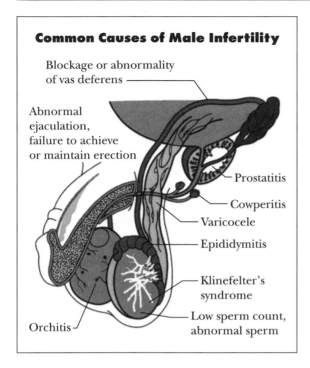

Common Causes of Male Infertility

Blockage or abnormality of vas deferens

Abnormal ejaculation, failure to achieve or maintain erection

Prostatitis

Cowperitis

Varicocele

Epididymitis

Klinefelter's syndrome

Orchitis

Low sperm count, abnormal sperm

Female infertility may be caused by an irregular menstrual cycle, blocked Fallopian tubes, or birth defects in the reproductive system. One genetic cause of infertility in females is Turner's syndrome. Most females with Turner's syndrome lack all or part of one of their X chromosomes. The disorder may result from an error that occurs during division of the parent's sex cells. Infertility and short stature are associated with Turner's syndrome. Other genetic disorders in females include trisomy X, tetrasomy X, and pentasomy. These syndromes are the female counterparts of Klinefelter's syndrome and are often associated with mental retardation.

At least 60 percent of miscarriages or pregnancy losses are caused by chromosomal abnormalities. Most babies with these abnormalities would not survive even if they were born. Chromosomal problems are more common if the mother is older and has a history of requiring longer than a year to conceive. Men who are older or who have a history of being subfertile can also contribute to genetic abnormalities. After the age of thirty-five, the structure within a woman's eggs is more likely to become damaged. Men over the age of forty-five have

an increased risk of damage to the structure of the chromosomes in their sperm.

Scientists believe that as their understanding of the genetic basis of infertility problems increases, new therapies will be developed to treat them. Most infertility cases are treated with drugs or surgery to repair the reproductive organs. No treatment is available to correct sex chromosomal abnormalities such as Turner's syndrome. However, some women with Turner's syndrome can have children. For women who cannot conceive, possible procedures include in vitro fertilization (fertilizing a woman's egg with sperm outside the body) and embryo transfer (moving the fertilized egg into a woman's uterus). Adoption is another option for infertile men and women.

—*Fred Buchstein*

See Also: Genetic Testing; Klinefelter's Syndrome; Turner's Syndrome.

Further Reading: *Does It Run in the Family? A Consumer's Guide to DNA Testing for Genetic Disorders* (1997), by Doris Teichler Zallen, provides an excellent introduction to genetics and genetic testing. For informative overviews of infertility, see *Dr. Richard Marrs' Fertility Book* (1997), by Richard Marrs et al., and *The Fertility Sourcebook: Everything You Need to Know* (1995), by M. Sara Rosenthal. To learn about medical genetics, see *Human Genetics: Concepts and Applications* (1994), by Ricki Lewis.

Insurance

Field of study: Human genetics

Significance: *Many social policy analysts and public health advocates worry that as genetic screening to identify individuals at risk for developing inherited diseases becomes more widely available, high-risk people may be denied health and life insurance coverage.*

Key terms

ALZHEIMER'S DISEASE: a degenerative brain disorder usually found among the elderly; sufferers gradually lose cognitive function and become unable to function independently

CHRONIC ILLNESS: an ongoing condition such as diabetes or hypertension

HIGH RISK: characterized by being likely to someday suffer from a particular disease or disabling condition

PREEXISTING CONDITION: a disease or disorder that is diagnosed prior to a person's application for insurance coverage

High-Risk Individuals and Preexisting Conditions

As researchers in human genetics improve screening techniques and discover more genetic markers for a variety of medical conditions, concern has grown among both health experts and the general public that genetic screening could lead to the denial of health insurance coverage to individuals with various genetic conditions. The insurance industry has always been reluctant to insure people identified as being at high risk or who suffer from preexisting conditions, a reluctance that has intensified as health-care costs have increased. For example, people with a family medical history of coronary artery disease have long been considered a higher risk than members of the general population. As a consequence, based on information provided through disclosures of family histories, these people have occasionally been denied health insurance coverage or required to pay higher premiums.

Similarly, people who suffer from conditions such as diabetes or hypertension and who change jobs or insurance carriers occasionally discover that their new medical insurance will not pay for any treatment for medical conditions that had been diagnosed prior to obtaining the new insurance. Such "preexisting" conditions are considered ineligible for payment of benefits. While some insurance companies will put a time limit on the restrictions for coverage of preexisting conditions of a few months or a year, providing there are no active occurrences of the disorder, other insurers may exclude making any payments related to a preexisting condition for an indefinite period of time. A person with a chronic condition such as diabetes may discover that while a new insurer will pay for conditions unrelated to the diabetes, such as a broken leg, the individual will be solely responsible for any diabetes-related expenses for the remainder of his or her life. Alternately, the sufferer of a chronic condition may discover that health insurance is available, but such insurance can be obtained only at a much higher premium than that paid by people who have not been diagnosed with preexisting conditions.

Insurance and Genetic Screening

Based on past practices by health insurance companies that restricted coverage of people identified as high risk or with preexisting conditions, it is not surprising that health-care policy analysts worry that genetic screening could serve primarily to allow insurers to eliminate potentially costly clients from the insurance pool. Insurance underwriting is essentially a gamble in which the insurer bets that the amount of money paid into the pool in the form of premiums will exceed the amount of money withdrawn in the form of benefits paid to doctors or patients. The health insurer is wagering that most people purchasing health insurance will live long, reasonably healthy lives with a minimal number of claims filed against the company. In contrast, the insured person is, in effect, betting that he or she will have frequent occasions to visit the doctor. As people live longer and medical treatments become more technologically complex, the profit margin for insurance companies shrinks.

By identifying more people who are at high risk of developing disorders such as breast cancer or Alzheimer's disease, insurers could reduce the number of clients who would be filing repeated and expensive claims for service. In the case of degenerative disorders such as Alzheimer's disease or Huntington's chorea, for example, patients may live for many years following the initial diagnosis of the disease while they become progressively more helpless and eventually require extended hospitalization or custodial care. An insurance company that wrote plans to cover nursing home care could decide to exclude people identified as carrying a gene putting them at risk of developing Alzheimer's disease. The insurance company's reasoning would be that because Alzheimer's sufferers may require many more years of custodial care than the average nursing home resident, it would be unprofitable to insure known future Alzheimer's sufferers. Such people would be seen as simply being too high risk.

In addition, if the insurer decided that a genetic marker constituted a preexisting condition, many people could be prevented from obtaining insurance coverage for the health problems most likely to affect them. Insurers could argue that even though the person with the genetic marker for breast cancer or Alzheimer's disease appears otherwise healthy at the time the screening is done, the very presence of the gene renders that person as already suffering from the disease and thus ineligible for coverage by their insurance pool if and when the disease actually appears.

A number of geneticists and other analysts have suggested that another inherent difficulty with genetic screening is that it opens the door for possible restriction of access to health insurance while not holding out any hope of a treatment or cure for the patient. It is now possible to detect the genetic markers for many conditions for which no effective preventive treatment exists. Alzheimer's disease provides a particularly poignant example. As of the late 1990's, the connection between genes identified as appearing in some early-onset Alzheimer's disease patients and the disease itself was still unclear. People who underwent genetic screening to discover if they carried that particular genetic marker could spend many decades worrying needlessly about their own risk of developing Alzheimer's disease while knowing that there was no way to prevent it. At the same time, the identification of the genetic marker would have identified the patient as a high risk for medical insurance.

On the other hand, in some cases the benefits of genetic screening may outweigh its potential costs. For example, certain cancers have long been recognized as running in some families. Doctors routinely counsel women with a family history of breast cancer to have annual mammograms and even, in cases where the risk seems particularly high, to undergo prophylactic mastectomy (removal of the breast). The discovery of a genetic marker for breast cancer suggests that women who are concerned that they are at higher-than-average risk for the dis-

ease can allay their fears through genetic screening rather than subjecting themselves to disfiguring surgery. Still, the very act of screening could become a double-edged sword. A positive test would not only confirm a woman's worst fears but could also result in her being denied high insurance coverage. Many patients with a high-risk family profile fear that even if the screening turns out negative, simply requesting the test will serve as a flag to health insurers, and they, too, will be assessed higher premiums or denied coverage based on their family histories.

In a climate of rising medical costs and efforts by both traditional insurance providers and health maintenance organizations to reduce expenses, many people feel there is good reason to fear that genetic screening will serve primarily as a tool to restrict access to health insurance. That is, rather than improving the quality of life of patients with genetic disorders, without government regulation to ensure equal access to health care benefits, genetic screening will primarily benefit insurance companies by allowing them to more effectively identify and exclude high-risk patients.

—*Nancy Farm Mannikko*

See Also: Cancer; Cystic Fibrosis; Genetic Testing; Hereditary Diseases.

Further Reading: Geneticist Doris Zallen's *Does It Run in the Family?* (1998) provides readers with the knowledge they need to make decisions regarding genetic testing and does so in an easy-to-understand way. In *Health Insurance: How to Get It, Keep It, or Improve What You've Got* (1996), Robert Enten provides practical advice to high-risk individuals on how to obtain insurance if they have been denied in the past. Jeremy Rifkin's *The Biotech Century: Harnessing the Gene and Remaking the World* (1998) discusses a variety of concerns regarding biotechnology and shows how genetic screening fits into a much wider area of debate in modern science. Finally, an anthology, *Code of Codes: Scientific and Social Issues in the Human Genome Project* (1993), edited by Daniel J. Kevles and Leoy Hood, includes several thought-provoking essays addressing questions regarding genetic screening and discrimination in both employment and insurance coverage.

Intelligence

Field of study: Human genetics

Significance: *The study of the genetic basis of intelligence is one of the most controversial areas in human genetics. Researchers generally agree that mental abilities are genetically transmitted to some extent, but there is disagreement over the degree to which mental abilities are products of genes and the degree to which they are products of environments. There is also disagreement over whether different mental abilities are products of a single ability known as "intelligence" and disagreement over how to measure intelligence.*

Key terms

MONOZYGOUS: developed from a single ovum (egg); identical twins are monozygous because they originate in the womb from a single fertilized ovum that splits in two

DIZYGOUS: developed from two separate ova; fraternal twins are dizygous

PSYCHOMETRICIAN: one who measures intellectual abilities or other psychological traits

INTELLIGENCE QUOTIENT (IQ): the most common measure of intelligence; it is based on the view that there is a single capacity for complex mental work and that this capacity can be measured by testing

Evidence for Genetic Links to Intelligence

Much of the research into the connection between genes and intelligence has focused on attempting to determine to what extent intellectual ability is a product of biological inheritance rather than a product of social influence. Studies concerned with exploring genetic links to mental abilities have usually approached the problem using four methods: associations of parental intelligence with the intelligence of offspring, associations of the intelligence of siblings (brothers and sisters), comparisons of dizygous (fraternal) twins and monozygous (identical) twins, and adoption studies.

To the extent that mental qualities are inherited, one should expect blood relatives to share these qualities with each other more than with nonrelatives. In an article published in 1981 in the journal *Science*, T. J. Bouchard, Jr., and Matt McGue examined studies that looked at statistical relationships of intellectual abilities among family members. These studies did reveal strong associations between mental capacities of parents and children and strong associations among the mental capacities of siblings. Further, if genes are involved in establishing mental abilities, one should expect that the more genes related people share, the more similar they will be in intelligence. Studies have indicated that fraternal twins are only slightly more similar to each other than nontwin siblings are to each other. Identical twins, developing from a single egg with identical genetic material, have even more in common. Bouchard and McGue found that there was an overlap of about 74 percent in the intellectual abilities of identical twins and an overlap of about 36 percent in the intellectual abilities of fraternal twins.

Family members may be similar because they live in similar circumstances, and identical twins may be similar because they receive nearly identical treatment. However, studies of adopted children show that the intellectual abilities of these children were more closely related to those of their biological parents than to those of their adoptive parents. Studies of identical twins who were adopted and raised apart from each other indicate that these twins have about 62 percent of their intellectual abilities in common.

Twin studies, in particular, have helped to establish that heredity is involved in a number of intellectual traits. Memory, number ability, perceptual skills, psychomotor skills, fluency in language use, and proficiency in spelling are only a few of the traits in which people from common genetic backgrounds tend to be similar to each other. However, psychometricians have not reached agreement on the extent to which mental abilities are products of genes rather than of environmental factors such as upbringing and opportunity. Some researchers estimate that only 40 percent of intellectual ability is genetic; others set the estimate as high as 80 percent.

It is important to keep in mind that even if most differences among human beings in mental abilities were caused by genetics, members of families would still show varied abilities. If, for example, there is a gene for high mathematical ability (gene *A*) and a gene for low mathematical ability (gene *a*), it is quite possible that a woman who has inherited each gene (*Aa*) from her parents will marry a man who has inherited each gene (*Aa*) from his parents. In this case, there is a 1 in 4 probability that they will have a child who is mathematically gifted (*AA*) and a 1 in 4 probability that they will have a child who is mathematically slow (*aa*).

The Problem of Defining and Measuring Intelligence

Debates over genetic links to intelligence are complicated by the problem of precisely defin-

ing and accurately measuring intelligence. It may be that abilities to build houses, draw, play music, or understand complex mathematical procedures are inherited as well as learned. Which of these abilities, however, constitute intelligence? Because of this debate, some people, such as Harvard psychologist Howard Gardner, have argued that there is no single quality of intelligence but rather multiple forms of intelligence.

If there is no single ability that can be labeled "intelligence," this means that one cannot measure intelligence or determine the extent

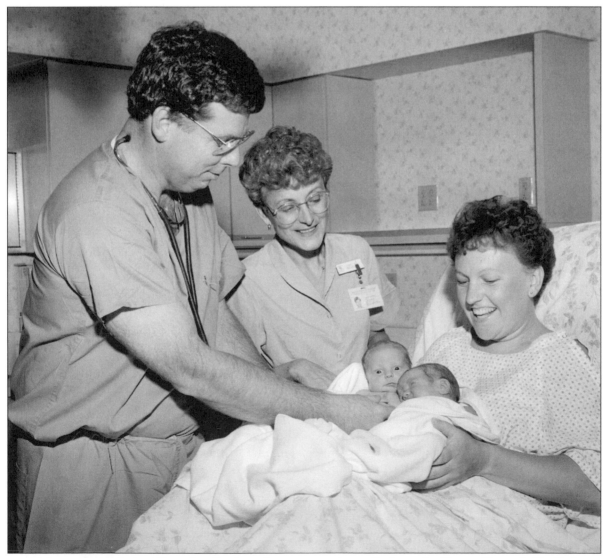

Twins are frequently the subjects of studies that seek to assess genetic contributions to intelligence. (James L. Shaffer)

to which general intellectual ability may be genetic in character. Intelligence quotient (IQ), the measure of intelligence most commonly used to study genetic links to intellectual ability, is based on the view that there is a great deal of overlap among various mental traits. Although a given individual may be skilled at music or writing and poor at mathematics, on the average, people who are proficient in one area also tend to be talented in other areas. Proponents of IQ measures argue that this overlap exists because there is a single, underlying, general intelligence that affects how people score on tests of various kinds of mental abilities. The opponents of IQ measures counter that even if one can speak of intelligence rather than "intelligences," it is too complex to be reduced to one number.

Impact and Applications

The passing of mental abilities from parents to children by genetic inheritance is a politically controversial issue because genetic theories of intelligence may be used to justify existing social inequalities. Social and economic inequalities among racial groups, for example, have been explained as differences among groups in inherited intelligence levels. During the nineteenth century, defenders of slavery claimed that black slaves were by nature less intelligent than the white people who held them in slavery. After World War I, the Princeton University psychologist C. C. Brigham concluded from results of army IQ tests that southern European immigrants had lower levels of inherited intelligence than native-born Americans and that blacks had even more limited intelligence. White supremacists and segregationists used Brigham's results to justify limiting the access of blacks to higher education and other opportunities for advancement. In 1969, Berkeley psychologist Arthur R. Jensen touched off a storm of debate when he published an article that suggested that differences between black and white children in educational success were caused in part by genetic variations in mental ability.

Wealth and poverty, even within racial and ethnic groups, have been explained as consequences of inherited intelligence. Harvard psychologist Richard Herrnstein and social critic Charles Murray have argued that American society has become a competitive, information-based society in which intellectual ability is the primary basis of upward mobility. They have maintained, furthermore, that much of intellectual ability is genetic in character and that people tend to marry and reproduce within their own social classes. Therefore, in their view, social classes also tend to be intellectual classes: a cognitive elite at the top of the American social system and a genetically limited lower class at the bottom.

Scientific truth cannot be established by accusing theories of being inconvenient for social policies of equal opportunity. Nevertheless, it is not clear that genetic differences in intelligence are necessarily connected to social status. Even those who believe that inherited intelligence affects social position generally recognize that social status is affected by many other factors such as parental wealth, educational opportunity, and cultural attitudes.

It seems evident that there are genetic links to mental ability. At the same time, however, the extent to which genes shape intellectual capacities, whether these capacities should be combined into one dimension called intelligence, and the validity of measures of intelligence remain matters of debate. The scientific debate, moreover, is difficult to separate from social and political debates.

—*Carl L. Bankston III*

See Also: Behavior; Biological Determinism; Race; Twin Studies.

Further Reading: *The Mismeasure of Man* (1996), by Stephen Jay Gould, is an influential criticism of IQ as a measure of intelligence and of the idea that intellectual abilities are inherited. At the other extreme, in *The Bell Curve: Intelligence and American Class Structure* (1994), Richard J. Herrnstein and Charles Murray maintain that IQ is a valid measure of intelligence, that intelligence is largely a product of genetic background, and that differences in intelligence among social classes play a major part in shaping American society. Howard Gardner's *Frames of Mind: The Theory of Multiple Intelligences* (1983) argues that there is no single mental ability to be inherited.

Isolates and Genetic Disease

Field of study: Human genetics

Significance: *The study of human genetics is complicated by many factors. The late onset of sexual maturity and the usually random mating habits of most humans make the following of rare mutations in human populations especially difficult. Isolates (small, isolated communities in which mates are chosen only from within the population and not from surrounding populations) can serve as natural laboratories for the study of human genetics, especially in the area of human disease.*

Key Terms

ALLELES: genetic variants of a particular gene

CONSANGUINEOUS: sharing a common genetic ancestry; members of the same family are consanguineous to varying degrees

GENETIC DISEASE: an inherited condition in which the normal functioning of the organism is impaired in some way

The Importance of Isolates

When studying the genetics of the fruit fly or any other organism commonly used in the laboratory, an experimenter can choose the genotypes of the flies that will be mated and can observe the next few generations in a reasonable amount of time. Experimenters can also choose to mate offspring flies with their siblings or with their parents. As one might expect, this is not possible when studying the inheritance of human characteristics. Thus, progress in human genetics most often relies on the observation of the phenotypes of progeny that already exist and matings that have already occurred. Many genetic diseases only appear when a person is homozygous for two recessive alleles; in order to be homozygous, a person must inherit the same recessive allele from both parents. Since the majority of these recessive alleles are rather rare in the population, the chance that both parents in a mating carry the same recessive allele is quite small. This makes the study of these diseases very difficult. The chance that both parents carry the same recessive allele is increased whenever mating occurs between individuals who share some of the same genetic background. These consanguineous matings produce measurably higher numbers of offspring with genetic diseases, especially when the degree of consanguinity is at the level of second cousin or closer.

In small religious communities in which marriage outside the religion is forbidden, and in small, geographically isolated populations in which migration into the population from the outside is at or near zero, marriages often occur between two people who share some common ancestry; therefore, the level of consanguinity can be quite high. These communities thus serve as natural laboratories in which to study genetic diseases. Geographically isolated mountain and island communities are found in many areas of the world, including the Caucasus Mountains of Eurasia, the Appalachian Mountains of North America, and many areas in the South Pacific. Culturally isolated communities are also of worldwide distribution. Among the Druse, a small sect of Islam, first-cousin marriages approach 50 percent of all marriages. Amish, Hutterites, and Dunkers in the United States are each descended from small groups of original settlers who immigrated in the eighteenth and nineteenth centuries and rarely mated with people from outside their religions.

The Amish

There are many reasons why the Amish serve as a good example of an isolate. The original immigration of Amish to America consisted of approximately two hundred settlers. In subsequent generations, the available mates came from the descendants of the original settlers. With mate choice this limited, it is inevitable that some of the marriages will be consanguineous. Consanguinity increases as further marriages take place between the offspring of consanguineous marriages. Current estimates are that the average degree of consanguinity of Amish marriages in Lancaster County, Pennsylvania, is at the level of marriages between second cousins.

Other factors that make the Amish good subjects for genetic research are their high fertility and their high level of marital fidelity. Thus, if both parents happen to be heterozy-

Geographically or culturally isolated populations, such as the Amish of the northeastern United States, are prone to a variety of genetic disorders. (Archive Photos)

gous for a particular genetic disease, the chance that at least one of the offspring will show the disease is high. In families of two children, there is a 44 percent chance that at least one child will show the trait. This increases to 70 percent of the families with four children and to more than 91 percent of the families with eight children, a common number among the Amish. Because of the high marital fidelity among the Amish, researchers do not have to worry about illegitimacy when making these estimates.

Many genetic diseases that are nearly nonexistent in the general population are found among the Amish. The allele for a type of dwarfism known as the Ellis-van Creveld syndrome is found in less than 0.1 percent of the general population; among the Lancaster Amish, however, the allele exists in approximately 7 percent of the population. Other genetic diseases at higher levels among the Amish include cystic fibrosis, limb-girdle muscular dystrophy, pyruvate kinase-deficient hemolytic anemia, and several inherited psychological disorders. Having more families and individuals with these diseases to study helps geneticists and physicians discover ways to treat the problems and even prevent them from occurring.

—*Richard W. Cheney, Jr.*

See Also: Human Genetics; Inborn Errors of Metabolism; Inbreeding and Assortative Mating.

Further Reading: Harold Cross, "Population Studies of the Old Order Amish," *Nature* 262 (July, 1976), describes the advantages of isolates and some of the genetic characteristics seen in the Amish populations. Victor McKusick et al., "The Distribution of Certain Genes in the Old Order Amish," *Cold Spring Harbor Symposia on Quantitative Biology* 29 (1964), and Victor McKusick et al., "Medical Genetic Studies of the Amish with Comparison to Other Populations," *Population Structure and Genetic Disorder* (1981), edited by A. W. Eriksson et al., describe many of the inherited conditions seen in the Amish community.

Klinefelter's Syndrome

Field of study: Human genetics
Significance: *Klinefelter's syndrome is a chromosomal sex disorder. This syndrome affects only males and accounts for ten out of every one thousand institutionalized mentally retarded adults in industrialized nations. It is one of the more common chromosomal aberrations.*

Key terms

AZOOSPERMIA: the absence of spermatozoa from the semen

CYTOGENETICS: the study of chromosomes by light microscopy

CHROMOSOMES: microscopic bodies that develop from the nucleus of the cell and contain all genes and hereditary traits

GYNECOMASTIA: a condition characterized by abnormally large mammary glands in the male that sometimes secrete milk

HYPOGONADISM: defective internal secretion of the gonads resulting in smaller than normal testicles in males

KARYOTYPE: a systemic order or array of the chromosomes of a single cell

MOSAICISM: a condition in which an individual has two or more cell populations derived from the same fertilized ovum, or zygote, so that some cells contain the usual XY chromosome pattern and others contain extra X chromosomes

Definition and Diagnosis

Klinefelter's syndrome is a relatively common genetic abnormality named after Harry Klinefelter, Jr., an American physician. The fundamental chromosomal defect associated with the syndrome is the presence of one or more extra X chromosomes. The normal human male karyotype (array of chromosomes) consists of twenty-two pairs of chromosomes plus the XY pair. The female also has twenty-two pairs but with an XX pair in place of the XY pair. Klinefelter's syndrome affects 1 in every 500 to 600 men. The incidence is relatively high in the mentally retarded population.

Klinefelter's syndrome affects males and results from one or more extra X chromosomes. It is a common cause of male hypogonadism (small testes), which results from the expression of an abnormal karyotype (usually an extra X chromosome), so that the individual has a combination of XXY rather than the usual XY male chromosomes. Other forms of Klinefelter's syndrome may be expressed in mosaicism, with males having both normal (XY) chromosome patterns in some cells and abnormal chromosome patterns in others. It may also be expressed in different chromosome patterns such as XXYY, XXXY, or XX.

Klinefelter's syndrome can be identified by the use of cytogenetic technology. In the area of cytogenetics, chromosomal analyses are done by growing human cells in tissue cultures and staining them microscopically, then photographing, sorting, and counting the chromosomes. The display of the karyotype is the end result of the technical part of cytogenetics.

Signs and Symptoms

The classic type of Klinefelter's syndrome usually becomes apparent at puberty, when the secondary sex characteristics develop. The testes fail to mature, causing primary hypogonadism. In this classic type, degenerative testicular changes begin that eventually result in irreversible infertility. Gynecomastia is often present, and it is usually associated with learning disabilities, mental retardation, and violent, antisocial behavior. Other common symptoms include tall stature and abnormal body proportions (disproportionate height relative to arm span), small testes, chronic pulmonary disease, varicosities of the legs, and diabetes mellitus (which occurs in 8 percent of those afflicted with Klinefelter's). Another 18 percent exhibit impaired glucose tolerance. Most people affected also have azoospermia (no spermatozoa in the semen) and low testosterone levels. However, men with mosaicism may be fertile.

Congenital hypogonadism appears as delayed puberty. Men with hypogonadism experience decreased libido, erection dysfunction, hot sweats, and depression. Genetic testing and careful physical examination may reveal Klinefelter's syndrome to be the reason for the primary complaint of infertility. Mental retardation is a frequent symptom of congenital chromosomal aberrations such as Klinefelter's syndrome because of probable coincidental defective development of the central nervous system. Early spontaneous abortion is a common occurrence in fetal abnormalities.

Treatment and Psychosocial Implications

Depending on the severity of the syndrome, treatment may include mastectomy to correct gynecomastia. Supplementation with testoster-

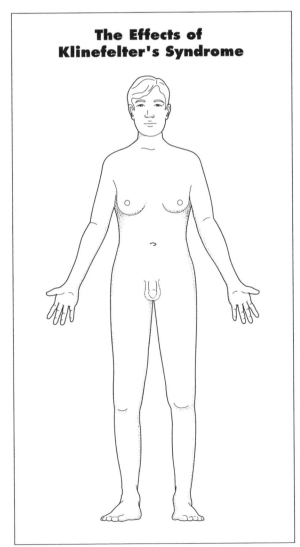

The effects of Klinefelter's syndrome include enlarged breasts, disproportionately long limbs, and a feminine body shape.

one may be necessary to induce the secondary sexual characteristics of puberty. The testicular changes that lead to infertility cannot be prevented. Any mental retardation present is irreversible. Psychotherapy with sexual counseling is appropriate when sexual dysfunction causes emotional problems. In people with the mosaic form of the syndrome who are fertile, genetic counseling is vital because they may pass on this chromosomal abnormality. Therapy should encourage people afflicted with this syndrome to discuss feelings of confusion and rejection that commonly accompany it. It should also attempt to reinforce their male identity. Hormonal therapy can provide some benefits, but both benefits and side effects of hormonal therapy should be made clearly understood to improve compliance. Some people afflicted with Klinefelter's syndrome are sociopathic; for this population, careful monitoring by probation officers or jail personnel can assist in identifying potential violent offenders, who can be offered psychological counseling.

—*Lisa Levin Sobczak*

See Also: Developmental Genetics; Fragile X Syndrome; Genetic Counseling; Human Genetics.

Further Reading: An overview of medical diagnosis and treatment is provided in *Current Medical Diagnosis and Treatment in 1997* (1997), edited by Lawrence M. Tierney, Jr., et al. The *Cecil Textbook of Medicine*, (1988), 18th ed., edited by James B. Wyngaarden and Lloyd H. Smith, Jr., and *Harrison's Principles of Internal Medicine* (1994), 13th ed., vol. 2, edited by Kurt J. Isselbacher et al., are classic sources used to educate physicians. Each of these sources provides a brief summary of Klinefelter's syndrome, along with many other ailments.

Knockout Genetics and Knockout Mice

Field of study: Molecular genetics
Significance: *In knockout methodology, a specific gene of an organism is inactivated, or "knocked out," allowing the consequences of its absence to be observed and its function to be deduced. The tech-*

nique, first and mostly applied to mice, permits the creation of animal models for inherited diseases and a better understanding of the molecular basis of physiology, immunology, behavior, and development. Knockout genetics is the study of the function and inheritance of genes using this technology.

Key terms

DEOXYRIBONUCLEIC ACID (DNA): the molecule that encodes genetic information in each living cell; it consists of two intertwined strands of units called nucleotides

EMBRYONIC STEM CELL: a cell derived from an early embryo that can replicate indefinitely in vitro and can differentiate into other cells of the developing embryo

GENE: the fundamental physical and functional unit of heredity; it consists of DNA and encodes, by its particular sequence of nucleotides, a specific functional product

GENOME: the total complement of genetic material for an organism

IN VITRO: a biological or biochemical process occurring outside a living organism, as in a test tube

IN VIVO: a biological or biochemical process occurring within a living organism

Knockout Methodology

The knockout mouse can be considered to be the first genetically engineered animal. Before this methodology, transgenic animals had been generated in which "foreign" deoxyribonucleic acid (DNA) was incorporated into their genomes in a largely haphazard fashion; such animals should more properly be referred to as "genetically modified." In contrast, knockout technology targets a particular gene to be altered. Prior to the creation of transgenic animals, any genetic change resulted from spontaneous and largely random mutations. Individual variability and inherited diseases are the results of this natural phenomenon—as are, on a longer time frame, the evolutionary changes responsible for the variety of living species on the earth. Spontaneously generated animal models of human inherited diseases have been helpful in understanding these conditions and developing treatments for them. However, these mutants were essentially gifts of nature, and their discovery was largely seren-

dipitous. In knockout mice, animal models are directly generated, expediting study of the pathology and treatment of inherited diseases.

In a knockout mouse, one of the animal's 100,000 genes has been selected to be inactivated in such a way that the defunct gene is reliably passed to its progeny. Developed independently by Mario Capecchi at the University of Utah and Oliver Smithies of the University of North Carolina, the process is formally termed "targeted gene inactivation," and, although simple in concept, it is operationally complex and technically demanding. It involves several steps in vitro: inactivating and tagging the selected gene, substituting the now-defunct gene for the functional gene in embryonic stem cells, and inserting the modified embryonic stem cells into an early embryo. The process then requires transfer of that embryo to a surrogate mother, which carries the embryo to term, and selection of offspring that are carrying the inactive gene. It may require several generations to verify that the genetic modification is being dependably transmitted. If the targeted gene has been previously isolated and characterized, the entire procedure requires up to a year and costs about $100,000.

Usefulness of Knockout Mice

Knockout mice are important because they permit the function of a specific gene to be established, and, since mice and humans share 99 percent of the same genes, the results can be widely applied to people. However, knockout mice are not perfect models in that some genes are specific to mice or humans, and similar genes can be expressed at different levels in the two species. Nevertheless, knockout mice are vastly superior to spontaneous mutants because the investigator selects the gene to be modified. Mice are predominantly used in this technology because of their short generation interval and small size; the short generation interval accelerates the breeding program necessary to establish pure strains, and the small size reduces the space and food needed to house and sustain them.

Knockout mice are, first of all, excellent animal models for inherited diseases, the study of which was the initial impetus for their crea-

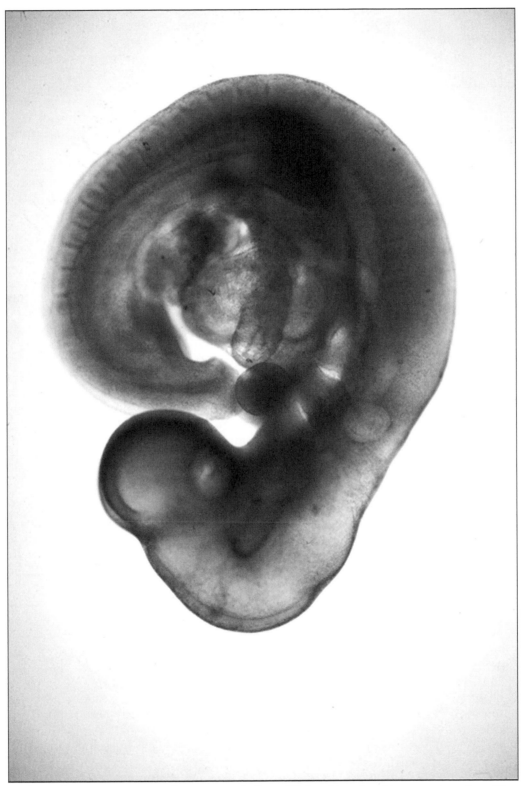

In the knockout process, mouse embryos are genetically modified and transferred to surrogate mothers. (Dan McCoy/Rainbow)

tion. The Lesch-Nyhan syndrome, a neurological disorder, was the focus of much of the early work with the knockout technology. The methodology has permitted the creation of previously unknown animal models for cystic fibrosis, Alzheimer's disease, and sickle-cell anemia, which will stimulate research into new therapies for these diseases. Knockout mice have also been developed to study atherosclerosis, cancer susceptibility, and obesity, as well as immunity, memory, learning, behavior, and developmental biology.

Knockout mice are particularly appropriate for studying the immune system because immune-compromised animals can survive if kept isolated from pathogens. More than fifty genes are responsible for the development and operation of B and T lymphocytes, the two main types of cells that protect the body from infection. Knockout technology permits a systematic examination of the role played by these genes. It has also proven useful in understanding memory, learning, and behavior, as knockout mice with abnormalities in these areas can also survive if human intervention can compensate for their deficiencies. Knockout mice have been created that cannot learn simple laboratory tests, cannot remember symbols or smells, lack nurturing behavior, or exhibit extreme aggression, which have implications for the fields of education, psychology, and psychiatry.

Developmental biology has also benefited from knockout technology. Animals with minor developmental abnormalities can be studied with relative ease, whereas those with highly deleterious mutations may be maintained in the heterozygous state, with homozygotes generated only as needed for study. The generation of conditional knockouts is facilitating study of the thirty-eight or more genes responsible for controlling the development of various tissues (lung, heart, skeleton, and muscle) during embryonic development. These genes can be explored methodically with knockout technology.

By 1997, over one thousand different knockout mice had been created worldwide. A primary repository for such animals is the nonprofit Jackson Laboratory in Bar Harbor, Maine, where over two hundred so-called induced mutant strains are available to investigators. Other strains are available from the scientists who first derived them or commercial entities licensed to generate and sell them.

Double Knockouts, Conditional Knockouts, and Reverse Knockouts

Redundancy is fairly common in gene function: Often, more than one gene has responsibility for the same or similar activity in vivo. Eliminating one redundant gene may have little consequence because another gene can fulfill its function. This has led to the creation of double knockout mice, in which two specific genes are eliminated. Double knockouts are generated by crossing two separate single knockout mice to produce double mutant offspring. Consequences of both mutations can then be examined simultaneously.

Some single knockout mice are deleteriously affected during embryonic development and do not survive to birth. This has led to the generation of conditional knockout mice, in which the gene is functional until a particular stage of life or tissue development triggers its inactivation. The approach is to generate animals with two mutations: The first is the addition of a new gene that causes a marked segment of a gene to be deleted in response to a temporal or tissue signal, and the second is to mark the gene that has been selected to be excised. In these animals, the latter gene remains functional until signaled to be removed.

Knockout methodology involves generation of loss-of-function or null mutations. Its reversal would permit the function of an inoperative gene to be restored. This reversal has been successfully accomplished in mice with the correction of the Lesch-Nyhan defect. Further experimentation may permit it to be applied to humans and other animals. Such targeted restoration of gene function would be the most direct way for gene therapy (the process of introducing a functional gene into an organism's cells) to cure inherited diseases.

—*James L. Robinson*

See Also: Biotechnology; Developmental Genetics; Gene Therapy; Genetic Engineering; Hereditary Diseases; Sickle-Cell Anemia.

Further Reading: Rosie Mestel, "The Mice Without Qualities," *Discover* 14 (March, 1993), briefly reports on the creation of knockout mice and their use in understanding the role of the missing gene. In "Targeted Gene Replacement," *Scientific American* 270 (March, 1994), one of the originators of the technology, Mario Capecchi, describes the steps involved and examples of its utility. *Developmental Biology* (1988), by Scott F. Gilbert, includes chapters on the methodology and its usefulness in a field with great impact. *Genetics* (1995), by Robert F. Weaver and Philip W. Hedrick, explains the technique and various applications, including understanding tumor suppressor genes.

Lactose Intolerance

Field of study: Human genetics

Significance: *Lactose intolerance is a common disorder associated with the digestion of milk sugar that affects a large portion of the human population and creates unpleasant intestinal effects. Its understanding has led to the commercial availability of alternative products that supplement the lack of dairy products in the diet.*

Key terms

ACQUIRED DEFICIENCY: a deficiency that is observed with aging

CONGENITAL DEFICIENCY: a deficiency that is attributed to genetic causes

GALACTOSEMIA: a disease attributed to the accumulation of galactose in the blood, caused by a lack of the enzyme that metabolizes galactose; galactosemia is not related to lactose intolerance, which is attributed to the lack of the lactase enzyme

LACTASE: an enzyme that breaks down lactose to the monosaccharides glucose and galactose in the small intestine during the metabolic process; its deficiency is responsible for the ill effects associated with lactose intolerance

LACTOSE: a sugar, also known as milk sugar, that constitutes 2 to 8 percent of milk content and makes up about 40 percent of an infant's diet

The Function of Lactose and Lactase

Milk is the primary source of nutrition for infants. One pint of cow's summer milk provides about 90 percent of the calcium, 30 to 40 percent of the riboflavin, 25 to 30 percent of the protein, and 10 to 20 percent of the calories needed daily. Lactose, also known as milk sugar, exists in the milk of humans, cows, and other mammals. About 7.5 percent of human milk consists of lactose, while cow's milk is about 4.5 percent lactose. This sugar is also one of the few carbohydrates exclusively associated with the animal kingdom since its biosynthesis takes place in the mammary tissue. It is produced commercially from whey, which is obtained as a by-product during the manufacture of cheese. Its so-called alpha form is used as an infant food. Its sweetness is about one-sixth that of sucrose (table sugar).

The metabolism (breaking down) of lactose to glucose and galactose takes place via a specific enzyme called lactase, which is produced by the mucosal cells of the small intestine. Because lactase activity is rate-limiting for lactose absorption, any deficiency in the enzyme is directly reflected in a diminished rate of the sugar absorption. This irregularity should not be confused with the intolerance to milk resulting from a sensitivity to milk proteins such as beta-lactoglobulin.

Consequences of Lactase Deficiency

There are three types of lactase deficiency: inherited deficiency, secondary low-lactase activity, and primary low-lactase activity. In inherited lactase deficiency, the symptoms of intolerance develop very soon after birth, as indicated by the presence of lactose in the urine. Patients are recommended a lactose-free diet as well as the consumption of live-culture yogurt, which provides the enzyme beta-galactosidase that attacks the small amounts of lactose that may be in the diet. Beta-galactosidase preparations are also commercially available.

Secondary low-lactase activity often occurs to normal humans and may be present during intestinal diseases such as colitis, gastroenteritis, kwashiorkor, and sprue. Post-peptic-ulcer operation effects may also indicate this type of disorder. In the very common case of primary low-lactase activity, humans develop this type of deficiency as they get older. Infants and small children have this enzyme in the active form that enables the metabolism. However, a large number of the adult population, estimated at almost 20 percent, gradually exhibit lactose intolerance, which arises from the gradual inability of their organism to synthesize the lactase enzyme in an active form. Susceptible individuals may start developing lactose intolerance as early as four years old.

As a result of lactose intolerance, relatively large quantities of the unhydrolyzed (unbroken) lactose pass into the large intestine, which causes the transfer of water from the interstitial fluid to the lumen by osmosis. At the same time, the intestinal bacteria produce organic acids as well as gases such as carbon dioxide, methane, and hydrogen, which lead to nausea and vomiting. The combined effect also produces cramps and abdominal pains. Although the condition is not life threatening, the results are quite unpleasant.

Definitive diagnosis of the condition is established by an assay for lactase content in the intestinal mucosa. Such a test requires that the individuals drink 50 grams of lactose in 200 milliliters of water. Blood specimens are then taken after 30, 60, and 120 minutes for glucose analysis. An increase of blood glucose by 30 milligrams per deciliter is considered normal, while a 20 to 30 milligram per deciliter increase is borderline. A lesser increase indicates lactate deficiency. This test, however, may still show deficiency results with individuals that have a normal lactase activity.

Lactase deficiency displays remarkable genetic variations. The condition is more prevalent among infants of Middle Eastern, Asian (especially Chinese and Thai), and African descent (such as the Ibo, Yoruba, and other tribes in Nigeria and the Hausa in Sudan). On the other hand, Europeans (especially northern) appear to be statistically less susceptible to the deficiency. Similarly, the Fula tribe in Sudan raises the fulani breed of cattle, and the Eastern African Tussi, who own cattle in Rundi, appear to be able to digest lactose. It is estimated that 10 to 20 percent of American Caucasians and about 75 percent of African Americans are affected.

The ill effects disappear as long as the diet excludes milk altogether. Louis Pasteur and other scientists of the late nineteenth century helped in the development of the microbiological and nutritional sciences. As a result, the production of cultured dairy products took over in an industrial and commercial form. Often people who exhibit partial lactose intolerance may still consume dairy products, including cheese and yogurt, if the food is processed or partially hydrolyzed. This may even be accomplished merely by heating or partially fermenting milk. Some commercial products, such as Lactaid, are designed for lactose-intolerant people because they include the active form of the lactase enzyme in either liquid or tablet form.

—*Soraya Ghayourmanesh*

See Also: Aging; Race.

Further Reading: A section on lactose intolerance is presented in John Hill et al., *An Introduction to General, Organic, and Biological Chemistry* (1993). Robert J. Ouellette, *Organic Chemistry* (1998), also contains a section on lactose metabolism. A thorough overview of lactose intolerance appears in H. A. Buller and R. J. Grant, "Lactose Intolerance," *Annual Reviews of Medicine* 141 (1990).

Lamarckianism

Field of study: Classical transmission genetics

Significance: *Although some aspects of Lamarckianism have been discredited, the basic premises of nineteenth century French biologist Jean Baptiste de Lamarck's philosophy have become widely accepted tenets of evolutionary theory. Lamarckianism became particularly intellectually suspect following fraudulent claims by the Soviet scientist Trofim Lysenko that he could manipulate the heredity of plants by changing their environment; by the 1990's, however, scientists had become more willing to acknowledge the influence of Lamarckianism in evolutionary biology.*

Key terms

ACQUIRED CHARACTERISTIC: a change in an organism brought about by its interaction with its environment

LYSENKOISM: a theory of transformation that denied the existence of genes

TRANSFORMIST: a nineteenth century term used to indicate a belief in evolutionary theory

Lamarckianism Defined

The term "Lamarckianism" has for many years been associated with intellectually disreputable ideas in evolutionary biology. Originally formulated by the early nineteenth century French scientist Jean-Baptiste Pierre Antoine de Monet, chevalier de Lamarck, Lamarckianism had two components that were often misinterpreted by scholars and scientists. The first was the transformist theory that animals gradually changed over time in response to their perceived needs. Many critics interpreted this to mean that species could adapt by wanting to change—in other words, that giraffes gradually evolved to have long necks because they wanted to reach the leaves higher in the trees or that pelicans developed pouched beaks because they wanted to carry more fish. Where Lamarck had suggested only that form followed function—for example, that birds that consistently relied on seeds for food gradually transformed to have beaks that worked best for eating seeds—critics saw the suggestion of active intent or desire.

The second component of Lamarckianism, that changes in one generation of a species could be passed on to the next, also led to misinterpretations and abuses of his ideas. In the most egregious cases, researchers in the late nineteenth and early twentieth centuries claimed that deliberate mutilations of animals could cause changes in succeeding generations—for example, they believed that if they cut the tails off a population of mice, succeeding generations would be born without tails. During the twentieth century, the Soviet agronomist Trofim Lysenko claimed to have achieved similar results in plants. Such claims have been thoroughly disproved.

Who Was Lamarck?

Such gross distortions of his natural philosophy would probably have appalled Lamarck. Essentially an eighteenth century intellectual, Lamarck was one of the last scientists who saw himself as a natural philosopher. He was born August 1, 1744, in Picardy, and as the youngest of eleven children was destined originally for the church. The death of his father in 1759 freed Lamarck to leave the seminary and enlist in the military, but an injury forced him to resign his commission in 1768. He sampled a variety of possible vocations before deciding to pursue a career in science.

His early scientific work was in botany. He devised a system of classification of plants and in 1778 published

Although many of Lamarck's theories have been disproved, his work exerted a strong influence on the science of heredity. (Library of Congress)

a guide to French flowers. In 1779, at the age of thirty-five, Lamarck was elected to the Académie des Sciences. Renowned naturalist Georges LeClerc, comte de Buffon, obtained a commission for Lamarck to travel in Europe as a botanist of the king. In 1789, Lamarck obtained a position at the Jardin du Roi as keeper of the herbarium. When the garden was reorganized as the Museum National d'Histoire Naturelle in 1794, twelve professorships were created; Lamarck became a professor of what would now be called invertebrate zoology.

Lamarck demonstrated through his lectures and published works that he modeled his career on that of his mentor, Buffon. He frequently went beyond the strictly technical aspects of natural science to discuss philosophical issues, and he was not afraid to use empirical data as a basis for hypothesizing. Thus, he often speculated freely on the transformation of species. *Philosophie Zoologique* (*Zoological Philosophy*), now considered his major published work, was issued in two volumes in 1809. In it, Lamarck elaborated upon his theories concerning the evolution of species through adaptation to changes in their environments. An essentially philosophical work, *Zoological Philosophy* is today remembered primarily for Lamarck's two laws:

First Law: In every animal which has not passed the limit of its development, a more frequent and continuous use of any organ gradually strengthens, develops and enlarges that organ and gives it a power proportional to the length of time it has been so used; while the permanent disuse of any organ imperceptibly weakens and deteriorates it, and progressively diminishes its functional capacity, until it finally disappears.

Second Law: All the acquisitions or losses wrought by nature on individuals, through the influence of the environment in which their race has long been placed, and hence through the influence of the predominant use or disuse of any organ; all these are preserved by reproduction to the new individuals which arise, provided that the acquired modifications are common to both sexes, or at least to the individuals which produce the young.

These two tenets constitute the heart of Lamarckianism.

During his lifetime, Lamarck's many books were widely read and discussed, particularly *Zoological Philosophy*. It is true Lamarck's ideas on the progression of life from simple forms to more complex forms in a great chain of being met with opposition, but that opposition was not universal. He was not the only "transformist" active in early nineteenth century science, and his influence extended beyond Paris. Whether or not Lamarck directly influenced Charles Darwin is a matter of debate, but it is known that geologist Charles Lyell read Lamarck, and Lyell in turn influenced Darwin.

Lamarckianism's fall into disrepute following Lamarck's death was prompted by social and political factors as well as scientific criteria. By the 1970's, after a century and a half of denigration, Lamarckianism began creeping back into evolutionary theory and scientific discourse. Researchers in microbiology have described processes that have been openly described as Lamarckian, while other scholars began to recognize that Lamarck's ideas did indeed serve as an important influence in developing theories about the influence of environment on both plants and animals.

—Nancy Farm Mannikko

See Also: Classical Transmission Genetics; Evolutionary Biology; Genetics, Historical Development of; Natural Selection.

Further Reading: Historians of science consider *The Spirit of System: Lamarck and Evolutionary Biology* (1977), by Richard W. Burkhardt, Jr., the most comprehensive examination of Lamarck and his time. Paul E. M. Fine's "Lamarckian Ironies in Contemporary Biology," *Lancet* (June 2, 1979), discusses how Lamarckianism has crept into twentieth century evolutionary theory even as some biologists continue to deny any Lamarckian influences. Hugh Elliot has translated Lamarck's *Zoological Philosophy: An Exposition with Regard to the Natural History of Animals* (1963) into English, making it available to readers curious about the origins of Lamarckianism. Uri Lanham's *Origins of Modern Biology* (1968) provides a good general history of biology.

Lethal Alleles

Field of study: Classical transmission genetics

Significance: *An allele is one possible form of a gene and produces a particular trait. Lethal alleles influence the viability of organisms, resulting in death if inherited in specific combinations. These alleles can exert their effects at various stages of life, depending on the allele, and are probably present in every organism, although they may not be detected in all cases.*

Key terms

HETEROZYGOTE: an individual who has different alleles of the same gene, such as *Aa*; such an individual would be considered hybrid for the trait controlled by that gene

HOMOZYGOTE: an individual who has two identical alleles for a particular gene, such as *AA* or *aa*; such an individual would be considered purebred for the trait controlled by that gene

WILD TYPE: a trait that is common in nature; alternatives to this trait are called mutants

Discovery of Lethal Alleles

A lethal allele is an allele that is capable of causing the death of an organism. Generally, lethal effects are observed when an organism inherits two copies of the allele (one copy from each parent); however, a single copy of the allele can produce serious effects on the phenotype of the organism as well. Lethal alleles were first characterized in 1904 by geneticist Lucien Cuenot. Cuenot was studying the inheritance of coat color in mice, particularly the inheritance pattern of a yellow coat color. The normal wild-type coat color of mice is often termed "agouti," which is a dark brown color. When Cuenot crossed yellow mice to wild-type mice, he obtained a ratio of one yellow to one wild-type mouse in the offspring. This type of ratio typically represents a testcross in which one of the parents is homozygous recessive for the gene in question, while the other parent in the cross is heterozygous for the gene. Knowing that wild-type mice always produced wild-type offspring when mated with each other (therefore being homozygous), Cuenot hypothesized that the yellow parent was the heterozygote in the cross.

However, in further crosses performed with yellow mice, a curious pattern began to develop. No matter which two yellow mice were crossed, the offspring always appeared in a ratio of two yellow to one wild-type mouse. This was an unexpected result for two reasons: First, if yellow mice were truly heterozygous, the expected ratio would be three yellow to one wild-type; second, at least some of the yellow mice would be expected to eventually be purebred, resulting in some crosses that produced nothing but yellow offspring. This latter result never occurred. Cuenot concluded that yellow was a trait that followed different rules from the majority of Mendelian traits discovered up until then.

An analysis of the crosses soon provided a possible explanation for the function of the yellow allele. The offspring from a simple Mendelian cross such as this one can be analyzed on two levels: according to either the physical characteristics of the offspring (phenotype) or the genetic characteristics of the offspring (genotype). Phenotypically, a cross between two yellow mice would be expected to produce an offspring ratio of $3/4$ yellow to $1/4$ wild-type if both parents were heterozygous. Genotypically, the same cross should produce a ratio of $1/4$ homozygous dominant (yellow), $1/2$ heterozygous (yellow), and $1/4$ homozygous recessive (wild-type), or a ratio of 1:2:1. When compared to the actual results of the experiment, Cuenot observed that one of the expected genotype groups, $1/4$ homozygous dominant yellow, did not appear to be present in the offspring, leaving only a 2:1 ratio. In fact, these offspring had died before birth, as an examination of the uterus of a pregnant yellow mouse revealed. Hence, the allele for yellow coat color is one that is lethal for mice in a double dose (homozygous); in a single dose, along with a wild-type allele (heterozygous), it does not result in death, instead producing mice with yellow coats, as shown in the following table (where *Y* stands for the yellow allele and *y* stands for the wild-type allele):

Genotype	Phenotype
YY	yellow (dead)
Yy	yellow (alive)
yy	wild-type

Another common example of a lethal allele is that carried by Manx cats. This breed of cats is naturally tailless because of spinal abnormalities that can be traced to a single allele. However, in embryos that receive two copies of this allele, the abnormalities become so severe that death of the embryo results.

Lethal Alleles in Humans

A wide variety of lethal alleles exist in humans. Many of these are not well defined, either because they occur in embryos that are not old enough to be detected before they die or because they cause spontaneous abortions that are never investigated genetically. Other human lethal alleles do not exert an effect until after birth. Two common examples of this type of lethal allele are cystic fibrosis (CF) and Huntington's chorea. CF is a genetic defect that, among other things, results in a disturbance in the salt and water transport and regulation in epithelial cells, especially those of the respiratory system. When this happens, the lungs and airways begin to accumulate heavy mucus that is difficult to dislodge, making it hard for the affected individual to breathe, resulting in an increased incidence of lung infections. Other organs affected by this disease include the pancreas and the reproductive organs. Although better treatments for CF continue to be developed, such as gene therapy for correcting the fundamental genetic defect, individuals with this disease still have a relatively limited life span and usually die before reaching adulthood.

Huntington's chorea is an example of a lethal allele with its effects in adulthood. This disease attacks the nervous system by causing degeneration of nervous tissue, eventually leading to convulsions and death. However, there is usually no sign of the presence of this allele until after childbearing age has passed; 50 percent or more of Huntington's chorea victims do not show any symptoms until after age forty, unfortunately ensuring that there will be a high risk of passing the allele to the offspring of the affected individual.

Lethal alleles are of pleiotropic genes. These are genes that have an effect on more than one trait. In the case of Cuenot's yellow mice, for example, the yellow allele had an effect on coat color and on the viability of embryos. In most cases, lethal alleles cause a single genetic defect that, in turn, sets off a cascade of events that eventually leads to the death of the organism. Cystic fibrosis is a good example of this. The only problem that can be directly explained genetically is the cellular problem with transport of salts and water. However, this defect then leads to other problems (accumulation of mucus) that have more progressive effects (severe and recurring lung infections) until the problems become so severe that death results.

Impact and Applications

Many lethal alleles are never detected because their adverse effects occur so early in the development of the embryo. Probably all people unknowingly carry one or more lethal alleles that, if passed on to their children in the appropriate combinations, would cause problems. Some of those alleles can be identified by genetic tests, enabling prospective parents to make informed decisions about having children. However, there are limitations to genetic testing: There are certain ethical problems inherent in the procedure, and testing can only be done for those lethal alleles that are known and well characterized.

Not all lethal alleles are lethal in every environment. There are many traits that, if present in the wild, would prove to be defective to the point of lethality; however, in more controlled environments, organisms with these traits may survive and be reproductive members of their species. Some of these alleles can play a role in various industries such as agriculture. Years of selective breeding have produced many varieties of plants and animals that would never survive in the wild but that thrive and produce when cultivated under carefully controlled conditions such as on a farm.

—*Randall K. Harris*

See Also: Classical Transmission Genetics; Cystic Fibrosis; Genetic Counseling; Huntington's Chorea; Monohybrid Inheritance.

Further Reading: For more information about cystic fibrosis, consult *Cystic Fibrosis: The Facts* (1995), by Ann Harris and Maurice Super. *Genetic Disorders Sourcebook* (1996), edited by Karen Bellenir, provides complete, nontechnical information about a variety of genetic defects, including some lethal alleles. Lois Wingerson, in *Mapping Our Genes: The Genome Project and the Future of Medicine* (1991), discusses prospects for increasing knowledge of genetic diseases by the application of new technology.

Linkage Maps

Field of study: Classical transmission genetics

Significance: *Linkage maps of the chromosomes of many species are used to predict the results of genetic crosses when the genes that are being examined are on the same chromosome.*

Key terms

ALLELES: different forms of the same gene

HOMOLOGOUS CHROMOSOMES: a pair of chromosomes (one from each parent) that are very similar to each other

MEIOSIS: cell division that reduces the chromosome number from two sets to one set, ultimately resulting in the formation of gametes (eggs, pollen, and sperm)

CROSSING-OVER: a meiotic event that exchanges reciprocal pieces of a pair of homologous chromosomes

DIHYBRID: an organism that is heterozygous for both of two different genes

Linkage and Crossing-Over

When Gregor Mendel examined inheritance of two traits at a time, he found that the dihybrid parent (*Aa* or *Bb*) produced offspring with the four possible combinations of these alleles at equal frequencies: $\frac{1}{4}AB$, $\frac{1}{4}Ab$, $\frac{1}{4}aB$, and $\frac{1}{4}ab$. He called this pattern "independent assortment." The discovery of meiosis explained the basis of independent assortment. If the *a* gene and the *b* gene are on nonhomologous chromosomes, then segregation of the alleles of one gene (*A* and *a*) will be independent of the segregation of the alleles of the other (*B* and *b*).

Even simple plants, animals, fungi, and protists have thousands of genes. The number of human genes is unknown, but it may be as high as 100,000. Human beings have forty-six chromosomes in each cell (twenty-three from the mother and twenty-three from the father): twenty-two pairs of autosomal chromosomes plus two sex chromosomes (two X chromosomes in females and an X and a Y chromosome in males). Since humans have only twenty-four kinds of chromosomes, there must be a few thousand genes on the average human chromosome.

If two genes fail to show independent assortment because they are on the same chromosome, they are said to be linked. For example, if the alleles *A* and *B* are on one chromosome and *a* and *b* are on the homologue of that chromosome, then the dihybrid (*AB/ab*) would form gametes with the combinations *AB* and *ab* more often than *Ab* and *aB*. How much more often? At one extreme, if there was no way to rearrange the pieces of two homologous chromosomes, then $\frac{1}{2}$w0 of the gametes would be *AB* and $\frac{1}{2}$ would be *ab*. At the other extreme, if the two genes were so far apart on a large chromosome that rearrangements occurred at

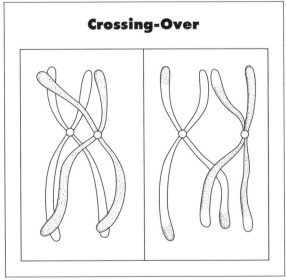

Crossing-Over

In the crossing-over process, chromosomes meet (left) and recombine (right).

almost every meiosis, they would assort independently. When two genes are on the same chromosome but show no linkage, they are said to be "syntenic."

Meiosis is usually accompanied by crossing-over. In the first stage of meiosis, each pair of homologous chromosomes pairs tightly (synapsis). At this stage of meiosis, each homologous chromosome is composed of two chromatids, so there are four complete DNA molecules (a tetrad) present in the paired homologous chromosomes. A reciprocal exchange of pieces of two paired homologous chromosomes produces new combinations of the linked genes (recombinants). For example, crossing-over in a dihybrid with *AB* on one chromosome and *ab* on its homologue could form *Ab* and *aB* recombinants. Each crossover forms one pair of recombinant chromosomes. The average number of crossovers during a meiotic division differs between species and sometimes between the sexes of a single species. For example, crossing-over does not occur in male fruit flies (*Drosophila melanogaster*), and it may occur slightly less often in human males than in females. Nevertheless, within a single sex of a single species, the number of crossovers during a meiotic division is fairly constant.

Linkage Maps

If two genes are very close together on the same chromosome, crossing-over between them is rare, and thus recombinant gametes are also rare. Conversely, crossing-over will occur more frequently between two genes that are farther apart on the same chromosome. This fact has been used to construct linkage maps (also called crossover maps or genetic maps) of the chromosomes of many species. The distances between genes on linkage maps are expressed as the percent crossing-over. One percent crossing-over is equal to one centiMorgan (cM). If two genes are 12 cM apart on a linkage map, a dihybrid will form twelve recombinant gametes for every eighty-eight nonrecombinant gametes. Linkage maps are made by combining data from many different experiments. For instance, suppose that a cross between a dihybrid *AB/ab* individual and a homozygous *ab/ab* individual produced 81 *AB/ab* + 83 *ab/ab*

progeny (non-crossover types) and 20 *Ab/ab* + 16 *aB/ab* progeny (crossover types). The linkage map distance between *a* and *b* equals $100(20 + 16)/(81 + 83 + 20 + 16) = 18$ cM.

The table shows the frequency of crossover gametes from test crosses of three different dihybrids, including the one already described:

gene pair	cM
a and *b*	18
a and *c*	7
b and *c*	11

It is clear that the *c* gene must be between *a* and *b* on the linkage map. The absolute order, *acb* or *bca*, is arbitrarily defined by the first person who constructs a linkage map of a species.

In this example, the linkage map is exactly additive. In real experiments, linkage map distances are seldom exactly additive.

Once a large number of genes on the same chromosome have been mapped, the linkage map is redrawn with map positions rather than map distances. For example, if many other experiments provided more information about linked genes, the following linkage map might emerge:

p	*q*	*a*	*c*	*b*	*r*	*s*
0	6	14	21	32	39	49

The *a* and *c* genes are still 7 cM apart ($21 - 14 = 7$), and the other distances on the first map are also still the same.

Very detailed linkage maps have been constructed for some plants, animals, fungi, and protists that are of particular value to medicine, agriculture, industry, or scientific research. Among them are *Zea mays* (maize), *Drosophila melanogaster* (fruit fly), and *Saccharomyces cerevisiae* (baker's yeast). The linkage map of *Homo sapiens* (humans) is not very detailed because it is unethical and socially impossible to arrange all of the desired crosses that would be necessary to construct one. Other techniques have allowed the construction of very detailed physical maps of human chromosomes.

Genetic Linkage Maps and the Structure of Chromosomes

It should be emphasized that the linkage map is not a scale model of the physical chromosome. It is generally true that the relative order of genes on the linkage map and the physical chromosome map are the same. However, the relative distances between genes on the linkage map may not be proportionately the same on the physical map. Consider three genes (*a*, *b*, and *c*) that are arranged in that order on the chromosome. Suppose that the *ab* distance on the physical map is exactly the same as the *bc* distance. If the frequency of crossing-over between *a* and *b* is higher than the cross-over frequency between *b* and *c*, then the *ab* linkage map distance will be larger than the *bc* linkage map distance. It is common to find small discrepancies between linkage maps and physical maps all along the chromosome. Large discrepancies are usually limited to genes that are close to the centromere. The frequency of crossing-over is generally very low near the centromere (an area rich in heterochromatin). If two genes are on opposite sides of the centromere, they will be fairly far apart on the physical map but very close together on the linkage map because of the low frequency of crossing-over between them.

—*James L. Farmer*

See Also: Chromosome Structure; Dihybrid Inheritance; DNA Structure and Function; Meiosis; Mendel, Gregor, and Mendelism.

Further Reading: Virtually every introductory biology textbook contains at least a superficial discussion of linkage maps. More advanced discussions can be found in Gordon Edlin, *Human Genetics* (1990), and Ursula Goodenough, *Genetics* (1984).

Meiosis

Field of study: Classical transmission genetics

Significance: *Meiosis is the process that reduces the number of sets of chromosomes per cell from two to one, often during the formation of eggs and sperm. It is the primary source of genetic individuality.*

Key terms

CHROMATID: one of the two conjoined copies of a chromosome that result from the duplication of the chromosomal material

CHROMOSOMES: one or more linear structures within cells each containing some or all of the genes of the cell

DIPLOID: a cell containing two sets of chromosomes, one derived from the father, one from the mother

GAMETE: a reproductive cell, usually either an egg or a sperm or their equivalents

HAPLOID: a cell containing one set of chromosomes

MEIOSIS: a special pair of cell divisions that reduces the chromosome content of a cell from two sets to one

NUCLEUS: a structure found in eukaryotic cells within which the chromosomes are located

The Role of Meiosis

During sexual reproduction of organisms with cells that have a distinct nucleus (eukaryotic cells), an egg and a sperm cell join to form the first cell of a new individual. Both the egg and the sperm contain a complete set of chromosomes, that is, a number of chromosomes characteristic to the particular organism. Consequently, the new cell formed by the union of egg and sperm has two sets of chromosomes. If this process continued generation after generation, the number of chromosomes per cell would double with each generation and soon become very large. This increase of chromosomes does not occur, however, because once in each life cycle the chromosome content of at least those cells destined to become gametes—eggs and sperm—undergo meiosis, a process that reduces the number of sets of chromosomes from two back to one per cell. Meiosis was discovered largely by Edouard

van Beneden, Theodor Boveri, and Oskar Hertwig in the 1880's.

Generally, the one set of chromosomes in each resulting gamete is a new assortment, with some derived from the father and some from the mother of the organism making the gamete. Furthermore, paternally and maternally derived chromosomes of the same kind often exchange pieces during meiosis and form chromosomes with a new assortment of genes. This new assortment of chromosomes and of genes within chromosomes, generally found in the gametes that start a new individual, means that the new organism will almost certainly have an assortment of genes quite different from that of either parent and possibly unique in the history of life.

The Timing of Meiosis

In vertebrates, most other animals, and many plants, meiosis occurs in direct connection with the formation of gametes. Since an individual results from the union of two gametes, each of its cells contains two sets of chromosomes. Such cells are called diploid. However, other patterns exist. For example, in many types of algae and protozoa, meiosis follows more or less immediately after the union of egg and sperm; the cells of the organism throughout almost the entire life cycle contain only one set of chromosomes. Such cells are called haploid. In still other cases, such as liverworts and mosses, specialized body cells undergo meiosis and then separate from the rest of the organism, forming independent individuals with only one set of chromosomes per cell. Certain cells of these individuals later may become egg or sperm cells without further change in number of sets of chromosomes. Thus, there is an alternation of generations, one diploid and one haploid, in the complete life cycle.

In humans and similar organisms, meiosis usually occurs immediately before the formation of sperm. However, it is not uncommon to find that meiosis in cells designated to become eggs starts long before the eggs are formed and might not be completed until after the eggs are released from the mother's ovary or perhaps not even until egg and sperm have joined.

The Process: First Division

Meiosis consists of two specialized cells divisions. In the first, sometimes called the reduction division, the number of chromosomes is cut in half. During the second, or equation, division, the two chromatids (conjoined copies) of each chromosome are separated. Each division has its own important peculiarities.

As is true of ordinary (mitotic) cell divisions, the chromosomal materials—including the DNA, the genes, and most of the protein components of the chromosomes—are replicated before the beginning of the division. After replication, each chromosome consists of two chromatids. The two chromatids of each chromosome are tightly attached to each other and will remain so until the second division of meiosis. At the start of meiosis, therefore, each chromosome consists of two chromatids. Each of the two chromatids of a chromosome carries the same genetic information; they are known as sister chromatids of each other.

Again as in mitotic cell divisions, at the start of the first meiotic division, chromosomes are long and very thin. As the process proceeds into its first phase, known as prophase I, they become shorter and wider. This part of pro-

phase I is known as the leptotene substage.

While the shortening and thickening of the chromosomes continues, a new phenomenon of great importance also begins. Since one of the two sets of chromosomes of each cell was originally contributed by the mother and one by the father, each set contains equivalent chromosomes, almost all carrying equivalent, although not necessarily identical, genetic information. A pair of chromosomes with equivalent information is known as a homologous pair, and the two chromosomes of a homologous pair are said to be homologs of each other. The new phenomenon consists of the coming together of the homologs and is called synapsis. During the time that the homologs find each other and join together, the cell is in the zygotene substage of prophase I. The four chromatids of each pair of chromosomes become entangled with one another in the next substage, known as pachytene.

During the diplotene substage of prophase I, it becomes visually evident under the microscope that each chromosome consists of two chromatids. The two chromatids of a single chromosome, the sister chromatids, are still attached to each other at a special place, the centromere. However, the rest of the sister chromatids are now separate and usually seen as separate. Since the homologs have synapsed by this time, each unit, or tetrad, consists of the

The Stages of Meiosis I

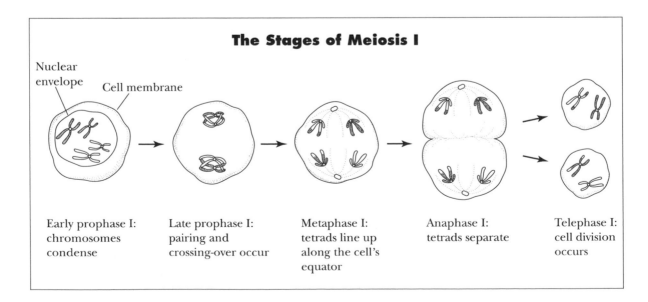

Early prophase I: chromosomes condense

Late prophase I: pairing and crossing-over occur

Metaphase I: tetrads line up along the cell's equator

Anaphase I: tetrads separate

Telephase I: cell division occurs

The Stages of Meiosis II

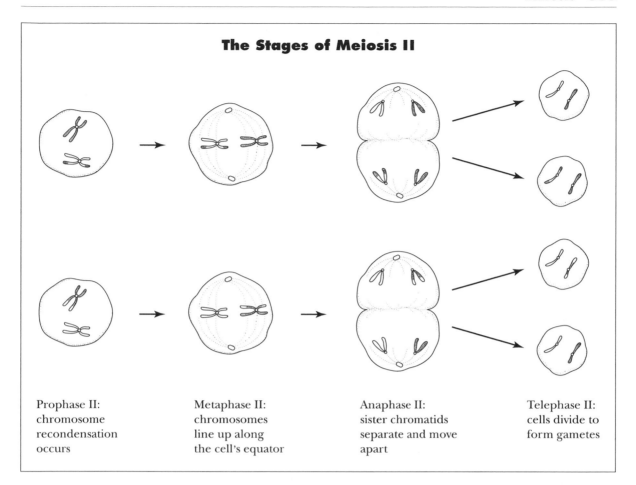

Prophase II:
chromosome
recondensation
occurs

Metaphase II:
chromosomes
line up along
the cell's equator

Anaphase II:
sister chromatids
separate and move
apart

Telephase II:
cells divide to
form gametes

four chromatids of the pair of homologous chromosomes.

During the last substage of prophase I, known as diakinesis, the chromosomes have become maximally short and thick. They are now ready for the next stage, metaphase I, in which the tetrads line up along the equator of the cell, which is the line at which the cell will soon divide into two.

Once the bundles of two homologous chromosomes or four chromatids are lined up, anaphase I begins. The homologous chromosomes of each tetrad separate, and one chromosome moves in one direction while the other moves directly opposite. These movements are coordinated by spindle fibers that extend from each pole of the cell to the centromeres of the chromosomes. The chromosomes move along, or are moved by, these fibers. Since each pair of sister chromatids still shares a single cen-

tromere, they must necessarily move together. Therefore, one chromosome of each homologous pair, consisting of two chromatids, moves toward one pole, and the other member of the pair moves toward the other pole.

Finally, in telophase I, one member of each pair of homologous chromosomes collects around one pole and the other around the other pole. New nuclear membranes then form around the chromosomes, and the cell divides into two cells, completing the first of the two meiotic divisions.

The Process: Second Division

The second meiotic division is quite straightforward, but no replication occurs between the two divisions. None is needed, as each chromosome still consists of two chromatids.

During prophase II, the chromosomes again become shorter and thicker and move toward the equators of each of the two cells. They arrive at the equators during metaphase II. Also in metaphase II, each centromere separates into two parts. Thus, the sister chromatids become completely separate structures and now make up a chromosome. Spindle fibers from opposite poles attach to the centromeres of each pair of newly formed chromosomes, setting the stage for anaphase II, during which the former sister chromatids travel in opposite directions toward the poles of the two future cells. This movement is completed during telophase II.

The chromosomes are now assembled at the two poles. New nuclear membranes are formed around them, and the cell divides into two parts. With the formation of a total of four cells, each containing one complete set of chromosomes, the process of meiosis is complete.

Genetic Consequences: Segregation

Gregor Mendel's principles of segregation and assortment are both largely the results of the events in meiosis, an idea developed mostly by Walter Sutton and Theodor Boveri in the first decade of the twentieth century. According to the principle of segregation, the two versions, or alleles, of each gene that are generally present in each diploid cell separate during the formation of gametes. One and only one of each pair goes to any one reproductive cell. Sutton and Boveri noticed that exactly the same can be said of the chromosomes of each homologous pair.

As discussed above, the chromosomes of each homologous pair separate toward the end of meiotic division I, and their sister chromatids, after becoming chromosomes, end up separately in each of the four cells resulting from meiosis. Since genes are constituents of these chromosomes, they, too, must separate and move individually to the prospective gametes. Consider a pair of homologous chromosomes. One carries a version, or allele, of the gene F dealing with some property of the organism, for example with the attachment of the ear lobe to the head. The other chromosome of the pair must, by the definition of "homologous pair," also carry a version of this gene. If one of the homologs carries the F allele of the

gene and the other the f one, then the organism in which the meiosis takes place will have the genetic arrangement or genotype Ff, but each of the gametes produced by that organism will contain either the F or the f version, not both.

Genetic Consequences: Assortment

Mendel's second principle, the principle of independent assortment, states that, with an important exception, the way that the pair of alleles of one gene segregates is independent of the way the pair of alleles of another gene segregates. (The exception applies to genes located fairly close to each other on the same pair of chromosomes.) This, too, is the result of meiosis. Suppose one observes the alleles A and a of the gene A, located on one pair of chromosomes, and also the alleles B and b of the gene B, located on another chromosome. The two chromosomes and four chromatids carrying the two versions of the A gene come together in synapsis during prophase I. Similarly, the pair of chromosomes carrying the alleles of gene B come together. Now, toward the end of prophase I, it may be that the chromosome with A disentangles itself on same side of the equator as does the chromosome with B of the B family. In that case, A and B will end up together in two of the four final cells and a and b together in the other two cells. However, it is equally likely that, after synapsis, the chromosome with A and the chromosome with b will be found on the same side of the equator and therefore will be together in two of the four final cells. Therefore, a and B will be together in the other two prospective gametes. If the chromosomes with, say, A and B originally came from the individual's mother and those with a and b from those of the father, then a reproductive cell with either Ab or aB would be a new combination.

This idea can be extended. If an organism has two pairs of chromosomes, then four types of gametes with respect to genes on these chromosomes are possible. In the above example, these are gametes with the gene complements AB, ab, Ab, and aB. If an organism has three pairs of chromosomes, then there are four different ways in which the chromosomes can move away from their synaptic union, resulting in eight possible types of gametes. If there are four pairs of chromosomes, there are eight possible chromosome arrangements in metaphase I, resulting in sixteen possible, different gametes. In general, when there are n pairs of chromosomes in a cell, there are 2^n possible types of gametes resulting from the sorting out of the synaptic unions of homologous chromosomes. For human beings, in which $n = 23$, that amounts to 8,388,608 kinds of gametes. Only two of these 8,388,608 combinations occurred in the parents of the individual producing the gametes.

This number of different gametes is possible in the production of both eggs and sperm. Therefore, there are 8,388,608 times 8,388,608, or about 7×10^{13}, combinations of chromosomes possible in the children of a single couple. This number is more than ten thousand times the entire population of the earth. It is, therefore, not surprising that each human being (except, to some extent, identical twins) is a genetically distinct individual. As discussed below, even this number understates the genetic variation among human beings.

Genetic Consequences: Crossing-Over

One additional process during meiosis increases even further the genetic difference between offspring and parent. During synapsis, the chromatids of the homologous chromosomes do not lie side by side; they lie across each other in a nearly chaotic pattern. At each place where the two chromatids cross, they may break and rejoin. If the two chromatids involved are sisters, it does not matter how the broken pieces rejoin; there is no genetic consequence. If, however, one chromatid belongs to one chromosome and the other to its homolog and each broken piece reattaches to the chromatid from which it did *not* come—that is, if they exchange places—then there usually are genetic consequences. Such an event is called "crossing-over."

To see why crossing-over has genetic consequences, consider a pair of homologs, one with the alleles *D* and *E* of the genes D and E, the other with the alleles *d* and *e*. Now suppose that crossing-over takes place on the chromosomes.

The original arrangement of *DE* and *de* will have become *De* and *dE*. Therefore, a different set of alleles of these genes will go to a gamete than would have been the case if no crossing-over had occurred. The assortment of genes received by the gametes and the offspring will be still more different from that in the parent. Since crossing-over is quite common, even more than 7×10^{13} genetic combinations are possible among the children of a single couple.

The many genetic combinations among the offspring of one pair of sexually reproducing organisms means a large variety of different gene combinations among the members of any sexually reproducing species. In the human case, for example, hardly any two people (except identical twins) have the same gene composition and therefore the same cell constituents. This fact makes exact matches for tissue or organ transplantation very unlikely and gives humans much of their biological individuality.

—*Werner G. Heim*

See Also: Chromosome Theory of Heredity; Classical Transmission Genetics; Linkage Maps; Nondisjunction and Aneuploidy.

Further Reading: Almost any modern textbook of cell biology or genetics will give a thorough, illustrated description of meiosis and its role in genetics. Good examples include Wayne M. Becker et al., *The World of the Cell*, 3d ed. (1996); Daniel L. Hartl and Elizabeth W. Jones, *Genetics: Principles and Analysis*, 4th ed. (1998); and Michael R. Cummings, *Human Heredity: Principles and Issues*, 4th ed. (1998). John A. Moore, *Science as a Way of Knowing* (1993), describes well the historical origin of knowledge about these matters.

Mendel, Gregor, and Mendelism

Field of study: Classical transmission genetics

Significance: *Gregor Mendel was a monk and a science teacher in Moravia when he wrote his famous paper about experimental crosses of pea plants. Little note was taken of it when it was* *published in 1866, but it provided concepts and methods that catalyzed the growth of modern genetics after 1900 and earned Mendel posthumous renown as the founder of the new science.*

Key terms

HYBRID: a plant form resulting from a cross between two distinct varieties

DOMINANCE: the appearance, in the hybrid, of a distinguishing trait from one parent to the exclusion of a contrasting trait from the other parent

GAMETES: reproductive cells that unite during fertilization to form an embryo; in plants, the pollen cells and egg cells are gametes

SEGREGATION: the process of separating a pair of Mendelian hereditary elements (genes), one from each parent, and distributing them at random into the gametes

INDEPENDENT ASSORTMENT: the segregation of two or more pairs of genes without any tendency for certain genes to stay together

Early Life

Born Johann Mendel on July 22, 1822, the future teacher, monk, abbot, botanist, and meteorologist grew up in a village in Moravia, a province of the Austrian Empire that later became part of Czechoslovakia (1918) and the Czech Republic (1993). His parents were peasant farmers and belonged to the large, German-speaking minority in this predominantly Czech province. Like most places in Moravia, Mendel's hometown had two names: Hynčice in Czech and Heinzendorf in German.

Johann Mendel was an exceptional pupil, but no local schooling was available for him beyond the age of ten. In 1833, he persuaded his parents to send him to town to continue his education. They were reluctant to let him go because they could ill afford to dispense with his help on the farm or finance his studies. In 1838, Mendel's father was partially disabled in a logging accident, and Johann, then sixteen and still at school, had to support himself. He earned just enough from tutoring to get by. At times, however, the pressure became too much for him. He suffered a breakdown in 1839 and returned home for several months to recuperate. He was to have several more of these stress-related illnesses, but no precise information is

available about their causes and symptoms.

In 1840, Mendel completed *Gymnasium*, as the elite secondary schools were called, and entered the University of Olomouc for the two-year program in philosophy that preceded higher university studies. He had trouble supporting himself in Olomouc, perhaps because there was less demand for German-speaking tutors, and his Czech was not good enough for teaching. He suffered another breakdown in 1841 and retreated to Hynčice during spring exams.

That summer, Mendel decided once more against staying and taking over the farm. Since his father could not work, the farm was sold to his elder sister's husband. Johann's share of the proceeds was not enough to see him through the Olomouc program, especially since he had to repeat a year because of the missed exams. However, his twelve-year-old sister sacrificed part of her future dowry so that he could continue. (He repaid her years later by putting her three sons through *Gymnasium* and university.)

Upon finishing at Olomouc in 1843, Mendel decided to enter the clergy. The priesthood filled his need for a secure position and held out possibilities for further learning and teaching, but Mendel does not seem to have felt called to it. Aided by a professor's recommendation, Mendel was accepted into the Augustinian monastery in Brno, the capital of Moravia, where he took the name Gregor. In 1847, after four years of preparation at the monastery, he was ordained a priest.

Priesthood and Teaching

The Brno monastery was active in the community and provided highly qualified instructors for *Gymnasia* and technical schools throughout Moravia. Several monks, including the abbot, were interested in science, and they had experimental gardens, a herbarium, a mineralogical collection, and an extensive library. Mendel found himself in learned company with opportunities for research in his spare time.

Unfortunately, Mendel's nerves failed him when he had to minister to the sick and dying. Assigned to a local hospital in 1848, he was so upset by it that he was bedridden himself within

Gregor Mendel, the "father of genetics," examining a flower in his laboratory. (Archive Photos)

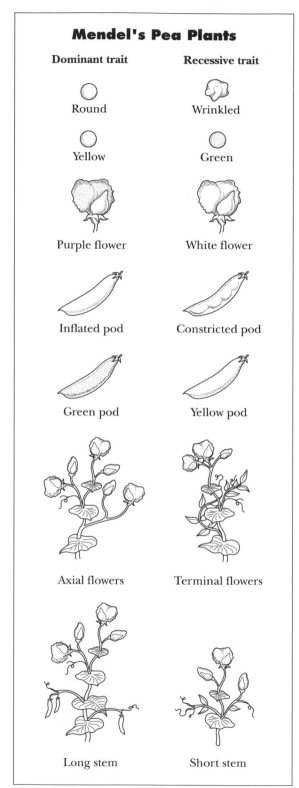

Mendel's Pea Plants

Dominant trait	Recessive trait
Round	Wrinkled
Yellow	Green
Purple flower	White flower
Inflated pod	Constricted pod
Green pod	Yellow pod
Axial flowers	Terminal flowers
Long stem	Short stem

Mendel evaluated the transmission of seven paired traits in his studies of garden peas.

five months. However, his abbot was sympathetic and let him switch to teaching. A letter survives in which the abbot explains this decision to the bishop: "[Mendel] leads a retiring, modest and virtuous religious life . . . and he devotes himself diligently to scholarly pursuits. For pastoral duties, however, he is less suited, because at the sick-bed or at the sight of the sick or suffering he is seized by an insurmountable dread, from which he has even fallen dangerously ill."

Mendel taught Latin and Greek, German literature, math, and science as a substitute at the *Gymnasium* and was found to be very good at teaching. Therefore, he was sent to Vienna in 1850 to take the licensing examinations so that he could be promoted to a regular position. These exams were very demanding and normally required more preparation than Mendel's two years at Olomouc. Mendel failed, but one examiner advised the abbot to let him try again after further study. The abbot took this advice and sent Mendel to study in Vienna for two years (1851-1853). There he took courses in biology, physics, and meteorology with some of the best-known scientists of his day, including physicist Christian Doppler and botanist Franz Unger.

For unknown reasons, Mendel returned to Moravia to resume substitute teaching and did not go to Vienna for the exams until 1856. This time he was too nervous to finish. After writing one essay, he fell ill and returned to Brno. Despite this failure, he was allowed to teach regular classes until 1868 even though he was technically only a substitute.

Scientific Work

During his teaching career, Mendel performed his famous experiments on peas in a garden at the monastery. He published the results in an 1866 article, which introduced fundamental concepts and methods of genetics. The first set of experiments involved fourteen varieties of pea plant, each with a single distinguishing trait. These traits made up seven contrasting pairs, such as seeds that were either round or wrinkled in outline or seed colors that were green or yellow. Upon crossing each pair, Mendel obtained hybrids identical to one parent variety. For example, the cross of round with wrinkled peas yielded only round peas; the cross of green with yellow peas yielded only yellow peas. He referred to traits that asserted themselves in the hybrids as "dominant." The others were "recessive" because they receded from view. The effect was the same regardless of whether he fertilized the wrinkled variety with pollen from the round or the round variety with pollen from the wrinkled. This indicated to Mendel that both pollen cells and egg cells contributed equally to heredity; this was a significant finding because the details of plant reproduction were still unclear.

Mendel next allowed the seven hybrids to pollinate themselves, and the recessive traits reappeared in the second generation. For instance, the round peas, which were hybrids of round and wrinkled peas, yielded not only more round peas but also some wrinkled ones. Moreover, the dominant forms outnumbered the recessives three to one. Mendel explained the 3:1 ratio as follows. He used the symbols A for the dominant form, a for the recessive, and Aa for the hybrid. A hybrid, he argued, could produce two types of pollen cell, one containing some sort of hereditary element corresponding to trait A and the other an element corresponding to trait a. Likewise, it could produce eggs containing either A or a elements. This process of dividing up the hereditary factors among the gametes became known as "segregation."

The gametes from the Aa hybrids could come together in any of four combinations: pollen A with egg A, pollen A with egg a, pollen a with egg A, and pollen a with egg a. The first three of these combinations all grew into plants with the dominant trait A; only the fourth produced the recessive a. Therefore, if all four combinations were equally common, one could expect an average of three plants exhibiting A for every one exhibiting a.

Allowing self-pollination to continue, Mendel found that the recessives always bred true. In other words, they only produced more plants with that same recessive trait; no dominant forms reappeared, not even in subsequent generations. Mendel's explanation was that the recessives could only have arisen from the pollen a and egg a combination, which excludes the A element. For similar reasons, plants with the dominant trait bred true one-third of the time, depending on whether they were the pure forms from the pollen A and egg A combination or the hybrids from the pollen A and egg a or pollen a and egg A combinations.

Mendel's hereditary elements sound like the modern geneticist's genes or alleles, and Mendel usually receives credit for introducing the gene concept. Like genes, Mendel's elements were material entities inherited from both parents and transmitted to the gametes. They also retained their integrity even when recessive in a hybrid. However, it is not clear whether he pictured two copies of each element in every

The Results of Mendel's Pea-Plant Experiments

Parental characteristics	First generation	Second generation	Second generation ratio
Round × wrinkled seeds	All round	5,474 round : 1,850 wrinkled	2.96 : 1
Yellow × green seeds	All yellow	6,022 yellow : 2,001 green	3.01 : 1
Gray × white seedcoats	All gray	705 gray : 224 white	3.15 : 1
Inflated × pinched pods	All inflated	882 inflated : 299 pinched	2.95 : 1
Green × yellow pods	All green	428 green : 152 yellow	2.82 : 1
Axial × terminal flowers	All axial	651 axial : 207 terminal	3.14 : 1
Long × short stems	All long	787 long : 277 short	2.84 : 1

cell, one copy from each parent, and he certainly did not associate them with chromosomes.

In a second set of experiments, Mendel tested combinations of traits to see whether they would segregate freely or tend to be inherited together. For example, he crossed round, yellow peas with wrinkled, green ones. That cross first yielded only round, yellow peas, as could be expected from the dominance relationships. Then, in the second generation, all four possible combinations of traits segregated out: not only the parental round yellow and wrinkled green peas but also new round green and wrinkled yellow ones. Mendel was able to explain the ratios as before, based on equally likely combinations of hereditary elements coming together at fertilization. The free regrouping of hereditary traits became known as "independent assortment." In the twentieth century, it was found not to occur universally because some genes are linked together on the same chromosome.

Mendel's paper did not reach many readers. As a *Gymnasium* teacher and a monk in Moravia without even a doctoral degree, Mendel could not command the same attention as a university professor in a major city. Also, it was not obvious that the behavior of these seven pea traits illustrated fundamental principles of heredity. Mendel wrote to several leading botanists in Germany and Austria about his findings, but only Carl von Nägeli at the University of Munich is known to have responded, and even he was skeptical of Mendel's conclusions. Mendel published only one more paper on heredity (in 1869) and did little else to follow up his experiments or gain wider attention from scientists.

Mendel pursued other scientific interests as well. He was active in local scientific societies and was an avid meteorologist. He set up a weather station at the monastery and sent reports to the Central Meteorological Institute in Vienna. He also helped organize a network of weather stations in Moravia. He envisioned telegraph connections among the stations and with Vienna that would make weather forecasting feasible. In his later years, Mendel studied sunspots and tested the idea that they affected the weather. He also monitored the water level in the monastery well in order to test a theory that changes in the water table were related to epidemics. A common thread that ran through these diverse research interests was that they all involved counting or measuring, with the goal of discovering scientific laws behind the numerical patterns. His one great success was in explaining the pea data with his concepts of dominance, segregation, and independent assortment.

Mendel felt pleased and honored to be elected abbot in 1868, even though he had to give up teaching and most of his research. He did not have the heart to say good-bye to his pupils. Instead, he asked the school director to announce his departure and give his last month's salary to the three neediest boys in the class. As abbot, Mendel had a reputation for generosity to the poor and to scientific and cultural institutions. He was also an efficient manager of the monastery and its extensive land holdings and a fierce defender of the monastery's interests. From 1874 on, he feuded with imperial authorities over a new tax on the monastery, which he refused to pay as long as he lived. Mendel's health failed gradually in the last years of his life. He had kidney problems and an abnormally fast heartbeat, the latter probably from nerves and nicotine. (A doctor recommended smoking to control his weight, and he developed a twenty-cigar-a-day habit.) He died January 6, 1884, of heart and kidney failure.

Impact and Applications

Years after Mendel's death, a scientific colleague remembered him saying, prophetically, "my time will come." It came in 1900, when papers by three different botanists reported experimental results similar to Mendel's and endorsed Mendel's long-overlooked explanations. This event became known as the rediscovery of Mendelism. By 1910, Mendel's theory had given rise to a whole new field of research, which was given the name "genetics." Mendel's hereditary elements were described more precisely as "genes" and were presumed to be located on the chromosomes. By the 1920's, the sex chromosomes were identified, the determination of sex was explained in Mendelian terms, and the arrangements of genes on chromosomes could be mapped.

The study of evolution was also transformed by Mendelian genetics, as Darwinians and anti-Darwinians alike had to take the new information about heredity into account. By 1930, it had been shown that natural selection could cause evolutionary change in a population by shifting the proportions of individuals with different genes. This principle of population genetics became a cornerstone of modern Darwinism.

Investigations of the material basis of heredity led to the discovery of the gene's deoxyribonucleic acid (DNA) structure in 1953. This breakthrough marked the beginning of molecular genetics, which studies how genes are copied, how mutations occur, and how genes exert their influence on cells. In short, all of modern genetics can trace its heritage back to the ideas and experiments of Gregor Mendel.

—*Sander Gliboff*

See Also: Classical Transmission Genetics; Dihybrid Inheritance; Monohybrid Inheritance.

Further Reading: The first biography of Gregor Mendel, Hugo Iltis's *Life of Mendel* (1932), is still among the best; for more recent accounts see *Gregor Mendel, the First Geneticist* (1996), by Víteslav Orel, or the shorter *Mendel* (1984), by the same author. In *The Origins of Mendelism* (1985), Robert Olby discusses the history of genetics from the 1700's through the rediscovery of Mendel.

Methane-Producing Bacteria

Field of study: Bacterial genetics
Significance: *Methane-producing bacteria, also known as methanogens, are unique, primitive organisms that convert simple molecules such as carbon dioxide, methanol, and hydrogen into methane. They can be found in sediments and biomasses and exist in a variety of shapes and growth patterns. They provide an excellent source of methane, which can serve as a clean fuel.*

Key terms

ANAEROBIC BACTERIA: bacteria that thrive in the absence of oxygen

EUKARYOTES: complex, unicellular or multicellular organisms with cells that have definite nuclei

PROKARYOTES: simple, unicellular organisms (such as bacteria and algae) with cells that lack clearly defined nuclei

METHANOGENESIS: the conversion of several substances to methane by microorganisms that reside in anaerobic habitats; such substances include hydrogen, carbon dioxide, formic acid, methanol, methyl amine, and acetate anion

REDUCTION: a process in which two hydrogen atoms (molecular hydrogen) or a pair of electrons are added to a substance

Characteristics of Methanogens

Methanogens are the only microorganisms that produce methane as a way of life, and they have crucial effects in various ecosystems. They are important in anoxic (oxygen-free) biogeochemical cycles that convert one-carbon compounds and acetate to methane via pathways that involve unusual coenzymes. The most common reduction pathways are those that involve the transformation of carbon dioxide to methane (an eight-electron process) and the transformation of a methyl group to methane (a two-electron step). Less popular pathways involve the reduction of dimethyl sulfide, carbon monoxide, and simple alcohols such as methanol, ethanol, and propanol.

Methanogens are found in anaerobic habitats that lack the presence of oxygen, such as sewage sludge, the intestinal tracts of animals, hot springs, and marsh soils. In this anaerobic microbial food chain, the final product, methane, is released, since it is water-insoluble and inert toward any type of further nonmicrobial oxidation. The ability to form methane, which is characterized by its so-called reducing potential, is achieved and maintained by the contribution of molecular hydrogen gas or hydrogen sulfide.

Methane is produced by hydrogenases of the fermentative anaerobes in freshwater sediments and the intestinal tracts of animals. In cows, for instance, two highly specialized digestive organs called the rumen and cecum block or delay the passage of cellulose fiber biomasses while the microorganisms produce methane gas from hydrogen and carbon dioxide or formic acid. As a result, a cow may be able to produce, via belching, twenty-five to thirty gallons of methane gas per day. In other methanogenic habitats, the reducing potential is reached by the presence of hydrogen sulfide, which is geochemically produced in hydrothermal vents. In sediments, hydrogen sulfide is produced by reduction of the sulfate anion.

Methanogens Versus Other Bacteria

Biologists have classified two forms of life: prokaryotes and eukaryotes. Prokaryotes are unicellular organisms, such as bacteria and algae, that lack a clearly defined nucleus. Eukaryotes are complex, unicellular or multicellular organisms that possess a clearly defined nucleus. There is also a distinct difference in the metabolic chemistry and genetic constitution of each type.

Methanogens, however, cannot be strictly defined as either eukaryotes or prokaryotes, as evidenced by at least three facts. First, the chemical composition of the cell walls is not complete in methanogens, which lack the peptidoglycan that common bacteria have. Second, the metabolic pathways in methanogenesis are unique. Third, the genetic code of methanogens differs from comparable materials of other organisms. It is therefore believed that although all cells had common ancestors, an evolutionary step took place about 3.4 billion years ago that separated eukaryotes from prokaryotes and, at the same time, created the new line of methanogens, which are also called archaebacteria (*archae* is the Greek term for "ancient"). Surprisingly, methanogens appear to adapt themselves to the present conditions of life despite the fact that their composition is as primitive as it was several billion years ago when the atmosphere contained no oxygen and when earth lacked the complex organic food materials that are currently in existence.

The isolation of methanogens is not simple since, in many cases, any contact with air leads to the destruction of its activity. As a result,

special bacteriological techniques must be employed. Thus, plating methanogens on solid agar in petri dishes requires anaerobic conditions, usually provided by anaerobic chambers. Because of the lack of peptidoglycan in the cell walls of methanogens, common antibiotics such as penicillin have no harmful effect. On the contrary, the growth of common bacteria at the expense of methanogens is avoided by the radical effect antibiotics have on the cell walls of the common bacteria. This technique serves as the "purification" step of methanogens, which can thus be isolated selectively from crude samples.

Several methanogens that differ in shape, source, and optimum growth temperature are currently known and are classified accordingly. *Methanococcus vannielii*, for instance, is isolated from marine sediment and grows at 20 to 40 degrees Celsius, while *Methanococcus jannaschii* is found in a deep-sea hydrothermal vent and has a growth temperature range of 48 to 94 degrees Celsius. Both are grain shaped. On the other hand, *Methanothermobacter thermoautotrophicus* is rod shaped, isolated from a sewage sludge digester, and has an optimum temperature range of 50 to 75 degrees Celsius. This microorganism, which can be routinely isolated in a petri dish under anaerobic conditions, has served in the understanding of the biochemistry involved in the reduction of carbon dioxide to methane. Other examples include *Methanospirillum bungateii*, which is found in sewage sludge and is spiral shaped, *Methanosaeta concilii*, which is shaped like a blunt rod and is encased in a sheath, and *Methanosphera stadtmaniae*, which is spherical.

Biodegradation of materials such as food grease is a very important subject that has been concerning scientists in the 1990's and will gain in significance as the earth's population increases. Methanogens appear to be of interest because of their ability to produce an important fuel, methane, via several biodegradation schemes. Two-thirds of the methane produced by many anaerobic habitats, including bioreactors that convert waste to methane, is derived from acetate. Nevertheless, acetate-utilizing methanogens are poorly characterized. The role of these bacteria is currently monitored by

Methane-producing bacteria flourish in hot springs and other oxygen-restricted habitats. (Michael J. Doolittle / Rainbow)

studying the effect of inhibitors, which have been found to retard the reduction of methyl coenzymes to methane.

—*Soraya Ghayourmanesh*

See Also: Archaebacteria; Bacterial Genetics and Bacteria Structure.

Further Reading: Extensive coverage of methane-producing bacteria is presented in *Methanogenesis: Ecology, Physiology, Biochemistry and Genetics* (1993), edited by J. G. Ferry. Methanogenesis is also discussed in a thorough manner in *Environmental Factors Influencing Methanogenesis in a Shallow Anoxic Aquifer* (1990), by R. E. Beeman, and *Methylotrophy and Methanogenesis*, by P. J. Laye (1983).

Miscegenation and Antimiscegenation Laws

Field of study: Human genetics

Significance: *Miscegenation is the crossing or hybridization of different races. As knowledge of the nature of human variability has expanded, finding useful definitions of the term "race" has become increasingly difficult. Limited understanding of the biological and genetic effects of race crossing played a major role in the development of the eugenics movement and the enactment of antimiscegenation laws in the first half of the twentieth century.*

Key terms

EUGENICS: the control of individual reproductive choices to improve the genetic quality of the human population

POSITIVE EUGENICS: selecting individuals to reproduce who have desirable genetic traits, as seen by those in control

NEGATIVE EUGENICS: preventing the reproduction of individuals who have undesirable genetic traits, as seen by those in control

HYBRIDIZATION: the crossing of two genetically distinct species, races, or types to produce mixed offspring

RACE: in the biological sense, a group of people who share certain genetically transmitted physical characteristics

What Is a Race?

Implicit in most biological definitions of race is the concept of shared physical characteristics that have come from a common ancestor. Humans have long recognized and attempted to classify and categorize different kinds of people. The father of modern systematics, Carolus Linnaeus, described, in his system of binomial nomenclature, four races of humans: Africans (black), Asians (dark), Europeans (white), and Native Americans (red). Skin color in humans has been, without doubt, the primary feature used to classify people, although there is no single trait that can be used to do this. Skin color is used because it makes it very easy to tell groups of people apart. However, there are thousands of human traits. What distinguishes races are differences in gene frequencies for a variety of traits. The great majority of genetic traits are found in similar frequencies in people of different skin color. There may not be a single genetic trait that is always associated with people of one skin color while not appearing at all in people of another skin color. It is possible for a person to differ more from another person of the same skin color than from a person of a different skin color.

Many scientists think that the word "race" is not useful in human biology research. Scientific and social organizations, including the American Association of Physical Anthropologists and the American Anthropological Association, have deemed that racial classifications are limited in their scope and utility and do not reflect the evolving concepts of human variability. It is of interest to note that subjects are frequently asked to identify their race in studies and surveys.

It is useful to point out the distinction between an "ethnic group" and a race. An ethnic group is a group of people who share a common social ancestry. Cultural practices may lead to a group's genetic isolation from other groups with a different cultural identity. Since members of different ethnicities may tend to marry only within their group, certain genetic traits may occur at different frequencies in the group than they do in other ethnic or racial groups, or the population at large.

Miscegenation

Sir Francis Galton, a cousin of Charles Darwin, is often regarded as the "father of eugenics." He asserted that humans could be selectively bred for favorable traits. In his 1869 book *Hereditary Genius,* he set out to prove that favorable traits were inborn in people and concluded

> that the average intellectual standard of the Negro race is some two grades below our own. That the average ability of the [ancient] Athenian race is, on the lowest possible estimate, very nearly two grades higher than our own—that is, about as much as our race is above that of the African Negro.

In spite of its scientific limitations, the work of Galton was widely accepted by political and scientific leaders. Bertrand Russell even suggested that the United Kingdom should issue color-coded "procreation tickets" to individuals based on their status in society: "Those who dared breed with holders of a different colored ticket would face a heavy fine." These "scientific" findings, combined with social and racial stereotypes, led to the eugenics movement and its development in many countries, including England, France, Germany, Sweden, Canada, and the United States.

Laws were passed to restrict the immigration of certain ethnic groups into the United States. Between 1907 and 1940, laws allowing forcible sterilization were passed in more than thirty states. Statutes prohibiting and punishing interracial marriages were passed in many states and, even as late as 1952, more than half the states still had antimiscegenation laws. The landmark decision against antimiscegenation laws occurred in 1967 when the U.S. Supreme Court declared the Virginia law unconstitutional. The decision, *Loving v. Virginia,* led to the erosion of the legal force of the antimiscegenation laws in the remaining states.

Impact and Applications

In spite of antimiscegenation laws and societal and cultural taboos, interracial matings have been a frequent occurrence. Many countries around the world, including the United States, are now racially heterogeneous societies. Genetic studies indicate that perhaps 20 to 30 percent of the genes in most African Americans are a result of admixture of white genes from mixed matings since the introduction of slavery to the Americas more than three hundred years ago. Miscegenation has been widespread throughout the world, and there may not even be such a thing as a "pure" race. No adverse biological effects can be attributed to miscegenation.

—*Donald J. Nash*

See Also: Eugenics; Eugenics: Nazi Germany; Race; Sterilization Laws.

Further Reading: *Coevolution: Genes, Culture, and Human Diversity* (1991), by W. Durham, examines the interactive roles of heredity and culture in bringing about human diversity. The ways in which race has been implicated with crime and intelligence is discussed in *Living with Our Genes* (1998), by Dean Hamer and Peter Copeland. Daniel J. Kevles provides a comprehensive account of the history of eugenics, sterilization laws, and antimiscegenation in *In the Name of Eugenics* (1995).

U.S. laws restricting interracial relationships were invalidated by a 1967 Supreme Court ruling. (Robert McClenaghan)

Mitochondrial Diseases

Field of study: Human genetics

Significance: *Mitochondrial genes are few in number but are necessary for animal cells to grow and survive. Mutations in these genes can result in age-related degenerative disorders and serious diseases of muscles and the central nervous system for which there is no generally effective treatment. Mitochondrial diseases are transmitted maternally and are usually associated with heteroplasmy, a state in which more than one type of gene arrangement, or genotype, occurs in the same individual.*

Key terms

MITOCHONDRIA: small structures enclosed by double membranes found in the cytoplasm of all higher cells that produce chemical power for the cells and harbor their own genetic material

MITOCHONDRIAL DEOXYRIBONUCLEIC ACID (MTDNA): the genetic material, composed of subunits called "nucleotides," found uniquely in mitochondria

MATERNAL INHERITANCE: the transmission pattern characteristically shown by mitochondrial diseases and mutations in mtDNA, where changes that occur in the mother's genetic material are inherited directly by children of both sexes without masking or interference by the mtDNA of the father

HETEROPLASMY: a mutation in which more than one set of mtDNA-coded gene products can be present in an individual organ or tissue type, a single cell, or a single mitochondrion

REPLICATIVE SEGREGATION: a mechanism by which individual mtDNAs carrying different mutations can come to predominate in any one mitochondrion

Mitochondrial Genetics and Disease

A tiny number of genes in animal cells are strictly inherited from the maternal parent and are found in the mitochondria of the cell's cytoplasm. The unique arrangement of subunits making up individual genes is highly mutable, and thousands of different arrangements, or genotypes, are cataloged in humans. Some variants in mitochondrial deoxyribonu-cleic acid (mtDNA) sequences can cause severe defects in sight, hearing, skeletal muscles, and the central nervous system. Symptoms of these diseases often include great fatigue. The diseases themselves are difficult to diagnose accurately, and they are currently impossible to treat effectively. New genetic screening methods based on polymerase chain reaction (PCR) technologies using muscle biopsies are essential for correct identification of these diseases.

A person normally inherits a single mtDNA type, but families are occasionally found in which multiple mtDNA sequences are present. This condition, called heteroplasmy, is often associated with mitochondrial disease. Heteroplasmy occurs in the major noncoding region of mtDNA without much impact, but if it exists in the genes that control the production of cellular energy, severe consequences result. Weak muscles and multiple organs are involved in most mitochondrial diseases, and there can be variable expression of a particular syndrome within the same family that may either increase or decrease with age. It is easiest to understand this problem by remembering that each cell contains a population of mitochondria, so there is the possibility that some mtDNAs will carry a particular mutation while others do not. Organs also require different amounts of adenosine triphosphate (ATP), the cell's energy source produced in mitochondria. If the population of mutated mitochondria grows to outnumber the unmutated forms, most cells in a particular organ may appear diseased. This process has been called "replicative segregation," and a mitochondrial disease is the result. Loss of mtDNA also occurs with increasing age, especially in the brain and heart.

Particular Mitochondrial Diseases

Mitochondrial diseases show a simple pattern of maternal inheritance. The first mitochondrial disease identified was Leber's hereditary optic neuropathy (LHON), a condition associated with the sudden loss of vision when the optic nerve is damaged, usually occurring in a person's early twenties. The damage is not reversible. Biologists now know that LHON is caused by at least four specific mutations that alter the mitochondrial proteins ND1, ND4, and CytB. A second mitochondrial syndrome is myoclonic epilepsy with ragged-red fiber disease (MERRF), which affects the brain and muscles throughout the body. This disease, along with another syndrome called mitochondrial encephalomyopathy, lactic acidosis, and stroke-like episodes (MELAS), is associated with particular mutations in mitochondrial transfer ribonucleic acid (mtRNA) genes that help produce proteins coded for by mtDNA. Finally, deletions and duplications of mtDNA are associated with Kearns-Sayre disease (affecting the heart, other muscles, and the cerebellum), chronic progressive external ophthalmoplegia (CPEO; paralysis of the eye muscles), rare cases of diabetes, heart deficiencies, and certain types of deafness. Some of these conditions have been given specific names, but others have not.

Muscles are often affected by mitochondrial diseases because muscle cells are rich in mitochondria. New treatments for these diseases are based on stimulating undamaged mtDNA in certain muscle precursor cells, called "satellite cells," to fuse to damaged muscle cells and regenerate the muscle fibers. Others try to prevent damaged mtDNA genomes from replicating biochemically in order to increase the number of good mtDNAs in any one cell. This last set of experiments has worked on cells in tissue culture but has not been used on patients. These approaches aim to alter the competitive ability of undamaged genes to exist in a cellular environment that normally favors damaged genes. Further advances in treatment will also require better understanding of the natural ability of mtDNA to undergo genetic recombination and DNA repair.

—Rebecca Cann

See Also: DNA Repair; DNA Replication; Extrachromosomal Inheritance; Genetic Testing; Mitochondrial Genes; Mutation and Mutagenesis; Polymerase Chain Reaction; Protein Synthesis.

Further Reading: Information presented by Peter Raven and George Johnson in *Biology* (1996) helps clarify what mitochondria are and why they require interaction with the cell's nucleus. Lynn Jorde et al., *Medical Genetics* (1995), presents a simple discussion of these diseases in the context of other genetic syndromes that are sex linked or sex limited in their inheritance patterns.

Mitochondrial Genes

Field of study: Classical transmission genetics

Significance: *Mutations in mitochondrial genes have been shown to cause several human genetic diseases associated with a gradual loss of tissue function. Understanding the role that mitochondrial genes and their nuclear counterparts play may lead to the development of treatments for these debilitating diseases. Analysis of the mitochondrial DNA sequence of different human populations has also provided information relevant to the understanding of human evolution.*

Key terms

ADENOSINE TRIPHOSPHATE (ATP): the molecule that serves as the major source of energy for the cell

ATP SYNTHASE: the enzyme that synthesizes ATP

ELECTRON TRANSPORT CHAIN: a series of protein complexes that pump H^+ ions out of the mitochondria as a way of storing energy that is then used by ATP synthase to make ATP

CYTOCHROMES: proteins found in the electron transport chain

RIBOSOMES: organelles that function in protein synthesis and are made up of a large and a small subunit composed of proteins and ribosomal ribonucleic acid (RNA) molecules

Mitochondrial Structure and Function

Mitochondria are membrane-bound organelles that exist in the cytoplasm of eukaryotic cells. Structurally, they consist of an outer membrane and a highly folded inner membrane that separate the mitochondria into several compartments. Between the two membranes is the intermembrane space; the innermost compartment bounded by the inner membrane is referred to as the "matrix." In addition to enzymes involved in glucose metabolism, the matrix contains several copies of the mitochondrial chromosome as well as ribosomes, transfer ribonucleic acid (tRNA), and all the other necessary factors required for protein synthesis. Mitochondrial ribosomes are structurally different from the ribosomes located in the cytoplasm of the eukaryotic cell and, in fact, more closely resemble ribosomes from bacterial cells. This similarity led to the endosymbiont hypothesis developed by Lynn Margulis, which proposes that mitochondria arose from bacteria that took up residence in the cytoplasm of the eukaryotic ancestor.

Embedded in the inner mitochondrial membrane is a series of protein complexes that are known collectively as the "electron transport chain." These proteins participate in a defined series of reactions that begin when energy is released from the breakdown of glucose and end when oxygen combines with $2H^+$ ions to produce water. The net result of these reactions is the movement of H^+ ions (also called protons) from the matrix into the intermembrane space. This establishes a proton gradient in which the intermembrane space has a more positive charge and is more acidic than the matrix. Thus mitochondria act as tiny batteries that separate positive and negative charges in order to store energy. Another protein that is embedded in the inner mitochondrial membrane is an enzyme called adenosine triphosphate (ATP) synthase. This enzyme allows the H^+ ions to travel back into the matrix. When this happens, energy is released that is then used by the synthase enzyme to make ATP. Cells use ATP to provide energy for all of the biological work they perform, including movement and synthesis of other molecules. The concept of linking the production of a proton gradient to ATP synthesis was developed by Peter Mitchell in 1976 and is referred to as the chemiosmotic hypothesis.

Mitochondrial Genes

The mitochondrial chromosome is a circular deoxyribonucleic acid (DNA) molecule that varies in size from about 16,000 base pairs in humans to over 100,000 base pairs in certain species of plants. Despite these size differences, mitochondrial DNA (mtDNA) contains only a few genes that tend to be similar over a wide range of organisms. This discussion will focus on genes located on the human mitochondrial chromosome that has been completely sequenced. These genes fall into two broad categories: those that play a role in mitochondrial protein synthesis and those involved in electron transport and ATP synthesis.

Mitochondria have their own set of ribosomes that consist of a large and a small subunit. Each ribosomal subunit is a complex of ribosomal RNA (rRNA) and proteins. Genes that play a role in mitochondrial protein synthesis include two rRNA genes designated 16S rRNA and 12S rRNA, indicating the RNA for the large and small subunits respectively. Also in this first category are genes for mitochondrial transfer RNA. Transfer RNA (tRNA) is an *L*-shaped molecule that contains the RNA anticodon at one end and an amino acid attached to the other end. The tRNA anticodon pairs with the codon of the messenger RNA (mRNA) and brings the correct amino acid into position to be added to the growing protein chain. Thus the tRNA molecule serves as a bridge between the information in the mRNA molecule and the sequence of amino acids in the protein. Mitochondrial tRNAs are different from those involved in protein synthesis in the cytoplasm. In fact, cytoplasmic tRNAs would not be able to function on mitochondrial ribosomes, nor could mitochondrial tRNAs work with cytoplasmic ribosomes. Thus, mtDNA contains a complete set of twenty-two tRNA genes.

Genes involved in electron transport fall into the second category of mitochondrial genes. The electron transport chain is divided into a series of protein complexes, each of which consists of a number of different proteins, a few of which are encoded by mtDNA. The NADH dehydrogenase complex (called complex I) contains about twenty-two different proteins. In humans, only six of these proteins are encoded by genes located on the mitochondrial chromosome. Cytochrome *c* reductase (complex III) contains about nine proteins, including cytochrome *b*, which is the only one whose gene is located on mtDNA. Cytochrome oxidase (complex IV) contains seven proteins, three of which are encoded by mitochondrial genes. About sixteen different proteins combine to make up the mitochondrial ATP synthase, and only two of these are encoded by mtDNA.

All of the proteins not encoded by mitochondrial genes are encoded by genes located on nuclear chromosomes. In fact, over 90 percent of the proteins found in the mitochondria are encoded by nuclear genes. These genes must be transcribed into mRNA in the nucleus, then the mRNA must be translated into protein on cytoplasmic ribosomes. Finally, the proteins are transported into the mitochondria where they function. By contrast, genes located on mtDNA are transcribed in the mitochondria and translated on mitochondrial ribosomes.

Impact and Applications

Any mutation occurring in a mitochondrial gene has the potential to incapacitate the ability of the mitochondria to synthesize ATP. Because human cells are dependent upon mitochondria for their energy supply, the effects of these mutations can be wide-ranging and debilitating, if not fatal. If the mutation occurs in a gene that plays a role in mitochondrial protein synthesis, the ability of the mitochondria to perform protein synthesis is affected. Consequently, proteins that are translated on mitochondrial ribosomes such as cytochrome b or the NADH dehydrogenase subunits cannot be made, leading to defects in electron transport and ATP synthesis. Mutations in mitochondrial tRNA genes, for example, have been shown to be the cause of several degenerative neuromuscular disorders. Genes involved in electron transport and ATP synthesis have a more directly negative effect when mutated. Douglas C. Wallace and coworkers identified a mutation within the NADH dehydrogenase subunit 4 gene, for example, that was the cause of a maternally inherited form of blindness and was one of the first mitochondrial diseases to be identified.

Of further interest is the study of nuclear genes that contribute to mitochondrial function. It is estimated that over two hundred such genes exist, only a few of which have been identified. Included in this list of nuclear genes are those encoding proteins involved in mtDNA replication, repair, and recombination; enzymes involved in RNA transcription

and processing; and ribosomal proteins and the accessory factors required for translation. It is presumed that a mutation in any of these genes could have negative effects upon the ability of the mitochondria to function. Understanding how nuclear genes contribute to mitochondrial activity is an essential part of the search for effective treatments for mitochondrial diseases.

Human evolutionary studies have also been affected by the understanding of mitochondrial genes and their inheritance. Researchers Allan C. Wilson and Rebecca Cann, knowing that mitochondria are inherited exclusively through the female parent, hypothesized that a comparison of mitochondrial DNA sequences in several human populations would enable them to trace the origins of the ancestral human population. These studies led to the conclusion that a female living in Africa about 200,000 years ago was the common ancestor for all modern humans; she is referred to as "mitochondrial Eve."

—*Bonnie L. Seidel-Rogol*

See Also: Chloroplast Genes; Extrachromosomal Inheritance; Mitochondrial Diseases.

Further Reading: In "Mitochondrial DNA in Aging and Disease," *Scientific American* (August, 1997), Douglas C. Wallace gives a more detailed explanation of human mitochondrial diseases. Though somewhat dated, "Mitochondrial DNA," by Leslie A. Grivell, in *Scientific American* (March, 1983) provides a detailed discussion of some of the unique features of mitochondrial genes. Finally, Allan C. Wilson and Rebecca L. Cann, "The Recent African Genesis of Humans," *Scientific American* (April, 1992), describes how studies of mitochondrial genes have led to information about human origins.

Mitosis

Field of study: Classical transmission genetics

Significance: *All cells require genetic information to control their activities. In multicellular organisms, each cell of the body possesses all of the genetic information of that organism. Growth, develop-*

ment, organismal reproduction, wound repair, and normal cell turnover all require the replication of cells. Mitosis is the process that ensures that each cell receives a complete set of genetic information identical to that of its parent cell.

Key terms

DEOXYRIBONUCLEIC ACID (DNA): the genetic material of most organisms in the form of a double-helical structure composed of two chains of subunits called nucleotides

CENTROMERE: a physical constriction on a chromosome that is important in chromosome movement during mitosis

CHROMOSOMES: rodlike structures located in the cell nucleus that consist of DNA and protein

CHROMATIN: a long, extended form of a chromosome consisting of DNA and its associated proteins

NUCLEUS: a large, membrane-bound structure inside of a cell that contains the chromosomes; a cell containing a nucleus is called a eukaryotic cell

Merging of Genetics and Cell Biology

In the 1860's, Gregor Mendel developed a theory of heredity based on his studies of the segregation of traits in pea plants. Without any knowledge of cellular structures, he correlated physical traits and their inheritance patterns with particles he termed "elemens," which correspond to what are now known as genes. Chromosomes were observed by cell biologists in the latter half of the nineteenth century, and the basic process of mitosis was worked out by 1875. It was not until 1900 that botanists began to connect the segregation of chromosomes in mitosis with the inheritance of traits as described by Mendel. In 1944, the chemical nature of the genetic material was discovered to be deoxyribonucleic acid (DNA). Later work confirmed that chromosomes are composed of DNA intimately associated with a number of protein molecules.

All multicellular organisms begin life as a single cell. As the organism grows, the number of cells rapidly increases. Each cell of the organism possesses the same genetic information (that is, the same number and types of chromosomes). Growth, wound healing, and the replacement of cells that have lived out their life spans all require that parent cells are duplicated, including their DNA. Mitosis is the process by which this cellular replication occurs.

Cell Structure and Chromosome Organization

The nucleolus is a dark-staining body located inside the cell nucleus. The function of the nucleolus is the production of ribosomes (a cell component involved in protein production). There may be more than one nucleolus in the nucleus, depending upon the species of organism. The presence or absence of the nucleolus helps to define stages of mitosis. Centrioles are barrel-shaped structures located outside the nucleus in the cytoplasm of animal cells. The centrioles are involved in the organization of long, thin, hollow proteins called microtubules, which are involved in the movement of chromosomes during mitosis.

A single chromosome consists of one stretch of DNA that can be millions of nucleotides long. The DNA is wrapped around special proteins called histones. Other nonhistone proteins are also involved in coiling the DNA into a highly condensed package called the chromosome. During interphase (a segment of the cell cycle during which growth occurs), the DNA-protein complexes are very relaxed and not highly condensed. The chromosomes cannot be seen as discrete entities in the microscope. When in this highly relaxed form, the chromosomes are referred to as chromatin. When the cell enters mitosis, the chromatin condenses, and the chromosomes are readily visible with a light microscope as rod-shaped structures within the cell nucleus.

In eukaryotic cells (those with a nucleus), chromosomes occur in pairs. For example, humans have twenty-three pairs of chromosomes, for a total of forty-six. Each pair of chromosomes has the same type of genes at the same location along the length of the chromosome. (An exception to this is the sex-determining chromosomes in some organisms. For example, male humans have an X and Y chromosome that do not contain the same type of genes.) Chromosomes from the same pair are referred to as homologous chromosomes, or homologues.

After the DNA has replicated and the chromosomes have condensed at the beginning of mitosis, each chromosome is actually a double chromosome; the replicated chromosomes are held together at the centromere. Even though the chromosome has replicated, the two are attached at the centromere, and therefore the duplicated chromosome structure is still considered to be a single chromosome. Each replicated portion of that chromosome is referred to as a "sister chromatid." In essence, each of the chromosomes held together at the centromere is identical, and each is a sister chromatid. Chromosome numbers at any stage of mitosis can be determined by counting the number of intact centromeres. The chromosomes can be distinguished by staining and microscopic observation; size, position of the centromere, and banding patterns distinguish one chromosome and its homologue from the others.

The Cell Cycle

The life history of a single cell, from its origin as the product of the division of its parent cell to its own mitotic division, is called the cell cycle. There are two major segments of the cell cycle: interphase and mitosis. During interphase, the cell grows, replicates its own components (including DNA), and conducts the basic housekeeping activities necessary to keep it alive and fulfill its role in maintaining the life of the organism. Mitosis is the actual process of cell division. Each of these major segments can be subdivided into other portions. The length of time a particular cell spends in each of the phases is highly individual, depending on the species and the particular type of cell. Each cell spends the majority of its life in interphase, usually around 90 percent of its time as a single cell. During this part of the cell cycle, the nucleus is clearly visible in the cell when viewed through a microscope. The chromosomes are in their long, extended form of diffuse chromatin and are very difficult to visualize. Interphase can be subdivided into three distinct portions: gap 1 (G_1), synthesis (S), and gap 2 (G_2).

The G_1 phase is generally the longest portion of the cell cycle. During this first "gap," the cell is extremely metabolically active as it grows and prepares for subsequent cell cycle stages. In animal cells, the centrioles are replicated during G_1. Embryonic cells that are rapidly dividing only spend a few minutes in G_1, but most cells spend many hours here. The S phase immediately follows G_1. During this phase, the chromosomes, including the DNA, are replicated. Again, embryonic cells complete this phase very rapidly, but adult cells require hours to complete chromosomal replication. The second gap phase follows the S phase and is a time of increased metabolic activity, particularly of protein synthesis, in preparation for cell division in the next phase.

Biochemistry and Mechanics of Mitosis

In an adult human, about 50 million cells die every second and are immediately replaced by the mitotic division of parent cells. The exact duplication and distribution of the chromosomes containing the genetic material is essential for the normal functioning of the new cells and the survival of the organisms. Four stages of mitosis have been defined based on activities within the cell and behavior of the chromosomes. These four stages are prophase, metaphase, anaphase, and telophase.

Upon completion of interphase, the cell has replicated its chromosomes and centromeres and stockpiled enough cellular components to divide into two cells. The duplicated centromeres are arranged at right angles to each other, and both pairs reside near each other at one end of the cell. The centrioles are now associated with small segments of microtubules radiating out from them, giving them a flowery appearance referred to as an aster. The centrioles, microtubules, and other associated proteins are called the centrosomes.

The cell now enters the first stage of mitosis, called prophase. Prophase is heralded by the condensation of the chromatin into chromosomes, revealing each chromosome to appear double stranded and consist of two sister chromatids. The nuclear membrane dissolves, and the nucleoli disappear. In the cytoplasm, a structure called the mitotic spindle begins to

The Stages of Mitosis

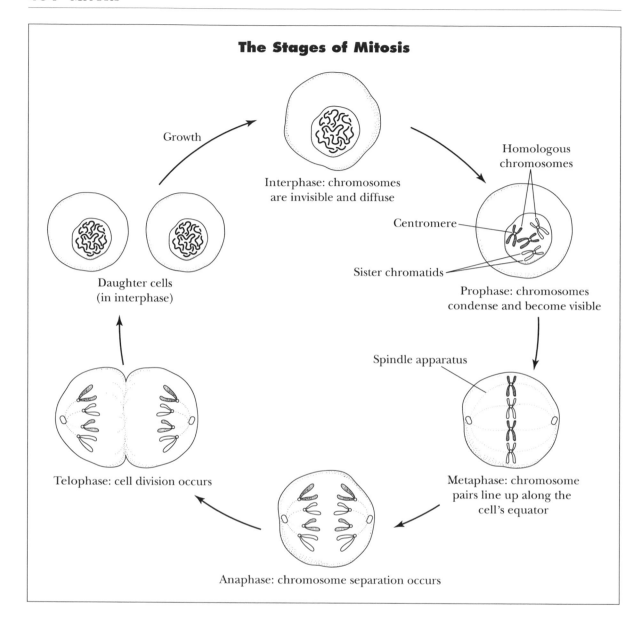

Growth

Interphase: chromosomes
are invisible and diffuse

Homologous
chromosomes

Centromere

Sister chromatids

Prophase: chromosomes
condense and become visible

Daughter cells
(in interphase)

Spindle apparatus

Telophase: cell division occurs

Metaphase: chromosome
pairs line up along the
cell's equator

Anaphase: chromosome separation occurs

form. Long fibers of microtubules and other proteins are assembled between the two centrosomes. As the microtubule bundles lengthen, the centrosomes are pushed apart.

During metaphase, the lengthening microtubule bundles have propelled the centrosomes to opposite poles of the cell. Bundles of microtubules extend from the poles toward the center of the cell. These microtubules bind to a specialized portion of the centromere of each chromosome called the kinetochore. Each sister chromatid has one kinetochore within the

single centromere. Movement of the microtubules cause the chromosomes to line up in a column at the equator of the cell. Kinetochores of each pair face opposite poles. The centrosome-microtubule structure is now called the spindle because of its distinctive shape. This equatorial position between the two poles of the mitotic spindle where the chromosomes have convened is referred to as the metaphase plate.

After the chromosomes are aligned on the metaphase plate, the centromeres connecting the sister chromatids divide, liberating the sister chromatids from each other. This centromeric division marks the beginning of anaphase. Each sister chromatid is now considered a single chromosome. The microtubules attached to the kinetochores begin to shorten by becoming disassembled within the centrosomes. This shortening action pulls the chromosomes to the poles of the cell. What were once sister chromatids now move toward opposite poles of the cell. At the end of anaphase, two poles have complete and equivalent collections of chromosomes.

At this point, the chromosomes and cell components must be subdivided into two new cells. Nuclear membranes begin to form around the polar clusters of chromosomes during telophase. Nucleoli reappear, and the chromosomes begin to relax, returning to their chromatin form. At this point, mitosis is technically complete, since the hereditary material has been replicated and divided into two new nuclei. To complete the process of cell division, cytokinesis (the division of the cytoplasm) occurs. In animal cells, a furrow is formed down the middle of the cell, which eventually pinches the single cell into two complete cells. Plant cells lay down a cell wall to divide the cell into the two new progeny cells. At the completion of mitosis, the two newly formed cells have the identical genetic information of the parent cell and of all cells within the organism. The new cells now begin their own cycle and begin the processes associated with interphase.

Impact and Applications

Propagation of the genetic material is one of the most profound biological phenomena. Without it, none of the cells within an organism that die of natural aging would be replaced, and the organism would rapidly die. If the chromosomal complements were not reproduced precisely in newly formed cells, the correct functions of those cells would not occur, leading to their deaths and again the demise of the organism. Most multicellular organisms begin their lives as single cells; without the ability to replicate the genetic material and duplicate cells, the organism would never grow and develop into a juvenile and finally an adult. In short, the continuation of any species would be impossible without mitosis.

Much contemporary research has been focused on understanding the underlying biochemical processes that control the cell cycle. Complex arrays of biochemicals, such as cyclin and mitosis promoting factor, are produced in increasing amounts, peak, fall, and continue the cycle in order to trigger each phase of the cell cycle. By understanding the underlying mechanisms that control the cell cycle and mitosis, it is hoped that some cell types that do not divide, such as nerve cells, will someday be able to be triggered to divide and proliferate, thus promoting the healing of brain and spinal cord injuries that currently cannot be mended.

The hallmark of cancer is the uncontrolled division of cells so that a mass called a tumor is produced. In cancerous cells, the cell cycle loses the innate timing present in a normal cell. Chromosome numbers in cancerous cells are virtually always incorrect. Extra chromosomes, missing chromosomes, and chromosome fragments can all be found in cancerous cells, revealing a wide range of errors occurring during mitosis. Many researchers are uncovering the genetic mechanisms that ultimately go awry in cancerous cells. Elucidation of both the genetics involved and of the proper control of the cell cycle may lead to cures for cancer by reestablishing the normal cell cycle and chromosome complement.

Aging has also been a subject of massive research efforts. It has been discovered that the ends of chromosomes, called telomeres, are composed of unique stretches of DNA. After each cell division, the length of the telomere shortens until, eventually, the chromosomes can no longer replicate and die. If the regulation of this telomere is deciphered, it may lead to treatments that slow the aging process. Interestingly, normal cells that are raised in laboratory flasks seem to be able to divide approximately fifty times before they die. Cancer cells in culture are immortal; therefore, the telomere phenomena does not apply to them. In this case, cancer cells may yield clues as to the regulation of telomere shortening, hence cell aging, which could help in finding treatments to slow cellular aging processes.

While each cell in a multicellular organism contains all of the genetic information of every cell, not all of the genes are expressed. For example, liver cells produce liver enzymes but not the neurotransmitters found in nerve cells. In theory, the chromosomes in the nucleus of any adult cell should be able to produce an entirely new individual. Removal of a nucleus from a single cell of an organism and growing an entirely new organism from it is called cloning. In 1997, the first mammal was cloned by placing the nucleus of an adult mammary cell of a sheep into an enucleated sheep egg (one without a nucleus) and implanting it in another sheep's uterus to develop. Dolly, as the world's first cloned mammal is known, is genetically identical to the donor of the mammary cell nucleus. The result of this cloning experiment has proven that under the proper conditions, the chromosomes in an adult cell can produce an entirely new individual. What started as a single egg with an adult nucleus grew, through mitosis, into an entirely new adult animal. The ethics of cloning and the possibilities of cloning humans are moral and ethical issues that are hotly debated.

—*Karen E. Kalumuck*

See Also: Chromatin Packaging; Chromosome Theory of Heredity; Classical Transmission Genetics; DNA Replication; Meiosis.

Further Reading: Neil Campbell's *Biology* (1996) provides a detailed but accessible overview for the general reader. *Essential Cell Biology: An Introduction to the Molecular Biology of the Cell* (1997), edited by Bruce Alberts, is a lavishly illustrated, clear, and concise overview geared toward the nonspecialist. A more advanced discussion can be found in *Molecular Biology of the Cell* (1995), by Bruce Alberts et al.

Molecular Genetics

Field of study: Molecular genetics
Significance: *Molecular genetics is the branch of genetics concerned with the central role that molecules, particularly the nucleic acids DNA and RNA, play in heredity. Understanding of molecular genetics is at the heart of biotechnology, which has had a tremendous impact on medicine, agriculture, forensics, and many other fields.*

Key terms

DEOXYRIBONUCLEIC ACID (DNA): a complex substance composed of two intertwined molecules, each composed of many nucleotides joined by chemical bonds, that serves as the genetic material for most organisms

RIBONUCLEIC ACID (RNA): a complex substance composed of many ribonucleotides joined by chemical bonds that serves as the informational intermediate between DNA and protein

GENOME: the assembly of the genetic information of an organism or one of its organelles

GENE: the genetic information that specifies a single protein or RNA product

REPLICATION: the process by which one DNA molecule is converted to two DNA molecules identical to the first

TRANSCRIPTION: the process of forming an RNA according to instructions contained in DNA

PROTEIN: a class of complex substances composed of many amino acids joined by chemical bonds

TRANSLATION: the process of forming proteins according to instructions contained in an RNA molecule

Identity and Structure of Genetic Material

Molecular genetics is the branch of genetics that deals with the identity of the molecules of heredity, their structure and organization, how these molecules are copied and transmitted, how the information encrypted in them is decoded, and how the information can change from generation to generation. In the late 1940's and early 1950's, scientists realized that the materials of heredity were nucleic acids. Deoxyribonucleic acid (DNA) was implicated as the substance extracted from a deadly strain of pneumococcal bacteria that could transform a mild strain into a lethal one and as the substance injected into bacteria by viruses as they start an infection. Ribonucleic acid (RNA) was shown to be the component of a virus that determined what kind of symptoms of infection appeared on tobacco leaves.

The nucleic acids are made up of nucleotides linked end to end to produce very long molecules. Each nucleotide has sugar and phosphate parts and a nitrogen-rich part called a base. Four bases are commonly found in each DNA and RNA. Three, adenine (A), guanine (G), and cytosine (C), are found in both DNA and RNA, while thymine (T) is normally found only in DNA and uracil (U) only in RNA. In double-helical DNA, two DNA strands are helically intertwined in opposite directions. The strands are held together in part by interactions specific to the bases. The A and T pair with one another, and G and C pair with one another. The RNAs adopt less regular structures. These may also require pairing between bases.

DNA and RNA, in various forms, serve as molecules of heredity. RNA is the genetic material that some viruses package in viral particles. One or several molecules of RNA may make up the viral information. The genetic material of most bacteria is a single circle of double-helical DNA, the circle consisting of from slightly more than 500,000 to about 5 million nucleotide pairs. In eukaryotes such as humans, the DNA genetic material is organized into multiple linear DNA molecules, each one the essence of a morphologically recognizable and genetically identifiable structure called a chromosome.

In each organism, the DNA is closely associated with proteins. Proteins are made of one or more polypeptides. Polypeptides are linear polymers, like nucleic acids, but the units linked end to end are amino acids rather than nucleotides. More than twenty kinds of amino acids make up polypeptides. Proteins are generally smaller than DNA molecules and assume a variety of shapes. Proteins contribute to the biological characteristics of an organism in many ways: They are major components of structures both inside (membranes and fibers) and outside (hair and nails) the cell; as enzymes, they initiate the thousands of chemical reactions that cells use to get energy and build new cells; and they regulate the activities of cells. Histone proteins pack eukaryotic nuclear DNA into tight bundles called "nucleosomes." Further coiling and looping of nucleosomes results in the compact structure of chromosomes. These can be seen with help of a micro-

scope. The complex of DNA and protein is called "chromatin."

The term "genome" denotes the roster of genes and other DNA of an organism. Most eukaryotes have more than one genome. The principal genome is the genome of the nucleus that controls most of the activities of cells. Two organelles, the mitochondria (which produce energy by oxidizing chemicals) and the plastids (which convert light to chemical energy in photosynthesis) have their own genomes. The organelle genomes have only some of the genes needed for their functioning. The others are present in the nuclear genome. Nuclear genomes have many copies of some genes. Some repeated sequences are organized tandemly, one after the other, while others are interspersed with unique sequences. Some repeated sequences are genes present in many copies, while others are DNAs of unknown function.

Copying and Transmission of Genetic Nucleic Acids

James Watson and Francis Crick's double-helical structure for DNA suggested to them how a faithful copy of a DNA could be made. The strands would pull apart. One by one, the new nucleotide units would then arrange themselves by pairing with the correct base on the exposed strands. When zipped together, the new units make a new strand of DNA. The process, called DNA replication, makes two double-helical DNAs from one original one. Each daughter double-helical DNA has one old and one new strand. This kind of replication, called semiconservative replication, has been confirmed by an experiment by Matthew Meselson and Franklin Stahl. Enzymes cannot copy DNA of eukaryotic chromosomes completely to each end of the DNA strands. This is not a problem for bacteria, whose circular genomes do not have ends. To keep the ends from getting shorter with each cycle of replication, eukaryotic chromosomes have special structures called "telomeres" at their ends that are targets of a special DNA synthesis enzyme.

When a cell divides, each daughter cell must get one and only one complete copy of the mother cell's DNA. In most bacterial chromo-

Matthew Meselson, who helped to clarify the process of DNA replication, operating an ultracentrifuge in 1958. (California Institute of Technology)

somes, this DNA synthesis starts at only one place, and that starting is controlled so that the number of starts equals the number of cell fissions. In eukaryotes, DNA synthesis begins at multiple sites, and each site, once it has begun synthesis, does not begin another round until after cell division. When DNA has been completely copied, the chromosomes line up for distribution to the daughter cells. Protein complexes called "kinetochores" bind to a special region of each chromosome's DNA called the "centromere." Kinetochores attach to microtubules, fibers that provide the tracks along which the chromosomes move during their segregation into daughter cells.

Gene Expression, Transcription, and Translation

DNA is often dubbed the blueprint of life. It is more accurate to describe DNA as the computer tape of life's instructions because the DNA information is a linear, one-dimensional series of units rather than a two-dimensional diagram. In the flow of information from the DNA tape to what is recognized as life, two steps require the decoding of nucleotide sequence information. The first step, the copying of the DNA information into RNA, is called "transcription," an analogy to medieval monks sitting in their cells copying, letter by letter, old Latin manuscripts. The letters and words in the new version are the same as in the old but are written with a different hand and thus have a slightly different appearance. The second step, in which amino acids are polymerized in response to the RNA information, is called "translation." Here, the monks take the Latin words and find English, German, or French equivalents. The product is not in the nucleotide language but in the language of polypeptide sequences. The RNAs that direct the order of amino acids are called messenger RNAs (mRNAs) because they bring instructions from the DNA to the ribosome, the site of translation.

Multicellular organisms consist of a variety of cells, each with a particular function. Cells also respond to changes in their environment. The differences among cell types and among cells in different environmental conditions are caused by the synthesis of different proteins. For the most part, regulation of which proteins are synthesized and which are not occurs by controlling the synthesis of the mRNAs for these proteins. Genes can have their transcription switched on or switched off by the binding of protein factors to a segment of the gene that determines whether transcription will start or not. An important part of this gene segment is the promoter. It tells the transcription apparatus to start RNA synthesis only at a particular point in the gene.

Not all RNAs are ready to function the moment their synthesis is over. Many RNA transcripts have alternating exon and intron segments. The intron segments are taken out with splicing of the end of one exon to the beginning of the next. Other transcripts are cut at several specific places so that several functional RNAs arise from one transcript. Eukaryotic mRNAs get poly-A tails (about two hundred nucleotide units in which every base is an A) added after transcription. A few RNAs are edited after transcription, some extensively by adding or removing U nucleotides in the middle of the RNA, others by changing specific bases.

Translation occurs on particles called ribosomes and converts the sequence of nucleotide residues in mRNA into the sequence of amino acid residues in a polypeptide. Since protein is created as a consequence of translation, the process is also called "protein synthesis." The mRNA carries the code for the order of insertion of amino acids in three nucleotide units called codons. Failure of the ribosome to read nucleotides three at a time leads to shifts in the frame of reading the mRNA message. The frame of reading mRNA is set by starting translation only at a special codon.

Transfer RNA (tRNA) molecules actually do the translating. There is at least one tRNA for each of the twenty common amino acids. Anticodon regions of the tRNAs each specifically pair with only a specific subset of mRNA codons. For each amino acid there is at least one enzyme that attaches the amino acid to the correct tRNA. These enzymes are thus at the center of translation, recognizing both amino acid and nucleotide residues.

The ribosomes have sites for binding of mRNA, tRNA, and a variety of protein factors. Ribosomes also catalyze the joining of amino acids to the growing polypeptide chain. The protein factors, usually loosely bound to ribosomes, assist in the proper initiation of polypeptide chains, in the binding of amino acid-bearing tRNA to the ribosome, and in moving the ribosome relative to the mRNA after each additional step. Three steps in translation use biochemical energy: attaching the amino acid to the tRNA, binding the amino acyl tRNA to the ribosome-mRNA complex, and moving the ribosome relative to the mRNA.

Protein Processing and DNA Mutation

The completed polypeptide chain is processed in one or more ways before it assumes its role as a mature protein. The linear string of amino acid units folds into a complex, three-dimensional structure, sometimes with the help of other proteins. Signals in some protein's amino acid sequences direct them to their proper destinations after they leave the ribosomes. Some signals are removable, while others remain part of the protein. Some newly synthesized proteins are called polyproteins because they are snipped at specific sites, giving several proteins from one translation product. Finally, individual amino acid units may get other groups attached to them or be modified in other ways.

The DNA information can be corrupted by reaction with certain chemicals, some of which are naturally occurring while others are present in the environment. Ultraviolet and ionizing radiation can also damage DNA. In addition, the apparatus that replicates DNA will make a mistake at low frequency and insert the wrong nucleotide. Collectively, these changes in DNA are called "DNA damage." When DNA damage goes unrepaired before the next round of copying of the DNA, mutations (inherited changes in nucleotide sequence) result. Mutations may be substitutions, in which one base replaces another. They may also be insertions or deletions of one or more nucleotides. Mutations may be beneficial, neutral, or harmful. They are the targets of the natural selection that drives evolution. Since some mutations are harmful, survival of the species requires that they be kept to a low level.

Systems that repair DNA are thus very important for the accurate transmission of the DNA information tape. Several kinds of systems have evolved to repair damaged DNA before it can be copied. In one, enzymes directly reverse the damage to DNA. In a second, the damaged base is removed, and the nucleotide chain is split to allow its repair by a limited resynthesis. In a third, a protein complex recognizes the DNA damage, which results in incisions in the DNA backbone on both sides of the damage. The segment containing the damage is removed, and the gap is filled by a limited resynthesis. In still another, mismatched base pairs, such as those that result from errors in replication, are recognized, and an incision is made some distance away from the mismatch. The entire stretch from the incision point to past the mismatch is then resynthesized. Finally, the molecular machinery that exchanges DNA segments, the recombination machinery, may be mobilized to repair damage that cannot be handled by the other systems.

Invasion and Amplification of Genes

Mutation is only one way that genomes change from generation to generation. Another way is via the invasion of an organism's genome by other genomes or genome segments. Bacteria have evolved restriction modification systems to protect themselves from such invasions. The gene for restriction encodes an enzyme that cleaves DNA whenever a particular short sequence of nucleotides is present. It does not recognize that sequence when it has been modified with a methyl group on one of its bases. The gene for modification encodes the enzyme that adds the methyl group. Thus the bacterium's own DNA is protected. However, DNA that enters the cell from outside, such as by phage infection or by direct DNA uptake, is not so protected and will be targeted for degradation by the restriction enzyme. Despite restriction, transfer of genes from one species to another (horizontal transfer) has occurred.

As far as is known, restriction modification systems are unique to bacteria. Gene transfer from bacteria to plants occurs naturally in diseases caused by bacteria of the *Agrobacterium* genus. As part of the infection process, these bacteria transfer a part of their DNA containing genes, only active in plants, into the plant genome. Studies with fungi and higher plants suggest that eukaryotes cope with gene invasion by inactivating the genes (gene silencing) or their transcripts (cosuppression).

Another way that genomes change is by duplications of gene-sized DNA segments. When the environment is such that the extra copy is advantageous, the cell with the duplication survives better than one without the duplication. Thus genes can be amplified under selective pressure. In some tissues, such as salivary glands of dipteran insects and parts of higher plant embryos, there is replication of large segments of chromosomes without cell division. Monster chromosomes result.

Genomes also change because of movable genetic elements. Inversions of genome segments occur in bacteria and eukaryotes. Other segments can move from one location in the genome to another. Some of these movements appear to be rare, random events. Others serve particular functions and are programmed to occur under certain conditions. One kind of mobile element, the retrotransposon, moves into new locations via an RNA intermediate. The element encodes an enzyme that makes a DNA copy of the element's RNA transcript. That copy inserts itself into other genome locations. The process is similar to that used by retroviruses to establish infection in cells. Other mobile elements, called transposons or transposable elements, encode a transposase enzyme that inserts the element sequence, or a copy of it, into a new location. When that new location is in or near a gene, normal functioning of that gene is disturbed.

The production of genes for antibodies (an important part of a human's immune defense system) is a biological function that requires gene rearrangements. Antibody molecules consist of two polypeptides called light and heavy chains. In most cells in the body, the genes for light chains are in two separated segments, and those for heavy chains are in three. During the maturation of cells that make antibodies, the genes are rearranged, bringing these segments together. The joining of segments is not precise. The imprecision contributes to the diversity of possible antibody molecules.

Cells of baker's or brewer's yeast have genes specifying their sex, or mating type, in three locations. The information at one location, the expression locus, is the one that determines the mating type of the cell. A copy of this information is in one of the other two sites, while the third has the information specifying the opposite mating type. Yeast cells switch mating types by replacing the information at the expression locus by information from a storage locus. Mating-type switching and antibody gene maturation are only two examples of programmed gene rearrangements known to occur in a variety of organisms.

Genetic Recombination

Recombination occurs when DNA information from one chromosome becomes attached to the DNA of another. When participating chromosomes are equivalent, the recombination is called "homologous." Homologous recombination in bacteria mainly serves a repair function for extreme DNA damage. In many eukaryotes, recombination is essential for the segregation of chromosomes into gamete cells during meiosis. Nevertheless, aspects of the process are common between bacteria and eukaryotes. Starting recombination requires a break in at least one strand of the double-helical DNA. In the well-studied yeast cells, a double-strand break is required. Free DNA ends generated by breaks invade the double-helical DNA of the homologous chromosome. Further invasion and DNA synthesis result in a structure in which the chromosomes are linked to one another. This structure, called a half-chiasma, is recognized and resolved by an enzyme system. Resolution can result in exchange so that one end of one chromosome is linked to the other end of the other chromosome and vice versa. Resolution can also result in restoration of the original linkage. In the latter case, the DNA around the exchange point may be that of the other DNA. This is known as gene conversion.

Impact and Applications

Molecular genetics is the heart of biotechnology. Its fundamental investigation of biological processes has provided tools for biotechnologists. Molecular cloning and gene manipulation in the test tube rely heavily on restriction enzymes, other nucleic-acid-modifying enzymes, and extrachromosomal DNA, all discovered during molecular genetic investigation. The development of nucleic acid hybridization, which allows the identification of specific molecular clones in a pool of others, required an understanding of DNA structure and dynamics. The widely used polymerase chain reaction (PCR), which can amplify minute quantities of DNA, would not have been possible without discoveries in DNA replication. Genetic mapping, a prelude to the isolation of many genes, was sped along by molecular markers detectable with restriction enzymes or the PCR. Transposable elements and the transferred DNA of *Agrobacterium*, because they often inactivate genes when they insert in them, were used to isolate the genes they inactivate. The inserted elements served as tags or handles by which the modified genes were pulled out of a collection of genes.

The knowledge of the molecular workings of genes gained by curious scientists has allowed other scientists to intervene in many disease situations, provide effective therapies, and improve biological production. Late twentieth century scientists rapidly developed an understanding of the infection process of the acquired immunodeficiency syndrome (AIDS) virus. The understanding, built on the skeleton of existing knowledge, has helped combat this debilitating disease. Molecular genetics has also led to the safe and less expensive production of proteins of industrial, agricultural, and pharmacological importance. The transfer of DNA from *Agrobacterium* to plants has been exploited in the creation of transgenic plants. These plants offer a new form of pest protection that provides an alternative to objectionable pesticidal sprays and protects against pathogens for which no other protection is available. Recombinant insulin and recombinant growth hormone are routinely given to those whose conditions demand them.

Through molecular genetics, doctors have diagnostic kits that can, with greater rapidity, greater specificity, and lower cost, determine whether a pathogen is present. Finally, molecular genetics has been used to identify genes responsible for many inherited diseases of humankind. Someday medicine may correct some of these diseases by providing a good copy of the gene, a strategy called gene therapy.

—*Ulrich Melcher*

See Also: Central Dogma of Molecular Biology; DNA Replication; DNA Structure and Function; Genetic Code; Mutation and Mutagenesis; Protein Synthesis; RNA Structure and Function.

Further Reading: *Molecular Biology Made Simple and Fun* (1997), by David Clark and Lonnie Russell, is a detailed and entertaining account of molecular genetics. Karl Drlica's *Understanding DNA and Gene Cloning* (1992) is a readable description of DNA analysis and molecular cloning, and his *Double-Edged Sword* (1994) discusses the impact of molecular genetics in medicine. Comprehensive treatments are found in *Molecular Biology of the Gene* (1988), by James Watson et al., and *Gene Structure and Expression* (1996), by John Hawkins.

Monohybrid Inheritance

Field of study: Classical transmission genetics

Significance: *Humans and other organisms show a number of different patterns in the inheritance and expression of traits. For many inherited characteristics in organisms, the pattern of transmission is governed by the principles of monohybrid inheritance, in which a trait is determined by one pair of genes. An understanding of monohybrid inheritance is critical for understanding how many medically significant traits in humans and economically significant traits in domestic plants and animals are passed from generation to generation.*

Key terms

GENE: a unit of inheritance at a given location on a chromosome

ALLELE: one of the pair of possible alternative

forms of a gene that occurs at a given site or locus on a chromosome

DOMINANT GENE: the controlling member of a pair of alleles that is expressed to the exclusion of the expression of the recessive member

RECESSIVE GENE: an allele that can only be expressed when the controlling or dominant allele is not present

Mendel and Monohybrid Inheritance

The basic genetic principles first worked out and described by Gregor Mendel in his classic experiments on the common garden pea have been found to apply to many inherited traits in all sexually reproducing organisms, including humans. Until the work of Mendel, plant and animal breeders tried to formulate laws of inheritance based upon the principle that characteristics of parents would be blended in their offspring. Mendel's success came about because he studied the inheritance of contrasting or alternative forms of one phenotypic trait at a time. The phenotype of any organism includes not only all of its external characteristics but also all of its internal structures, extending even into all of its chemical and metabolic functions. Human phenotypes would include characteristics such as eye color, hair color, skin color, hearing and visual abnormalities, blood disorders, susceptibility to various diseases, and muscular and skeletal disorders.

Mendel experimented with seven contrasting traits in peas: stem height (tall versus dwarf), seed form (smooth versus wrinkled), seed color (yellow versus green), pod form (inflated versus constricted), pod color (green versus yellow), flower color (red versus white), and flower position (axial versus terminal). Within each of the seven sets, there was no overlap between the traits and thus no problem in classifying a plant as one or the other. For example, although there was some variation in height among the tall plants and some variation among the dwarf plants, there was no overlap between the tall and dwarf plants.

Mendel's first experiments crossed parents that differed in only one trait. Matings of this type are known as monohybrid crosses, and the rules of inheritance derived from such matings

yield examples of monohybrid inheritance. These first experiments provided the evidence for the principle of segregation and the principle of dominance. The principle of segregation refers to the separation of members of a gene pair from each other during the formation of gametes (the reproductive cells: sperm in males and eggs in females). It was Mendel who first used the terms "dominant" and "recessive." It is of interest to examine his own words and to realize how appropriate his definitions are today: "Those characters which are transmitted entire, or almost unchanged by hybridization, and therefore in themselves constitute the characters of the hybrid, are termed the dominant and those which become latent in the process recessive." The terms dominant and recessive are used to describe the characteristics of a phenotype, and they may depend on the level at which a phenotype is described. A gene that acts as a recessive for a particular external trait may turn out not to be so when its effect is measured at the biochemical or molecular level.

An Example of Monohybrid Inheritance

The best way of describing monohybrid inheritance is by working through an example. Although any two people obviously differ in many genetic characteristics, it is possible, as Mendel did with his pea plants, to follow one trait governed by a single gene pair that is separate and independent of all other traits. In effect, by doing this, the investigator is working with the equivalent of a monohybrid cross. In selecting an example, it is best to choose a trait that does not produce a major health or clinical effect; otherwise, the clear-cut segregation ratios expected under monohybrid inheritance might not be seen in the matings.

Consider the trait of albinism, a phenotype caused by a recessive gene. Albinism is the absence of pigment in the hair, skin, and eyes. Similar albino genes have been found in many animals, including mice, buffalo, bats, frogs, and rattlesnakes. Since the albino gene is recessive, the gene may be designated with the symbol c and the gene for normal pigmentation as C. Thus a mating between a homozygous normal person (CC) and a homozygous albino person (*cc*) would be expected to produce children who are heterozygous (*Cc*) but phenotypically albino, since the normal gene is dominant to the albino gene. Only normal genes, *C*, would be passed on by the normally pigmented parent, and only albino genes, *c*, would be passed on by the albino parent. If there was a mating between two heterozygous people (*Cc* and *Cc*), the law of segregation would predict that each parent would produce two kinds of gametes: *C* and *c*. The resulting progeny would be expected to appear at a ratio of 1 *CC*: 2 *Cc*: 1 *cc*. Since *C* is dominant to *c*, ¾ of the progeny would be expected to have normal pigment, and ¼ would be expected to be albino. There are three genotypes (*CC*, *Cc*, and *cc*) and two phenotypes (normal pigmentation and albino). By following the law of segregation and taking account of the dominant gene, it is possible to determine the types of matings that might occur and to predict the types of children that would be expected:

Parents	Phenotypes	Offspring Expected
1. AA × AA	Normal × Normal	All AA (Normal)
2. AA × Aa	Normal × Normal	½ AA, ½ Aa (All Normal)
3. AA × aa	Normal × Albino	All Aa (Normal)
4. Aa × Aa	Normal × Normal	¼ AA, ½ Aa, ¼ aa (¾ Normal, ¼ Albino)
5. Aa × aa	Normal × Albino	½ Aa, ½ aa (½ Normal, ½ Albino)
6. aa × aa	Albino × Albino	All aa (Albino)

Because of dominance, it is not always possible to tell what type of mating has occurred. For example, in matings 1, 2, and 4 in the table, the parents are both normal in each case. Yet in mating 4, ¼ of the offspring are expected to be albino. A complication arises when it is realized that in mating 4 the couple might not produce any offspring that are *cc*; in that case, all offspring would be normal. Often, because of the small number of offspring in humans and other animals, the ratios of offspring expected under monohybrid inheritance might not be realized. Looking at the different matings and the progeny that are expected, it is easy to see how genetics can help to explain not only why children resemble their parents but also why children do not resemble their parents.

Modification of Basic Mendelian Inheritance

After Mendel's work was rediscovered early in the twentieth century, it soon became apparent that there were variations in monohybrid inheritance that apparently were not known to Mendel. Mendel studied seven pairs of contrasting traits, and in each case, one gene was dominant and one gene was recessive. For each trait, there were only two variants of the gene. It is now known that other possibilities exist. For example, other types of monohybrid inheritance include codominance (in which both genes are expressed in the heterozygote) and sex linkage (an association of a trait with a gene on the X chromosome). Nevertheless, the law of segregation operates in these cases as well, making it possible to understand inheritance of the traits.

Within a cell, genes are found on chromosomes in the nucleus. Humans have forty-six chromosomes. Each person receives half of the chromosomes from each parent, and it is convenient to think of the chromosomes in pairs. Examination of the chromosomes in males and females reveals an interesting difference. Both sexes have twenty-two pairs of what are termed "autosomes" or "body chromosomes." The difference in chromosomes between the two sexes occurs in the remaining two chromosomes. The two chromosomes are known as the sex chromosomes. Males have an unlike pair of sex chromosomes, one designated the X chromosome and the other, smaller one designated the Y chromosome. Females, on the other hand, have a pair of like sex chromosomes, and these are similar to the X chromosome of the male. Although the Y chromosome does not contain many genes, it is responsible for male development. A person without a Y chromosome would undergo female development. Since genes are located on chromosomes, the pattern of transmission of the genes demonstrates some striking differences from that of genes located on any of the autosomes. For practical purposes, "sex linked" usually refers to genes found on the X chromosome since the Y chromosome contains few genes. Although X-linked traits do not follow the simple pattern of transmission of simple monohybrid inheritance as first described by Mendel, they still conform to his law of segregation. Examination of a specific example is useful to understand the principle.

The red-green color-blind gene is an X-linked recessive since females must have the gene on both X chromosomes in order to exhibit the trait. For males, the terms recessive and dominant really do not apply since the male has only one X chromosome (the Y chromosome does not contain any corresponding genes) and will express the trait whether the gene is recessive or dominant. An important implication of this is that X-linked traits appear more often in males than in females. In general, the more severe the X-linked recessive trait is from a health point of view, the greater the proportion of affected males to affected females.

If the color-blind gene is designated cb and the normal gene Cb, the types of mating and offspring expected may be set up as they were for the autosomal recessive albino gene. In the present situation, the X and Y chromosomes will also be included, remembering that the Cb and cb genes will be found only on the X chromosome and that any genotype with a Y chromosome will result in a male.

"Carrier" females are heterozygous females who have normal vision but are expected to pass the gene to half their sons, who would be color blind. Presumably, the carrier female would have inherited the gene from her father, who would have been color blind. Thus, in some families the trait has a peculiar pattern of transmission in which the trait appears in a woman's father, but not her, and then may appear again in her sons.

Impact and Applications

The number of single genes known in humans has grown dramatically since Victor McKusick published the first *Mendelian Inheritance in Man* catalog in 1966. In the first catalog, there were 1,487 entries representing loci identified by Mendelizing phenotypes or by cellular and molecular genetic methods. In the 1994 catalog, the number of entries had grown to 6,459. Scarcely a day goes by without a news report or story in the media involving an example of monohybrid inheritance. Furthermore, genetic conditions or disorders regularly appear as the theme of a movie or play. An understanding of the principles of genetics and monohybrid inheritance provides a greater appreciation of what is taking place in the world, whether it is in the application of DNA fingerprinting in the courtroom, the introduction of disease-resistant genes in plants and animals, the use of genetics in paternity cases, or the description of new inherited diseases.

Perhaps it is in the area of genetic diseases

Parents	Phenotypes	Offspring Expected	
1. $X^{Cb}X^{Cb} \times X^{Cb}Y$	Normal × Normal	$X^{Cb}X^{Cb}$ normal female $X^{Cb}Y$ normal male	
2. $X^{Cb}X^{Cb} \times X^{cb}Y$	Normal × Color blind	$X^{Cb}X^{cb}$ normal female $X^{Cb}Y$ normal male	
3. $X^{Cb}X^{cb} \times X^{Cb}Y$	Normal × Normal	$X^{Cb}X^{Cb}$ $X^{Cb}X^{cb}$ ½ normal females, ½ carrier females $X^{Cb}Y$ $X^{cb}Y$ ½ normal males, ½ color-blind males	
4. $X^{Cb}X^{cb} \times X^{cb}Y$	Normal × Color blind	$X^{Cb}X^{cb}$ $X^{cb}X^{cb}$ ½ carrier females, ½ color-blind females $X^{Cb}Y$ $X^{cb}Y$ ½ normal males, ½ color-blind males	
5. $X^{cb}X^{cb} \times X^{Cb}Y$	Color blind × Normal	$X^{Cb}X^{cb}$ carrier females $X^{cb}Y$ color-blind males	
6. $X^{cb}X^{cb} \times X^{cb}Y$	Color blind × Color blind	$X^{cb}X^{cb}$ color-blind females $X^{cb}Y$ color-blind males	

that knowledge of monohybrid inheritance offers the most significant personal applications. Single-gene disorders usually fall into one of the four common modes of inheritance: autosomal dominant, autosomal recessive, sex-linked dominant, and sex-linked recessive. Examination of individual phenotypes and family histories allows geneticists to determine which mode of inheritance is likely to be present for a specific disorder. Once the mode of inheritance has been identified, it becomes possible to determine the likelihood or the risk of occurrence of the disorder in the children. Since the laws governing the transmission of Mendelian traits are so well known, it is possible to predict with great accuracy when a genetic condition will affect a specific family member. In many cases, testing may be done prenatally or in individuals before symptoms appear. As knowledge of the human genetic makeup increases, it will become even more essential for people to have a basic knowledge of how Mendelian traits are inherited.

—*Donald J. Nash*

See Also: Chromosome Theory of Heredity; Classical Transmission Genetics; Hereditary Diseases; Meiosis; Mendel, Gregor, and Mendelism.

Further Reading: *Mendelian Inheritance in Man* (1994), compiled by Victor A. McKusick, is a comprehensive catalog of Mendelian traits in humans. Although it is filled with medical terminology and clinical descriptions, there are interesting family histories and fascinating accounts of many of the traits. An introduction to the principles of heredity and a catalog of more than one hundred human traits is given in *The Family Genetic Sourcebook* (1990), by Benjamin A. Pierce. The principles of Mendelian inheritance are applied to the world of champion sled dogs in an article by Mark Derr in "The Making of a Marathon Mutt," *Natural History* (March, 1966).

Multiple Alleles

Field of study: Classical transmission genetics

Significance: *Alleles are alternate forms of genes located on the same sites of homologous chromo-somes. When three or more variations of a gene exist in a population, they are referred to as multiple alleles. The human blood groups (A, B, AB, and O) provide an example of multiple alleles.*

Key terms

BLOOD TYPE: one of the several groups into which blood can be classified based on the presence or absence of certain protein molecules on the red blood cells

CODOMINANT: two contrasting alleles that are both fully functional and both fully expressed when present in an individual

DOMINANT: an allele that masks the expression of another allele that is considered recessive to it

The Discovery of Alleles and Multiple Alleles

Although Gregor Mendel, considered to be the father of genetics, did not discover multiple alleles, an understanding of his work is necessary to understand their role in genetics. In the 1860's, Mendel formulated the earliest concepts of how traits or characteristics are passed from parents to their offspring. His work on pea plants led him to propose that there are two factors, since renamed "genes," that cause each trait that an individual possesses. A particular form of the gene, called the "dominant" form, will enable the characteristic to occur whether the offspring inherits one or two copies of that gene. The alternate form of the gene, or allele, will only be exhibited if two copies of this gene, called the "recessive," are present. For example, pea seeds will be yellow if two copies of the dominant, yellow-causing gene are present and will be green if two copies of the recessive gene are present. However, since yellow is dominant to green, an individual plant with one copy of each allele will be as yellow as a plant possessing two yellow genes. Mendel discovered only two alternate appearances, called "phenotypes," for each trait he studied. He found that violet is the allele dominant to white in causing flower color, while tall is the allele dominant to short in creating stem length.

Early in the twentieth century, examples of traits with more than an either/or phenotype caused by only two possible alleles began to be found in many types of living things. Coat color in rabbits is a well-documented example of multiple alleles. Not two but four alternative forms of the gene for coat color exist in rabbit populations, with different letters used to designate those colors. The gene producing color is labeled c; thus c^+ produces full, dark color, c^{ch} produces mixed colored and white hairs, c^h produces white on the body but black on the paws, and c creates a pure white rabbit.

It is important to note that although three or more alternative forms can exist in the whole population, each individual organism can still only possess two, acquiring only one from each of its two parents. What, then, of Mendel's principle of one allele dominant to the other? In the sample of rabbit color, apparently c^+ is dominant to the c^{ch}, which is dominant to c^h, with c, the gene for pure white, recessive to the other three. If mutation (the accidental change in the composition of a gene) can create four possible color alleles, is it not also possible that successive mutations might cause a much larger number of multiple alleles? This is apparently true, as the study over many years of eye color in *Drosophila melanogaster* (fruit fly) has revealed. There are apparently a vast number of alleles ranging from the dominant red gene through cherry, apricot, and numerous other shades to the recessive white.

Blood Types

One of the earliest examples of multiple alleles discovered in humans concerns the alternate forms of the genes that produce one's blood type. In 1900, the existence of four blood types (A, B, AB, and O) was discovered. By studying pedigrees (the family histories of many individuals), it was deduced by 1925 that these four blood types were caused by multiple alleles. The possible alleles one may inherit are named I^A, I^B, and I^O, or simply A, B, and O genes. Both A and B are apparently dominant to O. However, A and B are codominant to each other. Thus, if both are present, both are equally seen in the individual. A person with

The Relationship Between Genotype and Blood Type		
Genotype	Blood Type	Comments
AA	A	These two genotypes produce
AO	A	identical blood types.
BB	B	These two genotypes produce
BO	B	identical blood types.
AB	AB	Both dominant alleles are expressed.
OO	O	With no dominant alleles, the recessive allele is expressed.

two A genes or an A and an O has type A blood. Someone with two B genes or a B and an O has type B. Two O genes result in type O blood. Because A and B are codominant, the individual with one of each gene is said to have type AB blood.

To say people are "type A" means that they have a protein-sugar molecule of a particular combination embedded in the membrane of all red blood cells. These proteins are called "antigens." The presence of an A gene causes the production of an enzyme that transfers the sugar galactosamine to the protein. Since the B gene brings about the attachment of a different sugar called galactose, and since the O gene does not bring about the addition of any sugar, each blood type possesses different antigens on its red blood cells. Because of codominance, type AB people have both antigens interspersed on the red blood cells.

The danger of causing death by transfusing a different blood type than the patient possesses is brought about by the presence of the A or B antigens as well as the potential presence of A or B antibodies. Antibodies are chemical molecules in the plasma (the liquid portion of the blood). If, by error, type B blood, which contains anti-A antibodies, is given to a type A person, those antibodies will attach to the red blood cells, causing them to agglutinate, or form clumps. Since type AB possesses both kinds of antigen, either type A blood with its anti-B antibodies or type B with its anti-A antibodies poses a danger to the type AB person. Medical personnel must carefully check the blood type of both the recipient of a transfusion and the donated blood to avoid agglutination and subsequent death.

The use of blood types to establish paternity stems from the fact that a child's blood type can be used to determine what the parents' blood types could and could not be. Since a child receives one allele from each parent, certain people can be eliminated as that child's parents if the alleles they possess could not produce the combination found in the offspring. However, this only proves that a particular person could be the parent, as could millions of others who possess that blood type; it does not prove that a particular person alone is the parent. However, modern methods of analyzing the deoxyribonucleic acid (DNA) in many of the individual's genes now makes the establishment of paternity a very exact science.

Impact and Applications

The topic of multiple alleles has been found to have great relevance to many human disease conditions. One of these is cystic fibrosis (CF), the most common deadly inherited disease afflicting Caucasians. Characterized by a thick mucus buildup in lungs, pancreas, and intestines, it frequently brings about death by age twenty. Soon after the gene that causes CF was found in 1989, geneticists realized there may be as many as one hundred multiple alleles for this gene. The extent of the mutation in these alternate genes apparently causes the great variety in the severity of symptoms from one patient to another.

The successful transplantation of organs is also closely linked to the existence of multiple alleles. A transplanted organ has antigens on its cells that will be recognized as foreign and destroyed by the recipient's antibodies. The genes that build these cell-surface antigens, called human leukocyte antigen (HLA), occur in two main forms. HLA-A has nearly twenty different alleles, and HLA-B has more than thirty. Since any individual can only have two of each type, there is an enormous number of possible combinations in the population. Finding donors and recipients with the same or a very close combination of HLA alleles is a very difficult task for those arranging successful organ transplantation.

Geneticists are coming to suspect that multiple alleles, once thought to be the exception to the rule, may exist for the majority of human genes. If this is so, the study of multiple alleles for many disease-producing genes should shed much light on a great diversity of human genetic conditions and problems.

—*Grace D. Matzen*

See Also: Complete Dominance; Cystic Fibrosis; Mendel, Gregor, and Mendelism; Organ Transplants and HLA Genes.

Further Reading: William S. Klug supplies a solid explanation of multiple alleles in *Essentials of Genetics* (1996). In *An Introduction to the ABO Blood Group System* (1992), Catherine Newkirk explains blood types. James Cunningham provides useful information on cystic fibrosis in *An Introduction to Cystic Fibrosis for Patients and Families* (1994).

Mutation and Mutagenesis

Field of study: Molecular genetics

Significance: *A mutation is a heritable change in the structure or composition of DNA. Depending on the function of the altered DNA segment, the effect of a mutation can range from having no detectable influence on development to causing major deformities and even death. Mutation is a natural process by which new genetic diversity is produced. However, factors such as chemical pollutants and radiation that increase mutation rates can have a serious effect on health.*

Key terms

GENE POOL: all of the genes carried by all members of a population of organisms; the genetic diversity in the gene pool provides the variation that allows adaptation to new conditions

GERMINAL MUTATION: a mutation in gamete-forming (germinal) tissue; this can be passed from a parent to its offspring

MUTAGEN: a chemical or physical agent that causes an increased rate of mutation

MUTAGENESIS: the process of a heritable

change occurring in a gene, either spontaneously or in response to a mutagen

MUTATION RATE: the probability of a heritable change occurring in the genetic material over a given time period such as a cell division cycle or a generation

NUCLEOTIDES: the four different subunits of deoxyribonucleic acid (DNA); their order determines the kind of protein that will be produced by that gene

PHENOTYPE: the observable effects of a gene; phenotypes can be physical appearance, biochemical activity, cell function, or any other measurable factor

SOMATIC MUTATION: a mutation that occurs in a body cell and produces a group of mutant cells but is not transmitted to the next generation

WILD TYPE: the normal genetic makeup of an organism; a mutation alters the phenotype of a normal trait to produce a mutant phenotype

Mutation

A mutation is any change in the genetic material that can be inherited by the next generation of cells or progeny. A mutation can occur at any time in the life of any cell in the body. If a mutation occurs in the reproductive tissue, the change can be passed to an offspring in the egg or sperm. That new mutation may then affect the development of the offspring and be passed on to later generations. However, if the mutation occurs in cells of the skin, muscle, blood, or other body (somatic) tissue, the new mutation will only be passed on to other body cells when that cell divides. This can produce a mosaic patch of cells carrying the new genetic change. Most of these are undetectable and have no effect on the carrier. An important exception is a somatic mutation that causes the affected cell to lose control of the cell cycle so that it divides in an unlimited way, resulting in cancer. Many environmental chemicals and agents such as X rays and ultraviolet radiation that cause mutations therefore also cause cancer.

Mutation also has an important, beneficial role in natural populations of all organisms. The ability of a species to adapt to changes in its environment, combat new diseases, or respond to new competitors is dependent on genetic diversity in the population's gene pool. Without sufficient resources of variability, a species faced by a serious new stress can become extinct. The reduced population sizes in rare and endangered species will result in reduced genetic diversity and a loss of the capacity to respond to selection pressures. Zoo breeding programs often take data on genetic diversity into account when planning the captive breeding of endangered species. The creation of new agricultural crops or of animal breeds with economically desirable traits also depends on mutations that alter development in a useful way. Therefore, mutation can have both damaging and beneficial effects.

The Role of Mutations in Cell Activity and Development

The genetic information in a cell is encoded in the sequence of subunits, the nucleotides, that make up the deoxyribonucleic acid (DNA) molecule. A mutation is a change in the cell's genetic makeup, and it can range from changing just a single nucleotide in the DNA molecule to altering long pieces of DNA. To appreciate how such changes can affect an organism, it is important to understand how information is encoded in DNA and how it is translated to produce a specific protein. There are four different nucleotides in the DNA molecule: adenine (A), guanine (G), thymine (T), and cytosine (C). The DNA molecule is composed of two strands linked together by a sequence of base pairs (bp). An adenine on one strand pairs with a thymine on the other (A-T), and a guanine on one strand pairs with a cytosine on the other (G-C). When a gene is activated, one of the two strands is used as a model, or template, for the synthesis of a single-stranded molecule called messenger ribonucleic acid (mRNA). The completed mRNA molecule is then transported out of the nucleus, and it binds with ribosomes (small structures in the cytoplasm of the cell), where a protein is made using the mRNA's nucleotide sequence as its coded message. The nucleotides are read on the ribosome in triplets, with three adjacent nucleotides (called a codon) corresponding to one of the twenty amino acids found in protein.

Thus the sequence of nucleotides eventually determines the order of amino acids that are linked together to form a specific protein. The amino acid sequence in turn determines how the protein will work, either as a structural part of a cell or as an enzyme that will catalyze a specific biochemical reaction. A gene is often 1,000 bp or longer, so there are many points at which a genetic change can occur. If a mutation occurs in an important part of the gene, even the change of a single amino acid can cause a major change in protein function. Sickle-cell anemia is a good example of this. In sickle-cell anemia a base-pair substitution in the DNA causes the sixth codon in the mRNA to change from GAG to GUG. When this modified mRNA is used to create a protein, the amino acid valine is substituted for the normal glutamic acid in the sixth position in a string of 146 amino acids. This small change causes the protein to form crystals and thus deform cells when the amount of available oxygen is low. Since this protein is one of the parts of the oxygen-carrying hemoglobin molecule in red blood cells, this single DNA nucleotide change has potentially severe consequences for an affected individual.

Types of Mutation

Because they can be so diverse, one way to organize mutations is to describe the kind of molecular or structural change that has occurred. There are three broad classes of mutation. "Genomic mutations" are changes in the number of chromosomes in a cell. Inheriting an extra chromosome as in Down syndrome is an example of a genomic mutation. "Chromosome mutations" are changes in the structure of a chromosome and can include the loss, gain, or altered order of a series of genes. "Gene mutations" are genetic changes limited to an individual gene or the adjacent regions that control its activity during development. Thus, the amount of genetic information affected by a mutation can vary from a single gene to hundreds of them. Genes also vary in the severity of their effects. Some are undetectable in the carrier, some cause small defects or even beneficial changes in the function of a protein, while others can produce major changes in several different developmental processes at the same time.

Gene mutations are sometimes called point mutations because their genetic effects are limited to a single point, or gene, on a chromosome that can carry up to several thousand different genes. The simplest kind of point mutation is a base substitution, in which one base pair is replaced by another (for example, an A-T base pair at one point in the DNA molecule being replaced by a C-G base pair). This can change a codon triplet so that a different amino acid is placed in the protein at that point. This often changes the function of the protein, at least in minor ways. However, some base substitutions are silent. Since several different triplets can code for the same amino acid, not all base changes will result in an amino acid substitution.

Another common kind of gene mutation called a "frameshift" can have a much larger effect on protein structure. A frameshift mutation occurs when a nucleotide is added to, or lost from, the DNA strand when it is duplicated during cell division. Since translation of the mRNA is done by the ribosomes adding one amino acid to the growing protein for every three adjacent nucleotides, adding or deleting one nucleotide will effectively shift that reading frame so that all following triplets are different. By analogy, one can consider the following sentence of three-letter words: THE BIG DOG CAN RUN FAR. If a base (a letter, in this analogy) is added in the second triplet but the sentence is still read by three letters at a time, the meaning is completely altered. THE BIX GDO GCA NRU NFA R. In a cell, a nonfunctional protein is produced unless the frameshift is near the terminal end of the gene.

Environmental agents such as ultraviolet (UV) radiation can affect DNA and base pairing. Certain UV wavelengths, for example, cause some DNA nucleotides to pair abnormally. Gene mutations have also been traced to the movement of transposable DNA elements. Transposable elements were first discovered by Barbara McClintock while studying chromosome breakage and kernel traits in maize. Now they are known from many organisms, including humans. Transposable elements are small DNA segments that can become inserted into a chromosome and later excise and change their position. If one becomes inserted in the middle of a gene, it effectively separates the gene into two widely spaced fragments. In the fruit fly (*Drosophila melanogaster*), in which spontaneous mutations have been studied in detail at the DNA level, as many as half of the spontaneous mutations in certain genes have been traced to transposable elements.

There are four major kinds of chromosome mutations. A chromosome deletion or deficiency is produced when two breaks occur in the chromosome but are repaired by leaving out the middle section. For example, if the sections of a chromosome are labeled with the letters ABCDEFGHIJKLMN and chromosome breaks occur at F-G and at K-L, the broken chromosome can be erroneously repaired by enzymes that link the ABCDEF fragment to the LMN fragment. The genes in the unattached middle segment, GHIJK, will be lost from the chromosome. Losing these gene copies can affect many different developmental processes and even cause the death of the organism. Chromosome breaks and other processes can also cause some genes to be duplicated in the chromosome (for example, ABCDEFGHDE-FGHIJKLMN). A third kind of chromosome mutation, an inversion, changes the order of the genes when the segment between two chromosomal breaks is reattached backward (for example, ABCDJIHGFEKLMN). Finally, chromosome segments can be moved from one kind of chromosome to another in a structural change called a translocation. Some examples of heritable Down syndrome are caused by this type of chromosome mutation.

Genomic mutations are a large factor in the genetic damage that occurs in humans. Whole chromosomes can be lost or gained by errors during cell division. In animals, almost all examples of chromosome loss are so developmentally severe that the individual cannot survive to birth. On the other hand, since extra chromosomes provide an extra copy of each of their genes, the amount of each protein they code for is unusually high, and this, too, can create biochemical abnormalities for the or-

ganism. In humans, an interesting exception is changes in chromosome number that involve the sex-determining chromosomes, especially the X chromosome (the Y is relatively silent in development). Since normal males have one X and females have two, the cells in females inactivate one of the X chromosomes to balance gene dosage. This dosage compensation mechanism can, therefore, also come into operation when one of the X chromosomes is lost or an extra one is inherited because of an error in cell division. The resulting conditions, such as Turner's syndrome and Klinefelter's syndrome, are much less severe than the developmental problems associated with other changes in chromosome number.

Mutation Rate

A mutation is any heritable change in the genetic material, but there are several different ways one can look at genetic change. For example, errors can occur when the DNA molecule is being duplicated during cell division. In simple organisms such as bacteria, about one thousand nucleotides are added to the duplicating DNA molecule each second. The speed is slower in plants and animals, but errors still occur when mispairing between A and T or between C and G nucleotides occurs. DNA breaks are also common. These kinds of genetic change can be classified as "genetic damage." Some mutations are spontaneous, caused by changes that occur in the process of normal cell biochemistry. Other damage that can be traced to environmental factors changes bases, causes mispairing, or breaks DNA strands. Fortunately, almost all of this initial genetic damage is repaired by enzymes that recognize and correct errors in nucleotide pairing or DNA strand breaks. It is the unrepaired genetic damage that appears as new mutations. One of the first geneticists to design experiments to measure mutation rate was Hermann Müller, who received the Nobel Prize for his work on mutagenesis, including the discovery that X rays cause mutations.

The experiments by Müller provide a useful example of the kind of experimental design that can be used to measure mutation rates. Müller focused on new mutations (lethals) on the X chromosome of *Drosophila* that could cause the carrier to die. Since a male has only one X chromosome, a lethal mutation on that chromosome causes death. A living male *Drosophila* must, therefore, have no lethal mutations on his X chromosome. If a male *Drosophila* is treated with an agent such as X irradiation or certain chemicals, new lethal mutations can be detected when he is mated with special genetic strains of females. His X chromosomes are eventually passed on to male descendants. If a new lethal mutation exists on a specific X chromosome, all males that inherit that chromosome copy will die during development. Spontaneous mutation rates measured by this technique average about 1×10^{-5} for each gene. In other words, there is a probability of about 1 in 100,000 that a mutation will occur in a particular gene each generation. This is a very low probability for a specific gene, but when it is multiplied for all of the genes in an animal or plant, it is likely that a new mutation has occurred somewhere on the chromosomes of an organism each generation.

Spontaneous mutation rates vary to some extent from one gene to another and from one organism to another, but one major source of variation in mutation rate comes from external agents that act on the DNA to increase damage or inhibit repair. One of the most widely used techniques for measuring the mutagenic activity of a chemical was developed in the 1970's by Bruce Ames. The Ames test uses bacteria that have a mutation that makes them unable to produce the amino acid histidine. These bacteria cannot survive in culture unless they are given histidine in the medium. To test whether a chemical increases the mutation rate, it is mixed with a sample of these bacteria, and they are placed on a medium without histidine. Any colonies that survive represent bacteria in which a new mutation has occurred to reverse the original defect (a back-mutation). Since many chemicals that cause mutations also cause cancer, this quick and inexpensive test is now used worldwide to screen potential carcinogenic, or cancer-causing, agents.

Mutation rates in mice are measured by use of the specific-locus test. In this test, wild-type male mice are mated with females that are

Hermann Müller, who received a 1946 Nobel Prize for his investigations into the process of mutation. (Archive Photos)

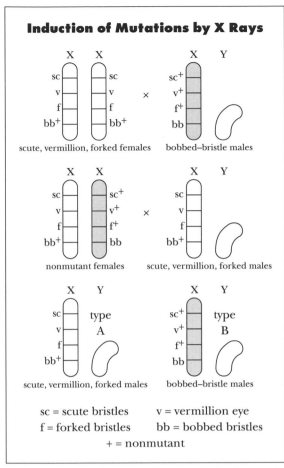

Induction of Mutations by X Rays

sc = scute bristles   v = vermillion eye
f = forked bristles   bb = bobbed bristles
                + = nonmutant

If, for a given fly and its descendants, an induced or spontaneous lethal mutation occurs in the paternal X chromosome (shaded), no third-generation males of type B will result. If a spontaneous lethal mutation occurs in an original maternal X chromosome, then no third-generation males of type A will result.

homozygous for up to seven visible, recessive mutations that cause changes in coat color, eye color, and shape of the ear. If no mutations occur in any of the seven genes in the germ cells of the male, the male offspring will all be wild type in appearance. However, a new mutation in any of the seven genes will yield a progeny with a mutant phenotype (for example, a new coat color). The same cross can also be used to identify new mutations in females. Since mice are mammals, they are a close model system to humans. Thus, results from mutation studies in mice have helped identify agents that are likely to be mutagenic in humans.

The Use of Mutations to Study Development

Mutations offer geneticists a powerful tool to analyze development. By understanding the way development is changed by a mutation, one can determine the role the normal gene plays. Although most people tend to think of mutations as causing some easily visible change in the appearance of a plant or animal (such as wrinkled pea seeds or white mouse fur), most mutations are actually lethal when present in two copies (homozygous). These lethal mutations affect some critical aspect of cell structure or other fundamental aspect of development or function. Genes turn on and off at specific times during development, and by studying the abnormalities that begin to show when a lethal mutation carrier dies, a geneticist can piece together a picture of the timing and role of important gene functions.

Another useful insight comes from mutations with effects that vary. For example, many mutations have phenotypic effects that depend on the conditions, such as temperature, in which the individual develops. An interesting example of such temperature sensitivity is the fur color of Siamese cats. The biochemical pathway for pigmentation is active in cool temperatures but is inactivated at warmer body temperature. For this reason, a Siamese cat will only be pigmented in the cooler parts such as the tips of the ears and tail. Gene interactions like this allow geneticists to study the conditions under which the protein coded by a mutant gene works.

It would be a mistake, however, to think that all mutations have large phenotypic effects. Many complex traits are produced by many genes working together and are affected by environmental variables such as temperature. These are called "quantitative" traits because they are measured on some kind of scale, such as size, number, or intensity. The mutations that affect quantitative traits are not different, except perhaps in the magnitude of their individual effects, from other kinds of gene mutation. Mutations in quantitative traits are a major source of heritable variation on which natural and artificial selection can act to change a phenotype.

Siamese cats are darker at their extremities as the result of a mutation that is affected by body temperature. (Yasmine Cordoba)

Impact and Applications

It will probably never be possible to eliminate all mutation events because many mutations are caused by small errors in normal DNA duplication when cells divide. Learning how mutations affect cell division and cell function can help one to understand processes such as cancer and birth defects that can often be traced to genetic change. Some explanations of processes such as aging have focused on mutation in somatic cells. Mutation is also the source of genetic variation in natural populations, and the long-term survival of a species depends on its ability to draw on this variation to adapt to new environmental conditions.

Two aspects of mutagenesis will continue to grow in importance. First, environmental and human-made mutagens will continue to be a source of concern as technological advances occur. Many scientists are working to monitor and correct potential mutagenic hazards. Second, geneticists are beginning to use molecular tools, such as transposable elements and the techniques of genetic engineering, to produce preplanned genetic changes. Directed mutagenesis of DNA offers a way to correct preexisting genetic defects or alter phenotypes in planned ways. Mutation is, therefore, both a source of problems and a source of promise.

—*James N. Thompson, Jr.*
—*R. C. Woodruff*

See Also: Biochemical Mutations; Cancer; Chemical Mutagens; Chromosome Structure; DNA Repair; DNA Structure and Function; Population Genetics.

Further Reading: Miroslav Radman and Robert Wagner, "The High Fidelity of DNA Duplication," *Scientific American* 259 (August, 1988), discusses the high degree of accuracy in the process of DNA duplication. Paul Howard-Flanders, "Inducible Repair of DNA," *Scientific American* 245 (November, 1981), explores the inducible repair of DNA. Radiation is an important environmental factor that can cause mutation, and Y. M. Shcherbak explores the consequences of the Chernobyl nuclear reactor accident in Russia in "Confronting the Nuclear Legacy: Ten Years of the Chernobyl Era," *Scientific American* (April, 1996). Many new techniques are being developed to study mutation, and J. Gossen and J. Vijg describe one of these, transgenic mice, in *Trends in Genetics* (1993). Transposable elements have recently been recognized as an important source of new mutations, and mobile genetic elements in humans are discussed in Gabriel E. Novick et al., "The Mobile Genetic Element ALU in the Human Genome," *BioScience* 46 (January, 1996). Mutations are also a common cause of cancer, and the September, 1996, issue of *Scientific American* is devoted to the topic "What You Need to Know About Cancer." Included in this issue is Robert A. Weinberg, "How Cancer Arises," a summary of the way cancer arises and some of the mutations and genes involved in human cancers.

Natural Selection

Field of study: Population genetics
Significance: *Natural selection is the mechanism proposed by Charles Darwin to account for biological evolutionary change. Using examples of artificial selection as analogies, he suggested that any heritable traits that allow an advantage in survival or reproduction to an individual organism would be "naturally selected" and increase in frequency until the entire population had the trait. Selection, along with other evolutionary forces, influences the changes in genetic and morphological variation that characterize biological evolution.*

Key terms

ADAPTATION: the evolution of a trait by natural selection, or a trait that has evolved as a result of natural selection

ARTIFICIAL SELECTION: selective breeding of desirable traits in organismal lineages

FITNESS: an individual's potential for natural selection as measured by the number of offspring of that individual relative to those of others

GROUP SELECTION: selection in which characteristics of a group not attributable to the individuals making up the group are favored

Natural Selection and Evolution

In 1859, English naturalist Charles Darwin published his magnum opus *On the Origin of Species*, in which he made two significant contributions to the field of biology: First, he proposed that biological evolution can occur by "descent with modification," with a succession of minor inherited changes in a lineage leading to significant change over many generations; and second, he proposed natural selection as the primary mechanism for such change. (This was also proposed independently by Alfred R. Wallace at about the same time.) Darwin reasoned that if an individual organism carried traits that allowed it to have some advantage in survival or reproduction, then those traits would be carried by its offspring, which would be better represented in future generations. In other words, the individuals carrying those traits would be "naturally selected" because of the advantages of the traits. For example, if a small mammal happened to have a color pattern that made it more difficult for predators to see, it would have a better chance of surviving and reproducing. The mammal's offspring would share the color pattern and the advantage over differently patterned members of the same species. Over many generations, the proportion of the species with the selected pattern would increase until it was present in every member of the species, and the species would be said to have evolved the color pattern trait.

Natural selection is commonly defined as "survival of the fittest," although this is often misinterpreted to mean that individuals who are somehow better than others will survive while the others will not. As long as the traits convey some advantage in reproduction so that the individual's offspring are better represented in the next generation, then natural selection is occurring. The advantage may be a better ability to survive, or it may be something else, such as the ability to produce more offspring.

For natural selection to lead to evolutionary change, the traits under selection must be heritable, and there must be some forms of the traits that have advantages over other forms (variation). If the trait is not inherited by offspring, it cannot persist and become more common in later generations. Darwin recognized this, even though in his time the mechanisms of heredity and the sources of new genetic variation were not understood. After the rediscovery of Gregor Mendel's principles of genetics in the early years of the twentieth century, there was not an immediate integration of genetics into evolutionary biology. In fact, it was suggested that genetic mutation might be the major mechanism of evolution. This belief, known as Mendelism, was at odds with Darwinism, in which natural selection was the primary force of evolution. However, with the "modern synthesis" of genetics and evolutionary theory in the 1940's and 1950's, Mendelian genetics was shown to be entirely compatible with Dar-

Alfred Wallace, whose investigations of natural selection paralleled the work of Charles Darwin. (National Library of Medicine)

winian evolution. With this recognition, the role of mutation in evolution was relegated to the source of variation in traits upon which natural selection can act.

The potential for natural selection of an organism is measured by its "fitness." In practice, the fitness of an individual is some measure of the representation of its own offspring in the next generation, often relative to other individuals. If a trait has evolved as a result of natural selection, it is said to be an "adaptation." The term "adaptation" can also refer to the process of natural selection driving the evolution of such a trait. There are several evolutionary forces in addition to selection (for example, genetic drift, migration, and mutation) that can influence the evolution of a trait, though the process is called adaptation as long as selection is involved.

Population Genetics and Natural Selection

Population geneticists explore the actual and theoretical changes in the genetic composition of natural or hypothetical populations. Not surprisingly, a large part of the theoretical and empirical work in the field has concentrated on the action of natural selection on genetic variation in a population. Ronald A. Fisher and J. B. S. Haldane were the primary architects of selection theory beginning in the 1930's, and Theodosius Dobzhansky was a pioneer in the detection of natural selection acting on genetic variants in populations of *Drosophila melanogaster* (fruit flies).

The most basic mathematical model of genes in a population led to the Hardy-Weinberg law, which predicts that there would be no change in the genetic composition of a population in the absence of any evolutionary forces such as natural selection. However, models that include selection show that it can have specific influences on a population's genetic variation. In such models, the fitness of an organism's genotype is represented by a fitness coefficient (or the related selection coefficient), in which the genotype with the highest fitness is assigned a value of 1, and the remaining genotypes are assigned values relative to the highest fitness. A fitness coefficient of 0 represents a lethal geno-

type (or, equivalently, one that is incapable of reproduction).

The simplest models of selection include the assumption that a genotype's fitness does not change with time or context and demonstrate three basic types of selection, defined by how selection acts on a distribution of varying forms of a trait (where extreme forms are rare and average forms are common). These three types are: directional selection (in which one extreme is favored), disruptive selection (in which both extremes are favored), and stabilizing selection (in which average forms are favored). The first two types (with the first probably being the most common) can lead to substantial genetic change and thus evolution, though in the process genetic variation is depleted. The third type maintains variation but does not result in much genetic change. These results create a problem: Natural populations generally have substantial genetic variation, but most selection is expected to deplete it. The problem has led population geneticists to explore the role of other forces working in place of, or in conjunction with, natural selection and to study more complex models of selection. Examples include models that allow a genotype to be more or less fit if it is more common (frequency-dependent selection) or that allow many genes to interact in determining a genotype's fitness (multilocus selection). Despite the role of other forces, selection is considered an important and perhaps complex mechanism of genetic change.

Detecting and Measuring Fitness

Although a great amount of theoretical work on the effects of selection has been done, it is also important to relate theoretical results to actual populations. Accordingly, there has been a substantial amount of research on natural and laboratory populations to measure the presence and strength of natural selection. In practice, selection must be fairly strong for it to be distinguished from the small random effects that are inherent in natural processes.

Ideally, a researcher would measure the total selection on organisms over their entire life cycles, but in some cases this may be too difficult or time-consuming. Also, a researcher may be interested in discovering what specific parts of the life cycle selection influences. For these reasons, many workers choose to measure components of fitness by breaking down the life cycle into phases and looking for fitness differences among individuals at some or all of them. These components can differ with different species but often include fertility selection (differences in the number of gametes produced), fecundity selection (differences in the number of offspring produced), viability selection (differences in the ability to survive to reproductive age), and mating success (differences in the ability to successfully reproduce). It is often found in such studies that total lifetime fitness is caused primarily by fitness in one of these components, but not all. In fact, it may be that genotypes can have a disadvantage in one component but still be selected with a higher overall fitness because of greater advantages in other components.

There are several empirical methods for detection and measurement of fitness. One relatively simple way is to observe changes in gene or genotype frequencies in a population and fit the data on the rate of change to a model of gene-frequency change under selection to yield an estimate of the fitness of the gene or genotype. The estimate is more accurate if the rate of mutation of the genes in question is taken into account. In the famous example of "industrial melanism," it was observed that melanic (dark-colored) individuals of the peppered moth *Biston betularia* became more common in Great Britain in the late nineteenth century, corresponding to the increase in pollution that came with the industrial revolution. It was suggested that the melanic moths were favored over the lighter moths because they were camouflaged on tree trunks where soot had killed the lichen and were therefore less conspicuous to bird predators. Although it is now known that the genetics of melanism are more complex, early experiments suggested that there was a single locus with a dominant melanic allele and a recessive light allele; the data from one hundred years of moth samples were used to infer that light moths have two-thirds the survival ability of melanic moths.

Later, a second method of fitness measure-ment was applied to the peppered moth using a mark-recapture experiment. In such an experiment, known quantities of marked genotypes are released into nature and collected again some time later. The change in the proportion of genotypes in the recaptured sample provides a way to estimate their relative fitnesses. In practice, this method has a number of difficulties associated with making accurate and complete collections of organisms in nature, but the fitness measure of melanic moths by this method was in general agreement with that of the first method. A third method of measuring fitness is to measure deviations from the genotype proportions expected if a population is in Hardy-Weinberg equilibrium. This method can be very unreliable if deviations are the result of something other than selection.

Units of Selection

Darwin envisioned evolution by selection on individual organisms, but he also considered the possibility that there could be forms of selection that would not favor the survival of the individual. He noted that in many sexual species, one sex often has traits that are seemingly disadvantageous but may provide some advantage in attracting or competing for mates. For instance, peacocks have a large, elaborately decorated tail that is energetically costly to grow and maintain and might be a burden when fleeing from predators. However, it seems to be necessary to attract and secure a mate. Darwin, and later Fisher, described how such a trait could evolve by sexual selection if the female evolves a preference for it, even if natural selection would tend to eliminate it.

Other researchers have suggested that in some cases selection may act on biological units other than the individual. Richard Dawkins's *The Selfish Gene* (1976) popularized the idea that selection may be acting directly on genes and only indirectly on the organisms that carry them. This distinction is perhaps only a philosophical one, but there are specific cases in which genes are favored over the organism, such as the "segregation distorter" allele in *Drosophila* that is overrepresented in offspring of heterozygotes but lethal in homozygous conditions.

The theory of kin selection was developed to explain the evolution of altruistic behavior such as self-sacrifice. In some bird species, for example, an individual will issue a warning call against predators and subsequently be targeted by the predator. Such behavior, while bad for the individual, can be favored if those benefiting from it are close relatives. While the individual may perish, relatives that carry the genes for the behavior survive and altruism can evolve. Kin selection is a specific type of group selection in which selection favors attributes of a group rather than an individual. It is not clear whether group selection is common in evolution or limited to altruistic behavior.

Impact and Applications

The development of theories of selection and the experimental investigation of selection have always been intertwined with the field of evolutionary biology and have led to a better understanding of the history of biological change in nature. More recently, there have been medical applications of this knowledge, particularly in epidemiology. The specific mode of action of a disease organism or other parasite is shaped by the selection pressures of the host it infects. Selection theory can aid in the understanding of cycles of diseases and the response of parasite populations to antibiotic or vaccination programs used to combat them.

Although the idea of natural selection as a mechanism of biological change was suggested in the nineteenth century, artificial selection in the form of domestication of plants and animals has been practiced by humans for many thousands of years. Early plant and animal breeders recognized that there was variation in many traits, with some variations being more desirable than others. Without a formal understanding of genetics, they found that by choosing and breeding individuals with the desired traits, they could gradually improve the lineage. Darwin used numerous examples of artificial selection to illustrate biological change and argued that natural selection, while not necessarily as strong or directed, would influence change in much the same way. It is important to make a clear distinction between the two processes: Breeders have clear, long-term goals

in mind in their breeding programs, but there are no such goals in nature. There is only the immediate advantage of the trait to the continuation of the lineage. The application of selection theory to more recent breeding programs has benefited human populations in the form of new and better food supplies.

—*Stephen T. Kilpatrick*

See Also: Artificial Selection; Evolutionary Biology; Hardy-Weinberg Law; Population Genetics.

Further Reading: Richard Lewontin et al., *Dobzhansky's Genetics of Natural Populations* (1981), contains many original classic studies of selection in natural populations. John Endler, *Natural Selection in the Wild* (1986), is a good reference on measuring fitness. Sewell Wright's four-volume *Evolution and Genetics of Populations* (1968-1978) is an excellent introduction to the theory of natural selection. H. Kettlewell, *The Evolution of Melanism* (1973), reviews a famous example of natural selection and fitness measurement in the peppered moth. G. C. Williams, *Adaptation and Natural Selection* (1966), is a good introduction to adaptation and units of selection.

Neural Tube Defects

Field of study: Human genetics

Significance: *Neural tube defects are a category of birth defects that usually result from the failure of the neural tube to close properly during gestational development. Many neural tube defects can be prevented through folic acid supplementation and avoidance of other risk factors. However, because the neural tube closes during the first gestational month, preventive measures must be instituted prior to pregnancy. Therefore, prevention of neural tube defects is contingent upon the planning or expectation of pregnancies and concomitantly initiating positive lifestyle changes.*

Key terms

NEURAL TUBE: the embryonic precursor to the spinal cord and brain that normally closes at small openings, or neuropores, by the twenty-eighth day of gestation

SPINA BIFIDA: a neural tube defect that usually

results from the failure of the posterior neuropore to close properly during gestation

ANENCEPHALUS: a neural tube defect characterized by the failure of the cerebral hemispheres of the brain and the cranium to develop normally

ETIOLOGY: the cause or causes of a disease or disorder

MULTIFACTORIAL: characterized by a complex interaction of genetic and environmental factors

Formation of the Neural Tube

Neural tube defects represent congenital anomalies that are both ancient and prevalent in human populations. Documented cases of anencephalus and spina bifida have been found among the skeletal remains of the ancient Egyptians and prehistoric Native Americans. In the contemporary world, spina bifida remains one of the most common birth defects. Yet despite their prevalence and antiquity, some questions concerning the causes of neu-ral tube defects remain unanswered.

Neural tube defects result from a disruption in the formation or closure of the neural tube, which, during embryonic development, differentiates into the brain and spinal cord. The neural tube develops first out of the neural plate. The borders of the neural plate are folded, forming the neural groove. The neural groove becomes progressively deeper, placing the two folds in opposition. Final development of the neural tube occurs as the dorsal folds fuse along the midline. Closure of this structure commences around the third gestational week, beginning at its midportion and ending at the anterior and posterior neuropores around the twenty-fifth and twenty-seventh gestational days, respectively.

Classification of Neural Tube Defects

Disruption in the formation and closure of the posterior neuropores is associated with spina bifida. Spina bifida occulta is generally unaccompanied by protrusion of the spinal cord or its coverings through the open, un-

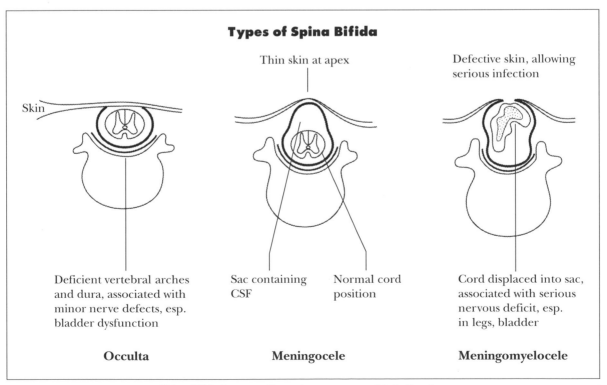

Types of Spina Bifida

Thin skin at apex

Defective skin, allowing serious infection

Skin

Deficient vertebral arches and dura, associated with minor nerve defects, esp. bladder dysfunction

Sac containing CSF

Normal cord position

Cord displaced into sac, associated with serious nervous deficit, esp. in legs, bladder

Occulta   **Meningocele**   **Meningomyelocele**

Spina bifida is among the most common neural tube disorders.

fused arches of the vertebrae. On the other hand, much more severe conditions, categorized as spina bifida cystica, result in the herniation of neural tissues and the formation of cystic swelling. One form of spina bifida cystica, meningomyelocele, is marked by the protrusion of both the meninges and the spinal cord through the unfused vertebral arch. In the most severe cases, portions of the spinal cord and nerve roots are encased in the walls of the sac, damaging and hindering normal neurological functioning and development. In such instances, the severity of the neurological dysfunction depends on the location of the lesion along the vertebral column, as nerves below the defect are adversely affected. Meningocele, a more moderate manifestation of spina bifida cystica, is encountered four to five times less frequently than meningomyelocele. Unlike the latter condition, the cystic sacs of meningoceles are made up solely of meninges and spinal fluid. This factor, coupled with the lack of involvement of the spinal cord, generally affords a more favorable prognosis, although some sensory and motor deficits may persist after surgery.

Anencephalus, which results from the disruption of the anterior neuropore, is the most devastating and severe of all the neural tube defects. Infants born with this birth defect are lacking significant areas of their brain and skull. The region normally occupied by the cerebral hemispheres consists of a formless mass of highly vascular connective tissue, while most of the bones of the cranial vault are simply absent. Many anencephalic infants are stillborn; most die soon after birth.

Encephalocele, like anencephalus, is believed to result from defective closure of the anterior neuropore. In these conditions, a saclike protrusion of neural tissue occurs through an opening along the midline of the skull. The prognosis and outcome of infants born with encephalocele depends upon the size of the lesion and the extent to which neural tissues are involved.

Prevalence-at-Birth Rates and Causes

Prevalence-at-birth rates of neural tube defects show substantial geographic and temporal variation. Historically, some of the highest prevalence-at-birth rates have been documented in the British Isles and range from as high as 4.5 in 1,000 births in Belfast, Ireland, to as low as approximately 1.5 in 1,000 births in London, England. In the United States, the highest rates of neural tube defects have historically occurred in northeastern states. Rates of neural tube defects are declining in most areas of the world, although regional outbreaks, marked by higher birth prevalence rates, have been reported and are generally unexplained. Typically, rates in the contemporary United States average around 1 to 2 in 1,000 births, and the risk of having an infant with a neural tube defect increases by about 2 percent if a couple has previously had a child with such a defect.

Among the most important risk factors are those relating to the diet and health status of prospective mothers. Also, there are indications that excessive elevation of a woman's body temperature during early pregnancy, through hot baths or recreational hot tubs, may increase her chances of having an infant with a neural tube defect. A number of studies have suggested that women who give birth to infants with neural tube defects have lower health status and poorer diets than women whose children lack these anomalies. Inadequate levels of folate appear to place women at a greater risk of having an infant with a neural tube defect. Folate supplementation of women who are planning a pregnancy is now recommended by many authorities; women who are planning to become pregnant should consult their doctors. Tests for alpha-fetaprotein in the mother's blood during the prenatal period can help detect the presence of a neural tube defect in the developing fetus.

—*Mary K. Sandford*

See Also: Congenital Defects; Prenatal Diagnosis.

Further Reading: Excellent, thorough overviews of all aspects of neural tube defects are found in *Epidemiology of Anencephalus and Spina Bifida* (1980), by J. M. Elwood and J. H. Elwood, and *Epidemiology and Control of Neural Tube Defects* (1993), by J. M. Elwood et al.

Nondisjunction and Aneuploidy

Field of study: Classical transmission genetics

Significance: *Nondisjunction is the faulty disjoining of replicated chromosomes during mitosis or meiosis, which causes an alteration in the normal number of chromosomes (aneuploidy). Nondisjunction is a major cause of Down syndrome and various sex chromosome anomalies. Understanding the mechanisms associated with cell division may provide new insight into the occurrence of these aneuploid conditions.*

Key terms

DEOXYRIBONUCLEIC ACID (DNA): the genetic material of most organisms, composed of nucleotide units whose arrangement carries genetic information

CHROMOSOME: a rodlike structure contained in the nucleus of cells that is composed of DNA and protein

MITOSIS: nuclear division of chromosomes, usually accompanied by cytoplasmic division; two daughter cells are formed with identical genetic material

MEIOSIS: a series of two nuclear divisions that occur in gamete formation in sexually reproducing organisms

GENE: a portion of a DNA molecule in a chromosome that contains specific genetic information

Background

In all living organisms, hereditary information is contained in a chemical called deoxyribonucleic acid (DNA). In multicellular organisms and some single-cell organisms, DNA is packaged in rodlike structures called chromosomes. Each cell in multicellular organisms contains all the hereditary information for that individual. A given species has a characteristic chromosome number. There are two of each kind of chromosome in sexually reproducing organisms; this is the diploid chromosome number for a species. In humans (*Homo sapiens*), there are forty-six chromosomes; in corn (*Zea mays*), there are twenty chromosomes. The haploid chromosome number is a set of chromosomes (one of each kind of chromosome) for a particular species. In humans, there are twenty-three different kinds of chromosomes; in corn, there are ten different kinds of chromosomes. One set of chromosomes is contributed to a new individual by each parent in sexual reproduction through the egg and sperm. Thus, a fertilized egg will contain two sets of chromosomes.

A karyotype is a chromosome map that displays the number and kinds of chromosomes a particular organism possesses. The human karyotype contains twenty-two pairs of autosomes (chromosomes that are not sex chromosomes) and one pair of sex chromosomes. Females normally possess two X chromosomes in their cells, one inherited from each parent. Males have a single X chromosome, inherited from the mother, and a Y chromosome, inherited from the father.

The many cells of a multicellular organism are created as the fertilized egg undergoes a series of cell divisions. In each cell division cycle, the chromosomes are replicated, and, subsequently, one copy of each chromosome is distributed to two daughter cells through a process called mitosis. When gametes (eggs or sperm) are produced in a mature organism, a modification of mitosis occurs called meiosis. Meiosis produces cells that become gametes. These cells contain one set of chromosomes instead of two sets of chromosomes. When two gametes join (when a sperm cell fertilizes an egg cell), the diploid chromosome number for the species is restored, and, potentially, a new individual will form with repeated cell divisions.

When replicated chromosomes are distributed to daughter cells during mitosis or meiosis, the replicated chromosomes are said to disjoin from each other (disjunction). Occasionally, this distribution of replicated chromosomes to daughter cells is faulty, and chromosomes do not disjoin. When faulty disjoining (nondisjunction) of replicated chromosomes occurs, a daughter cell may result with either one chromosome more than normal or one fewer than normal. This alteration in the normal number of chromosomes is called "aneuploidy." One chromosome more than normal is referred to as a "trisomy." For example, Down syndrome is caused by trisomy 21 in humans. One chromosome fewer than normal is called "monosomy." Turner's syndrome in humans is an example of monosomy. Turner's individuals are women who have only one X chromosome in their cells, whereas human females normally have two X chromosomes. When nondisjunction occurs in the dividing cells of a mature organism or a developing organism, a portion of the cells of the organism may be aneuploid. If nondisjunction occurs in meiosis during gamete formation, then a gamete will not have the correct haploid chromosome number. If that gamete is involved in a conception event, the resulting embryo will be aneuploid. Examples of human aneuploid conditions occurring in live births include Down syndrome (trisomy 21), Edwards' syndrome (trisomy 18), Patau syndrome (trisomy 13), triple X syndrome, Klinefelter's syndrome, and Turner's syndrome. Most aneuploid human conceptions do not survive to birth.

Causes of Nondisjunction

There are both environmental and genetic factors associated with nondisjunction in plants and animals. Environmental factors that may induce nondisjunction include physical factors such as heat, cold, maternal age, and ionizing radiation, in addition to a wide variety of chemical agents.

In humans, it is well established that increased maternal age is a cause of nondisjunction associated with the occurrence of Down syndrome. For mothers who are twenty years of age, the incidence of newborns with Down syndrome is 0.4 in 1,000 newborns. For mothers over forty-five years of age, the incidence of newborns with Down syndrome is 17 in 1,000 newborns. While it is clear that increased maternal age is linked to nondisjunction, it is not at all clear what specific physiological, cellular, or molecular mechanisms or processes are associated with this increased nondisjunction. While nondisjunction in maternal meiosis may be the major source of trisomy 21 in humans, paternal nondisjunction in sperm formation does occur and may result in aneuploidy. It is important to remember that there is no way to prevent nondisjunction, even in those who lead very healthy lives prior to, during, and after conception.

In a study conducted by Karl Sperling and colleagues published in the *British Medical Journal* (July 16, 1994), low-dose radiation in the form of radioactive fallout from the Chernobyl nuclear accident (April, 1986) was linked to a significant increase in trisomy 21 in West Berlin in January, 1987: twelve births of trisomy 21 compared to the expected two or three births. This study suggests that, at least under certain circumstances, ionizing radiation may affect the occurrence of nondisjunction. Researchers have shown that ethanol (the alcohol in alcoholic beverages) causes nondisjunction in mouse-egg formation, suggesting a similar possibility in humans. Other researchers have found that human cells in tissue culture (cells growing on nutrient media) had an increased occurrence of nondisjunction if the media was deficient in folic acid. This implies that folic acid may be necessary for normal chromosome segregation or distribution during cell division.

Scientists know from genetics research that mutations (changes in specific genes) in the fruit fly result in the occurrence of nondisjunction. This genetic component of nondisjunction is further supported by the observation that an occasional family gives birth to more than one child with an aneuploid condition. In these instances, it is likely that genetic factors are contributing to repeated nondisjunction.

Impact and Applications

There are several reasons scientists are devoting research efforts to understanding the consequences of nondisjunction and aneuploidy. First, at least 15 to 20 percent of all recognized human pregnancies end in spontaneous abortions. Of these aborted fetuses, between 50 and 60 percent are aneuploid. Second, of live births, 1 in 700 is an individual with Down syndrome. Mental retardation is a major symptom in individuals with Down syndrome. Thus, nondisjunction is one cause of mental retardation. Finally, aneuploidy is common in cancerous cells. Scientists do not know whether nondisjunction is part of the multistep process of tumor formation or whether aneuploidy is a

The reactor explosion at the Chernobyl nuclear plant in 1986 has been linked to an increase in cases of nondisjunction in babies born in central Europe. (AP/Wide World Photos)

consequence of tumor growth. Continued research into the mechanics of cell division and the various factors that influence that process will increase the understanding of the consequences of nondisjunction and possibly provide the means to prevent its occurrence.

—*Jennifer Spies Davis*

See Also: Down Syndrome; Klinefelter's Syndrome; Meiosis; Mitosis; Turner's Syndrome.

Further Reading: "The Causes of Down Syndrome," *Scientific American* 257 (August, 1987), discusses nondisjunction and the ongoing research into what genes occur on chromosome 21 and how they contribute to Down syndrome. Dr. Henry C. Heins, "What Causes Down Syndrome," *American Baby* 53 (April, 1991), gives an informative explanation of the causes of Down syndrome and other kinds of chromosome abnormalities. Several support organizations are listed. *Down's Syndrome: An Introduction for Parents* (1988), by Cliff Cunningham, provides in-depth advice and information for parents with Down syndrome children.

Oncogenes

Field of study: Molecular genetics

Significance: *Oncogenes are a group of genes originally identified in RNA tumor viruses and later identified in many other types of human malignancy. The discovery of oncogenes has revolutionized the understanding of cancer genetics and contributed to the development of a model of cancer as a multistage genetic disorder. The identification of these abnormally functioning genes in many types of human cancer has also provided new molecular targets for therapeutic intervention.*

Key terms

PROTO-ONCOGENES: cellular genes that carry out specific steps in the process of cellular proliferation; as a consequence of mutation or deregulation, these genes may be converted into cancer-causing genes

MUTATION: any alteration of gene structure; mutations may result in the production of abnormal proteins with altered or absent functional activity or may cause changes in the level of gene expression

The Discovery of Oncogenes

The discovery of oncogenes has been closely linked to the study of the role of a group of ribonucleic acid (RNA) tumor viruses (*Retroviridae*) in the etiology of many animal cancers. In the early part of the twentieth century, Peyton Rous identified a virus (called Rous sarcoma virus after its discoverer) capable of inducing tumors called sarcomas in chickens. Many other RNA tumor viruses capable of causing tumor formation in animals or experimental systems were also discovered, which led to a search for specific viral genes responsible for the cancer-causing properties of these viruses.

The identification of these cancer-causing genes (oncogenes) awaited developments in the area of recombinant deoxyribonucleic acid (DNA) technology and molecular genetics, which ultimately facilitated the molecular analysis of this group of genes. These analyses revealed that viral oncogenes were actually cellular genes that were incorporated into the genetic material of the RNA tumor virus during the process of infection. The acquisition of these host-cell genes was responsible for the cancer-causing properties of these viruses. The first oncogene discovered was the *src* gene of the Rous sarcoma virus. Subsequently, at least thirty different oncogenes were discovered in avian and mammalian RNA tumor viruses. Each of these oncogenes has a cellular counterpart that is the origin of the viral gene; with the exception of the Rous sarcoma virus, the incorporation of the host-cell gene into the virus, involving a process called transduction, results in the loss of viral genes, generating a defective virus.

In addition to the oncogenes originally identified in viruses, approximately fifty oncogenes have been identified in malignant tumors as part of chromosomal rearrangements or the amplification or mutations of specific genes. The first genetic rearrangement linked to a specific type of human malignancy involved the discovery of the "Philadelphia" chromosome in patients with chronic myelogenous leukemia (CML). This chromosome represents a shortened version of chromosome 22, which results from an exchange of genetic material between chromosomes 9 and 22 (called a reciprocal translocation). Subsequent molecular analyses showed that the oncogene *abl*, originally identified in a mammalian RNA tumor virus, was translocated to chromosome 22 in CML patients as a consequence of the reciprocal translocation event. Additional human malignancies involving translocated oncogenes previously identified in RNA tumor viruses have been identified, notably the oncogene *myc* in patients with Burkitt's lymphoma, a disease largely endemic to parts of Africa.

Additional genetic rearrangements involving oncogenes have been identified in the form of gene amplifications (segments of genetic material duplicated many times in genetically unstable tumor cells). These gene amplifications may be associated with the presence of multiple copies of genetic segments along a chromosome, designated as homogeneously staining regions (HSRs), or may appear in the form of

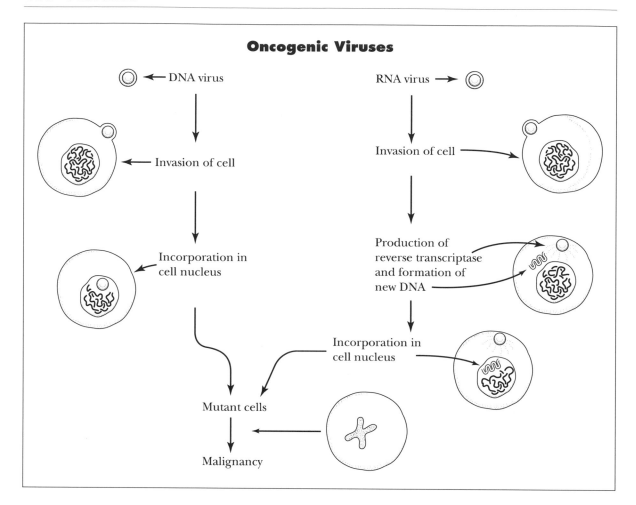

minichromosomes containing the amplified genes, termed double-minutes (DMs). For example, late-stage neuroblastomas often contain numerous double-minute chromosomes containing amplified copies of the *N-myc* gene.

The Properties of Oncogenes

Studies of viral oncogenes provided the first evidence of their role in malignancy. Dramatic evidence was provided by studies of the *sis* oncogene of simian sarcoma virus, which showed that it represents an altered form of the mammalian growth factor called platelet derived growth factor (PDGF). Growth factors are proteins that bind to receptors on sensitive cells to initiate an intracellular signaling cascade, which results in cellular proliferation. This seminal discovery of the identity of the *sis* oncogene suggested a model explaining the prop-

erties of oncogenes. The proto-oncogene model states that oncogenes are derived from normal host genes called proto-oncogenes, which encode gene products involved in cellular proliferative events. If proto-oncogenes are altered as a consequence of mutation or deregulation of gene activity, they may subvert normal cell-division processes, resulting in unregulated cellular proliferation, a hallmark of malignancy.

Subsequent analyses of oncogene activities and the structure and function of the cellular proto-oncogenes from which they are derived have provided strong evidence for this model. Viral and cellular oncogenes with functions affecting every stage of the cell-cycle proliferative pathway have been identified. In addition to altered growth factors such as *sis*, researchers have also identified altered growth factor receptors such as the epidermal growth factor receptor (*erb-b*), elements of the intracellular signal cascade (*src* and *ras*), nuclear transcriptional activators (*myc*), cell-cycle regulators called cyclin-dependent kinases (*cdks*), and cell death inhibitors (*bcl2*) in human tumors of diverse tissue origin. Each of these oncogenic gene products represents an altered form of normal cellular genes that participate in cell-division pathways. Numerous mutations in proto-oncogenes have been identified, including single-base changes (point mutations), gene truncations, gene amplifications, and gene rearrangements resulting from exchanges between different chromosomes called translocations.

One of the most dramatic discoveries involved a comparative analysis of the structures of the normal form of the cellular *ras* proto-oncogene and the oncogene isolated from human bladder carcinomas. Surprisingly, a single-base change was sufficient to convert a normal cellular gene to a cancer-causing gene. The observed mutations were localized to regulatory regions of the *ras* gene product, resulting in its permanent activation. Molecular analyses of many other oncogenes have shown that the observed mutations fall into several categories: nucleotide base changes that result in gene products whose functions are not subject to normal inhibitory processes, overproduction of gene products caused by gene amplification or translocation, and loss of regulatory components caused by gene translocation or truncation. The generalized consequence of these mutations is to convert normal cellular gene products important in cell division to dominant, dysregulated gene products that cause the inappropriate stimulation of cellular proliferation.

Interestingly, most tumors analyzed show the involvement of multiple oncogenes and tumor suppressor genes. Studies of tumor development in human colorectal carcinomas in which it is possible to identify discrete stages of tumor development have indicated a progressive increase in the number and types of cellular oncogenes implicated at successive stages of tumorigenesis. From these studies, a model of oncogenesis has emerged in the form of a multistage disorder characterized by the successive accumulation of mutations in specific cellular oncogenes and tumor suppressor genes, which results in the inability to regulate cellular proliferation.

Impact and Applications

The identification of oncogenes has provided enormous amounts of information on the cellular mechanisms responsible for the loss of growth control in cancer cells. In addition, these dysfunctional gene products represent potential targets for therapeutic applications. Research studies have been directed toward the design of pharmacological inhibitors of specific oncogenes such as *ras* and *erb-b* in order to block dysregulated, proliferative pathways in malignant cells. Additional molecular targets include overexpressed oncogenes that stimulate cellular proliferation or blood vessel formation (angiogenesis), processes critical to tumor establishment. The advantages of these approaches include enhanced cytotoxicity as well as a potential decrease in side effects compared to conventional chemotherapeutic drugs that do not target tumor cell dysfunction. Structural abnormalities in oncogene products may be used in the development of monoclonal antibodies directed against these dysfunctional proteins. Toxins may also be linked to the antibodies to generate immunotoxins whose cell-killing activities directly target malignant cells. Malignant melanoma (skin cancer) has been the focus of many of these targeted approaches directed against specific abnormal gene products. Successful clinical applications will most likely combine approaches involving cytotoxic drugs and inhibitors targeting multiple sites of oncogene dysfunction in the cancer cell.

—*Sarah Crawford Martinelli*

See Also: Cancer; Gene Therapy; Tumor-Suppressor Genes.

Further Reading: *Oncogenes* (1995), by Geoffrey Cooper, provides the reader with a general overview. Harold Varmus, "The Molecular Genetics of Cellular Oncogenes," *Annual Review of Genetics* 18 (1994), details the structure and function of specific oncogenes. Leland Hartwell et al., "Cell Cycle Control and

Cancer," *Science* 266 (1994), provides a clear description of the role of oncogenes in cell-cycle dysregulation.

One Gene-One Enzyme Hypothesis

Field of study: Molecular genetics
Significance: *The immediately apparent functions of a gene are replication and transcription, but the net effect of these processes is genetic control of the organism's metabolism. The "one gene-one enzyme" hypothesis explains the biochemical or molecular function of a gene and postulates that each gene controls the synthesis of a single enzyme. The current modified concept is "one gene-one polypeptide." The manner in which different genes may affect related steps in the development of a particular trait has implications in the understanding of mutations and the treatment of disorders in plants, animals, and humans.*

Key terms

PEPTIDE: two amino acids that are joined by a peptide bond; a linear arrangement of many amino acids is called a "polypeptide"

ENZYME: an organic, catalytic agent (a protein) that accelerates a specific chemical reaction in a living system; an enzyme may consist of a single polypeptide chain or multiple polypeptide chains

History of Biochemical Genetics

The one gene-one enzyme hypothesis is the foundation of biochemical genetics. In 1902, English physician Archibald Garrod proposed that the disease alkaptonuria (an arthritic condition) in humans was a Mendelian recessive trait. Alkaptonuria, caused by a defect in a liver enzyme (homogentisic acid oxidase), is characterized by the excretion of wine-colored urine. The disease is associated with a failure of the breakdown of the benzene ring in the degradation of phenylalanine and tyrosine because of a gene defect, with associated absence of an enzyme that cleaves the ring of alkapton; alkapton is excreted in urine instead of being degraded to fumaric and acetoacetic acids. The report was published in 1902 in *Lancet* and in Garrod's book *Inborn Errors of Metabolism* in 1909.

Garrod's discovery, like Gregor Mendel's work, was not fully appreciated for more than thirty years. William Bateson referred to the disease in his 1909 book *Mendel's Principles of Heredity*. One of Bateson's students and Leonard Troland attempted to integrate biochemistry with genetics. Troland's article, published in *American Naturalist*, suggested that "the germ-cell contains 'determiners' for the production of enzymes.... On the supposition that the actual Mendelian factors are enzymes, nearly all . . . general difficulties instantly vanish, and I am not acquainted with any evidence which is inconsistent with this supposition." In 1922, Hermann Müller discussed the possible role of enzymes in chemical reactions but rejected the notion that genes were enzymes.

Among biochemists, Sir Frederick Gowland Hopkins came close to appreciating Garrod's work when he wrote in the 1913 *Report of the British Association*:

> Extraordinarily profitable have been the observations made upon individuals suffering from those errors of metabolism which Dr. Garrod calls "metabolic sports, the chemical analogues of structural malformations." In these individuals Nature has taken the first essential step in an experiment by omitting from their chemical structure a special catalyst which at one point in the procession of metabolic chemical events is essential to its continuance. At this point there is arrest, and intermediate products come to light.

Sewall Wright, in 1941, and J. B. S. Haldane, in 1942, brought Garrod's work to George Beadle's attention. Beadle and Boris Ephrussi had worked on eye pigmentation genes in *Drosophila melanogaster* in 1936 and suggested that the synthesis of the brown eye pigment arises by the following metabolic chain:

Precursor	→	Substance I	→	Substance II	→	Brown Pigment
(Tryptophan)		(Formylkynurenin)		(Hydroxykynurenin)		
		(Vermilion)		(Cinnabar)		

George Beadle (left) and Linus Pauling circa 1952. (California Institute of Technology)

Working with the bread mold *Neurospora*, Beadle and Edward Tatum published their definitive work on biochemical genetics in 1941. They exposed spores of the fungus to X rays and other agents that cause mutations (changes in the deoxyribonucleic acid, or DNA, sequence). Genetic crosses between the wild type (standard genotype) and mutants revealed that because of a single gene mutation, most of the mutants carried a block at a single step in the reaction sequence leading up to the synthesis of the amino acid, vitamin, purine, or pyrimidine required for growth. This led them to propose the one gene-one enzyme theory, which paved the way for molecular genetics.

One Gene-One Polypeptide Hypothesis

The one gene-one enzyme concept was broadened to one gene-one protein, since many genes code for proteins that are not enzymes. Many proteins have multiple polypeptide chains; in some, all the chains are identical (coded by the same gene), while in others, two or more different polypeptide chains, each with a distinctive amino acid sequence, exist. Thus a further modification of the concept was needed. The enzyme hexosaminidase is a good example of a two-gene control. Two different genes in human adults are known to control the structure of the hemoglobin molecule. Hemoglobin is responsible for the oxygen-carrying capacity of the individual's blood. Mutations affecting hemoglobin production in humans may involve a substitution of one amino acid by another. For example, sickle-cell anemia occurs in individuals who are homozygous for a recessive gene. The alpha chains of sickle-cell hemoglobin have the same structure as in normal hemoglobin, but in the beta chains, one amino acid is replaced by another (valine in the sickle-cell hemoglobin and glutamic acid in the normal) at one particular point in the chain. One gene controls the structure of the alpha chain, and the other gene controls the structure of the beta chain. Thus the correctly modified hypothesis is one gene-one polypeptide.

The work of Charles Yanofsky and others in the 1960's clearly established that the gene and the polypeptide chain are collinear; that is, the order of bases (nucleotides) in the DNA is the same as the order of respective encoded amino acids in the polypeptide chain. The linear sequence of nucleotides in a gene specifies the linear sequence of amino acids in a protein.

Impact and Application

The minimum size of genes can be estimated from the knowledge of the number of amino acids involved in a protein. Since each amino acid of a polypeptide chain is coded by a sequence of three consecutive nucleotides (codon) in a DNA strand and since the codons are arranged sequentially (corresponding to the sequence of amino acids in the polypeptide), the length of a gene (nucleotides) can be estimated. For example, an average polypeptide chain of four hundred amino acids would correspond to twelve hundred base pairs or nucleotides in DNA.

In humans, there are numerous genetic diseases that are caused by a defective protein. Phenylketonuria is caused by defective activity of the enzyme phenylalanine hydroxylase, which normally converts phenylalanine to tyrosine. Large amounts of phenylalanine appear in the blood, which leads to the disease. Patients with Tay-Sachs disease lack hexosaminidase A enzyme. They may be normal at birth but soon experience acute reaction to sounds. Very few patients with this disease survive beyond two years of age. In hereditary fructose intolerance, a mutation eliminates the enzyme fructose-1-phosphate aldolase. Infants suffer adverse reactions and may die if fructose is fed to them. Argininemia is caused by a defective enzyme, arginase, which allows arginine to accumulate in the blood. One treatment involves infecting patients with Shope rabbit papilloma virus, which stimulates arginase activity.

Gene therapy, in which an abnormal gene is repaired or replaced with a normal gene, is, in general, in experimental stages for humans. Transfer of genes such as growth hormones has already been achieved in animals. In 1990, French Anderson and colleagues treated a four-year-old girl suffering from the adenosine deaminase (ADA) disorder with genes coding for ADA. The genes were inserted into blood cells taken from the patient, and the modified cells were transfused back into her body. Her immune system was restored to a normal state as the new genes began to produce ADA enzyme. Gene therapies for cystic fibrosis (CF) and some cancers have also been employed in patients. For CF correction, the gene must be delivered to the cells lining the lungs. Scientists are using adenovirus (a common cold virus) to insert the gene. CF patients inhale adenovirus that carries a normal CF gene.

Gene therapy is currently applicable to somatic cells but not to germ cells; it is therefore effective only for the treated individual. The changes cannot be passed on to the next generation. Obstacles to be overcome include putting a gene into an exact spot on a chromosome and getting the gene to express in the tissue. Synthetic genes have been introduced into and expressed in prokaryotes. Eventually, synthetic genes will play a major role in gene therapy in eukaryotes as well. Protein engineering via targeted alterations in DNA to produce a desired trait (for example, stress tolerance) and site-directed mutagenesis are important applications of gene-protein relationship. The knowledge of genes and the enzymes for which they code has significant implications for gene therapy and medical treatment of genetic disorders.

—*Manjit S. Kang*

See Also: DNA Structure and Function; Inborn Errors of Metabolism; Sickle-Cell Anemia.

Further Reading: Jeffrey H. Miller, *Discovering Molecular Genetics* (1996), provides key papers of pioneering scientists whose research led to the establishment of the one gene-one polypeptide hypothesis. *Heritage from Mendel* (1967), edited by R. Alexander Brink, contains an excellent chapter on "Mendelism, 1965," by George Beadle. George P. Rédei, *Genetics* (1982), also presents an account of the hypothesis.

Organ Transplants and HLA Genes

Field of study: Immunogenetics
Significance: *Organ transplantation has saved the lives of countless people. Although the success rate*

for organ transplantation continues to improve, many barriers remain, including an inadequate supply of donor organs and the phenomenon of transplant rejection. Transplant rejection is caused by an immune response by the recipient to molecules on the transplanted organs that are coded for by the human leukocyte antigen (HLA) gene complex.

Key terms
ANTIGENS: molecules recognized as foreign to the body by the immune system, including molecules associated with disease-causing organisms (pathogens)
HISTOCOMPATIBILITY ANTIGENS: molecules expressed on transplanted tissues that are recognized as foreign by the immune system, causing rejection of the transplant: the most important histocompatibility antigens in vertebrates are coded for by a cluster of genes called the major histocompatibility complex (MHC)
LOCUS (LOCI): the location of a gene on a chromosome
ALLELES: the two alternate forms of a gene at the same locus on a pair of homologous chromosomes
POLYMORPHISM: the presence of many different alleles for a particular locus in individuals of the same species

Transplantation
The replacement of damaged organs by transplantation was one of the great success stories of modern medicine in the latter decades of the twentieth century. During the 1980's, the success rates for heart and kidney transplants showed marked improvement and, most notably, the one-year survival for pancreas and liver transplants rose from 20 percent and 30 percent to 70 percent and 75 percent, respectively. These increases in organ survival were largely attributable to improvements in two aspects of the transplantation protocol that directly reduced tissue rejection: the development of more accurate methods of tissue typing that allowed better tissue matching of donor and recipient, and the discovery of more effective and less toxic antirejection drugs. In fact, these changes helped make transplantation procedures so common by the 1990's that the

low number of donor organs became a major limiting factor in the number of lives saved by this procedure.

Rejection and the Immune Response

The rejection of transplanted tissues is associated with genetic differences between the donor and recipient. Transplants of tissue within the same individual, called autografts, are never rejected. Thus the grafting of blood vessels transplanted from the leg to an individual's heart during bypass operations are never in danger of being rejected. On the other hand, organs transplanted between genetically distinct humans tend to undergo clinical rejection within a few days to a few weeks after the procedure. During the rejection process, the transplanted tissue is gradually destroyed and loses its function. When examined under the microscope, tissue undergoing rejection is observed to be infiltrated with a variety of cells, causing its destruction. These infiltrating cells are part of the recipient's immune system, which recognizes molecules on the transplant as foreign to the body and responds to them as they would to a disease-causing, pathogenic organism.

The human immune response is a complex system of cells and secreted proteins that has evolved to protect the body from invasion by

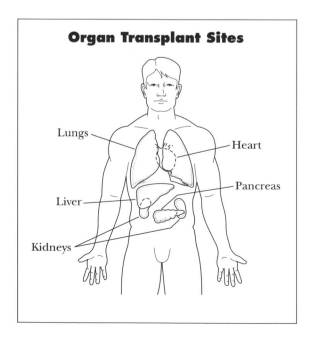

Organ Transplant Sites

Lungs

Heart

Pancreas

Liver

Kidneys

pathogens. Immune mechanisms are directed against molecules or parts of molecules called antigens. The ultimate function of the immune response is to recognize pathogen-associated antigens as foreign to the body and to eliminate and destroy the organism, thus resolving the disease. On the other hand, the immune response is prevented, under most circumstances, from attacking the antigens expressed on the tissues of the body in which they originate. The ability to distinguish between self and foreign antigens is critical to protecting the body from pathogens and to the maintenance of good health.

A negative consequence of the ability of the immune system to discriminate between self and foreign antigens is the recognition and destruction of transplants. The antigens associated with transplants are recognized as foreign in the same fashion as pathogen-associated antigens, and many of the same immune mechanisms used to kill pathogens are responsible for the destruction of the transplant. The molecules on the transplanted tissues recognized by the immune system are called histocompatibility antigens. The term "histocompatibility" refers to the fact that transplanted organs are often not compatible with the body of a genetically distinct recipient. All vertebrate animals have a cluster of genes that code for the most important histocompatibility antigens, called the major histocompatibility complex (MHC).

MHC Polymorphism, HLA Genes, and Tissue Typing

Each MHC locus is highly polymorphic, meaning that many different alleles exist within a population (members of a species sharing a habitat). The explanation for the polymorphism of histocompatibility antigens is related to the actual function of these molecules within the body. Clearly, histocompatibility molecules did not evolve to induce the rejection of transplants, despite the fact that this characteristic led to their discovery and name.

Histocompatibility molecules function by regulating immunity against foreign antigens. Each allele codes for a protein that allows the immune response to recognize a different set of antigens. Many pathogens, including the viruses associated with influenza and acquired immunodeficiency syndrome (AIDS), undergo genetic mutations that lead to changes in their antigens, making it more difficult for the body to make an immune response to the virus. The existence of multiple MHC alleles in a population, therefore, ensures that some individuals will have MHC alleles allowing them to mount an immune response against a particular pathogen. If an entire population lacked these alleles, their inability to respond to certain pathogens could threaten the very existence of the species. The downside of MHC polymorphism, however, is the immune response to the donor's histocompatibility antigens that causes organ rejection.

The human leukocyte antigen (HLA) gene complex is located on chromosome 6 of humans. Six important histocompatibility antigens are coded for by the HLA complex: the *A*, *B*, *C*, *DR*, *DP*, and *DQ* alleles. Differences in HLA antigens between the donor and recipient are determined by tissue typing. For many years, tissue typing was performed using antibodies specific to different HLA alleles. Antibodies are proteins secreted by the cells of the immune system that are used in the laboratory to identify specific antigens. As scientists began to clone the genes for the most common HLA alleles in the 1980's and 1990's, however, it appeared that direct genetic analysis would eventually replace or at least supplement these procedures.

Fewer differences in these antigens between donor organ and recipient mean a better prognosis for transplant survival. Therefore, closely related individuals who share many of their histocompatibility alleles are usually preferred as donors. When a family member is not available, the process of finding a donor is problematic. Worldwide computer databases are used to match potential donors with recipients, who are placed on a waiting list based on the severity of their disease.

Immunosuppressive Antirejection Drugs

Perhaps the most important medical breakthrough responsible for the increased success of organ transplantation occurred in the last two decades of the twentieth century. This breakthrough involved the discovery and successful use of antirejection drugs, most of which act by suppressing the immune response to the transplanted tissue. Immunosuppressive drugs are usually given in high doses for the first few weeks after transplantation or during a rejection crisis, but the dosage of these drugs is usually reduced thereafter to avoid their toxic effects.

Cyclosporine is by far the most effective of these drugs and has largely been responsible for the increased efficacy of liver, pancreas, lung, and heart transplantation procedures. In spite of its successes, cyclosporine has limitations in that it can cause kidney damage when given in high doses. Azathioprine, associated with bone marrow toxicity, was largely supplanted by the introduction of the less toxic cyclosporine. However, azathioprine has been used as part of a combined cyclosporine-azathioprine regimen. This practice allows the reduction of both the cyclosporine and azathioprine dosages, reducing the toxicity of both drugs. The search for more effective and less toxic antirejection drugs continues. Individuals receiving immunosuppressive therapy have other concerns in addition to the toxicity of the drugs themselves. As these individuals will have an impaired ability to mount an immune response to pathogens, their susceptibility to a variety of diseases will be increased. Thus transplant recipients must take special precautions to avoid exposure to potential pathogens, especially when receiving high doses of the drugs.

—*James A. Wise*

See Also: Antibodies; Genetic Screening; Immunogenetics; Multiple Alleles.

Further Reading: "Life, Death, and the Immune System," *Scientific American* 269 (September, 1993), is a special issue that provides an excellent overview of the immune system. Charles Janeway and Paul Travers provide an excellent review of the HLA complex in *Immunobiology: The Immune System in Health and Disease* (1994). Robert I. Lechler et al., "The Molecular Basis of Alloreactivity," *Immunology Today* 11 (March, 1990), discusses the molecular basis of transplantation rejection.

Parthenogenesis

Field of study: Classical transmission genetics

Significance: *Parthenogenesis is the development of unfertilized eggs, which produces individuals that are genetically alike and allows rapid expansion of a population of well-adapted individuals into a rich environment. This clonal reproduction strategy is used by a number of species for rapid reproduction under very favorable conditions, and it appears to offer a selective advantage to individuals living in disturbed habitats.*

Key terms

ADAPTIVE ADVANTAGE: increased fertility in offspring as a result of passing on favorable genetic information

DIPLOID: having two sets of genetic information

FERTILIZATION: the fusion of two cells (egg and sperm) in sexual reproduction

HAPLOID: having one set of genetic information

MEIOSIS: cell division that reduces the chromosome number from diploid to haploid in the production of the sperm and the egg

ZYGOTE: the product of fertilization in sexually reproducing organisms

The Nature of Parthenogenesis

Parthenogenesis is derived from two Greek words that mean "virgin" (*parthenos*) and "origin" (*genesis*) and describes a form of reproduction in which females lay diploid eggs (containing two sets of chromosomes) that develop into new individuals without fertilization—there is no fusion of a sperm nucleus with the egg nucleus to produce the new diploid individual. This is a form of clonal reproduction because all of the individuals are genetically identical to the mother and to each other. Parthenogenesis mechanisms do not show any single pattern and have evolved independently in different organisms. In some organisms, such as rotifers and aphids, parthenogenesis alternates with normal sexual reproduction. When there is a rich food source, such as new rose bushes emerging in the early spring, aphids reproduce by parthenogenesis; late in the summer, however, as the food source is decreasing, sexually reproducing females appear. The same pattern has been observed in rotifers, in which a decrease in the quality of the food supply leads to the appearance of females that produce haploid eggs by normal meiosis that require fertilization for development. The strategy appears to involve the clonal production of large numbers of genetically identical individuals that are well-suited to the environment when the conditions are favorable and the production of a variety of different types, by the recombination that occurs in a normal meiosis and the mixing of alleles from two individuals in sexual reproduction, when the conditions are less favorable. In social insects, such as bees, wasps, and ants, parthenogenesis is a major factor in sex determination, although it may not be the only factor. In these insects, eggs that develop by parthenogenesis remain haploid and develop into males, while fertilized eggs develop into diploid, sexually reproducing females.

In algae and some forms of plants, parthenogenesis also allows rapid reproduction when conditions are favorable. In citrus, seed development by parthenogenesis maintains the favorable characteristics of each plant. For this reason, most commercial citrus plants are propagated by asexual means, such as grafting. Parthenogenesis has also been induced in organisms that do not show the process in natural populations. In sea urchins, for example, development can be induced by mechanical stimulation of the egg or by changes in the chemistry of the medium. Even some vertebrate eggs have shown signs of early development when artificially stimulated, but haploid vertebrate cells lack all of the information required for normal development, so such "zygotes" cease development very early.

Parthenogenesis in Vertebrates

Parthenogenesis has been observed in vertebrates such as fish, frogs, and lizards. In these parthenogenetic populations, all the individuals are females, so reproduction of the clone is restricted to parthenogenesis. Parthenogenetic fish often occur in a population along

Parthenogenesis has been observed in frogs, lizards, and fish as well as in many invertebrate and plant species. (Ben Klaffke)

with sexually reproducing individuals. The parthenogenetic forms produce diploid eggs that develop without fertilization; in rare cases, however, fertilization of a parthenogenetic egg gives rise to a triploid individual that has three sets of chromosomes rather than the normal two sets (two from the diploid egg and one from the sperm). In some groups, penetration of a sperm is necessary to activate development of the zygote, but the sperm nucleus is not incorporated into the zygote.

Evidence indicates that in each of these vertebrate situations, the parthenogenetic populations have resulted from a hybridization between two different species. The parthenogenetic forms always occur in regions where the two parental species overlap in their distribution, often an area that is not the most favorable habitat for either species. The hybrid origin has been confirmed by the demonstration that the animals have two different forms of an enzyme that have been derived from the two different species in the region. Genetic identity has also

been confirmed using skin graft studies. In unrelated organisms, skin grafts are quickly rejected because of genetic incompatibilities; clonal animals, on the other hand, readily accept grafts from related donors. Parthenogenetic fish from the same clone accept grafts that confirm their genetic identify, but rejection of grafts by other parthenogenetic forms from different populations shows that they are different clones and must have a different origin. This makes it possible to better understand the structure of the populations and helps in the study of the origins of parthenogenesis within those populations. Comparisons of nuclear genetic content and mitochondrial genetic content also allow the determination of both species origin and the maternal species of the parthenogenetic form since the mitochondria are almost exclusively transmitted through the vertebrate egg. Within the hybrid, a mechanism has originated that allows the egg to develop without fertilization, although, as already noted, penetration by a sperm may be required to activate development in some of the species.

The advantage of parthenogenesis appears to be the production of individuals that are genetically identical. Since the parthenogenetic form may, at least in vertebrates, be a hybrid, it is heterozygous at most of its genetic loci. This provides variation in the form of enzymes present for each function that may provide the animal with a greater range of responses to the environment. Maintaining this heterozygous genotype may give the animals a selective advantage in environments where the parental species are not able to reproduce successfully and may be a major reason for the persistence of this form of reproduction. The vertebrate species are found in disturbed habitats, so their unique genetic composition may allow for adaptation to these unusual conditions.

Mechanisms of Development

The mechanisms of diploid egg development are as diverse as the organisms in which this form of reproduction is found. In normal meiosis, the like chromosomes of each pair separate at the first division and the copies of each chromosome separate at the second division (producing four haploid cells). During the meiotic process in the egg, three small cells (the polar bodies), each with one set of chromosomes, are produced, and one set of chromosomes remains as the egg nucleus. In parthenogenetic organisms, some modification of this process occurs that results in an egg nucleus with two sets of chromosomes—the diploid state. In some forms, the first meiotic division does not occur, so two chromosome sets remain in the egg following the second division. In other forms, one of the polar bodies fuses back into the cell so that there are two sets of chromosomes in the final egg. In another variation, there is a replication of chromosomes after the first division, but no second division takes place in the egg, so the chromosome number is again diploid. In all of these mechanisms, the genetic content of the egg is derived from the mother's genetic content, and there is no contribution to the genetic content from male material.

The situation may be even more complex, however, because some hybrid individuals may retain the chromosomal identity of one species by a selective loss of the chromosomes of the other species during meiosis. The eggs may carry the chromosomes of one species but the mitochondria of the other species. The haploid eggs must be fertilized, so these individuals are not parthenogenetic, but their presence in the population shows how complex reproductive strategies can be and how important it is to study the entire population in order to fully understand its dynamics: A single population may contain individuals of the two sexual species, true parthenogenetic individuals, and triploid individuals resulting from fertilization of a diploid egg.

—*D. B. Benner*

See Also: Classical Transmission Genetics; Extrachromosomal Inheritance; Meiosis; Natural Selection.

Further Reading: Most "modern" genetics textbooks say very little about parthenogenesis. An excellent older source is *Mechanisms of Mendelian Heredity* (1923), by four of the most influential individuals in American genetics: Thomas Hunt Morgan, Alfred Sturtevant, Hermann Muller, and Calvin Bridges. K. A. Goddard et al., *Journal of Heredity* 89, (March/April, 1998), is an informative research article on a fish population, with a good review of mechanisms and problems. Extensive coverage is found in *Evolution and Ecology of Unisexual Vertebrates* (1980), by R. M. Dawley and J. P. Bogart. A treatment of the parthenogenic lizards can be found in *The Biology of Reptiles*, vol. 15 (1985), edited by C. Gans and F. Billet.

Paternity Tests

Field of study: Human genetics

Significance: *Establishing paternity is important for many people beyond the immediate family as it has become a matter of policy on a wide range of issues such as child support, health insurance, veteran's and social security benefits, and legal access to medical records. It may also affect a child's future as it relates to inherited diseases.*

Key terms

DEOXYRIBONUCLEIC ACID (DNA) FINGERPRINT: the largely individual banding pattern produced when a sample of DNA is broken down with a restriction enzyme

FORENSIC GENETICS: the use of genetic tests and principles to resolve legal questions

HUMAN LEUKOCYTE ANTIGENS (HLA): antigens produced by a cluster of genes that play a critical role in the outcome of transplants; because they are made up of a large number of genes, they are used in individual identification and the matching of parents and offspring

PATERNITY EXCLUSION: the indication, through genetic testing, that a particular man is not the biological father of a particular child

Genetic Principles Involved in Paternity Testing

The basic genetic principles utilized in paternity testing have remained the same from the first applications of ABO blood groups to applications of deoxyribonucleic acid (DNA) fingerprinting. Available tests may positively exclude a man from being a child's biological father. Evidence supporting paternity, however, cannot be considered conclusive. Ultimately, a court must decide whether a man is determined to be the legal father based on all lines of evidence.

The genetic principles can be illustrated with a very simple example that uses ABO blood types. The four blood groups (A, B, AB, and O) are controlled by three pairs of genes. In the example, however, only three of the blood groups will be used to demonstrate the range of matings with the possible children for each of them (see table below).

Example 1: A man is not excluded.

Mother: A
Child: A
Putative Father: AB

It can be seen that the mothers in matings 1 and 4 satisfy the condition of the mother being A and possibly having a child being A. Mating 4 satisfies the condition of a father being AB, the mother A, and a possible child being A. Results indicate that the putative father could be the father. He is not excluded.

Example 2: A man is excluded.

Mother: A
Child: A
Putative Father: B

Again, it is seen that the mothers in matings 1, 4, and 7 satisfy the condition of the mother

Mating Number	Genes of Parents		Blood Type of Parents		Possible Children	
	Father	Mother	Father	Mother	Genes	Blood Type
1	AA	AA	A	A	AA	A
2	AA	AB	A	AB	AA or AB	A or AB
3	AA	BB	A	B	AB	AB
4	AB	AA	AB	A	AA or AB	A or AB
5	AB	AB	AB	AB	AA, AB, or BB	A, AB, or B
6	AB	BB	AB	B	AB or BB	AB or B
7	BB	AA	B	A	AB	AB
8	BB	AB	B	AB	AB or BB	AB or B
9	BB	BB	B	B	BB	B

being A and possibly having a child being A. Mating 7 satisfies the condition of a father being B and the mother A, but mating 7 cannot produce a child being A. The putative father cannot be the father, and he is excluded.

DNA Fingerprinting

After the initial use of ABO blood groups in paternity testing, it became apparent that there were many cases in which the ABO phenotypes did not permit exclusion. Other blood group systems were also used, including the MN and Rh groups. The more blood groups utilized, the more the probability of exclusion (or non-exclusion) increased. Paternity tests were not restricted to blood groups alone; they also used tissue types and serum enzymes.

The most powerful tool developed has been DNA testing. DNA fingerprinting was developed in England by Sir Alec Jeffreys. DNA is extracted from white blood cells and broken down into fragments by bacterial enzymes (restriction endonucleases). The fragments are separated by size, and specific fragments are identified. Each individual has different DNA profiles, but the profiles of parents and children have similarities in greater proportion than those between unrelated people. Also, frequencies of different fragments tend to vary among ethnic groups. It is possible not only to exclude someone who is not the biological father, but also to determine actual paternity with a probability greater than 99.9 percent.

Impact and Applications

The personal, social, and economic implications involved in paternity testing have far-reaching consequences. Blood-group analysis is cheaper but less consistent than DNA testing. Paternity can often be excluded but rarely proven as positive with the same degree of accuracy that DNA testing provides. Human leukocyte antigen (HLA) testing can also be used but suffers from many of the same problems as blood-group analysis. The development of DNA testing after 1984 revolutionized the field of paternity testing. DNA fingerprinting has made decisions on paternity assignments virtually 100 percent accurate. The same technique has also been applied in cases of individual identification, and results have helped to release people who have been falsely imprisoned as well as convict other people with the analysis of trace evidence.

—*Donald J. Nash*

See Also: Classical Transmission Genetics; DNA Fingerprinting; Forensic Genetics.

Further Reading: The applications and social implications of genetic testing are covered by Doris Teichler Zallen in *Does It Run in the Family?* (1997). Warren Cohen, "Kid Looks Like the Mailman? Gentic Labs Boom as the Nation Wonders Who's Daddy," *U.S. News and World Report* 122 (January 27, 1997), discusses paternity testing at genetic laboratories. Extensions of DNA testing are described by Marilyn S. Pollack in "Prenatal Paternity Testing: Why, When, How?" *Medical Laboratory Observer* 29 (June, 1997).

Polymerase Chain Reaction

Field of study: Genetic engineering and biotechnology

Significance: *Polymerase chain reaction (PCR) is the in vitro (in the test tube) amplification of specific nucleic acid sequences. In a few hours, a single piece of DNA can be copied one billion times. Because this technique is simple, rapid, and very sensitive, it is used in a very wide range of applications, including forensics, disease diagnosis, molecular genetics, and nucleic acid sequencing.*

Key terms

DEOXYRIBONUCLEIC ACID (DNA): the genetic material for most organisms; a long-chain macromolecule made of units called "nucleotides" that can be double- or single-stranded

MOLECULAR CLONING: the process of combining a piece of DNA into a plasmid, virus, or phage vector to obtain many identical copies of that DNA

ENZYME: a molecule, typically a protein, that aids or accelerates a cellular reaction without itself being altered by the reaction

DNA POLYMERASE: an enyzme that copies or replicates DNA; it uses a single-stranded DNA as a template for synthesis of a complementary new strand and requires a small

Kary B. Mullis accepts his 1993 Nobel Prize for invention of the polymerase chain reaction technique. (Reuters/Pressens Bild/Archive Photos)

section of double-stranded DNA to initiate synthesis

The Development of Polymerase Chain Reaction

Polymerase chain reaction (PCR) was developed by Kary B. Mullis in the mid-1980's. The technique revolutionized molecular genetics and the study of genes. One of the difficulties in studying genes is that a specific gene is one of as many as 100,000 genes in a complex genome. To obtain the number of copies of a specific gene required for accurate analysis requires the time-consuming techniques of molecular cloning and detection of specific deoxyribonucleic acid (DNA) sequences. The polymerase chain reaction changed the science of molecular genetics by allowing huge numbers of copies of a specific DNA sequence to be produced without the use of molecular clon-

ing. The tremendous significance of this discovery was recognized by the awarding of the 1993 Nobel Prize in Chemistry to Mullis for the invention of the PCR method.

How Polymerase Chain Reaction Works

PCR begins with the creation of a single-stranded DNA template to be copied. This is done by heating double-stranded DNA to temperatures near boiling (about 94 to 99 degrees Celsius). This is followed by the annealing (binding of a complementary sequence) of pairs of oligonucleotides (short nucleic acid molecules about fifteen to twenty nucleotides long) called "primers." Because DNA polymerase requires a double-stranded region to prime (initiate) DNA synthesis, the starting point for DNA synthesis is specified by the location at which the primer anneals to the template. The primers are chosen to flank the

DNA to be amplified. This annealing is done at a lower temperature (about 30 to 65 degrees Celsius). The final step is the synthesis by DNA polymerase of a new strand of DNA complementary to the template starting from the primers. This step is carried out at temperatures about 65 to 75 degrees Celsius. These three steps are repeated many times (for many cycles) to amplify the template DNA. The time for each of the three steps is typically one to two minutes. If, in each cycle, one copy is made of each of the strands of the template, the number of DNA molecules produced doubles each cycle. Because of this doubling, more than one million copies of the template DNA are made at the end of twenty cycles.

The PCR reaction is made more efficient by the use of heat-stable DNA polymerases that are isolated from bacteria that live at very high temperatures in hot springs or deep-sea vents and by the use of a programmable water bath (called a thermal cycler) to rapidly change the temperatures of samples to each of the temperatures needed in each of the steps of a cycle.

Impact and Applications

PCR is extremely rapid. One billion copies of a specific DNA can be made in a few hours. It is also extremely sensitive. It is possible to copy a single DNA molecule. Great care must be used to avoid contamination, for even trace contaminants could readily be amplified by this method.

PCR is a useful tool for many different applications. It is used in basic research to obtain DNA for sequencing and other analyses. PCR is used in disease diagnosis, in prenatal diagnosis, and to match donor and recipient tissues for organ transplants. Because a specific sequence can be amplified greatly, much less clinical material is needed to make a diagnosis. The assay is also rapid, so results are available sooner. PCR is used to detect pathogens, such as the causative agents for Lyme disease or for acquired immunodeficiency syndrome (AIDS), that are difficult to culture. PCR can even be used to amplify DNA from ancient sources such as mummies, bones, and other museum specimens. PCR is an important tool in forensic investigations. Target DNA from trace amounts of biological material such as semen, blood, and hair roots can be amplified. There are probes for regions of human DNA that show hypervariability in the population and therefore make good markers to identify the source of the DNA. PCR can therefore be used to evaluate evidence at the scene of a crime, help identify missing people, and resolve paternity cases.

—*Susan J. Karcher*

See Also: DNA Fingerprinting; DNA Replication; DNA Structure and Function; Forensic Genetics; Paternity Tests.

Further Reading: The inventor of the polymerase chain reaction, Kary B. Mullis, describes the initial development of the technique in his article, "The Unusual Origin of the Polymerase Chain Reaction," *Scientific American* 262 (April, 1990). A discussion of applications of the technique is presented in Norman Arnheim et al., "Application of the Polymerase Chain Reaction to Organismal and Population Biology," *Bioscience* 40 (March, 1990) and in Philip E. Ross, "Eloquent Remains," *Scientific American* 266 (May, 1992). The importance of polymerase chain reaction is presented in Ruth L. Guyer and Daniel E. Koshland, "The Molecule of the Year," *Science* 246 (December 22, 1989). James D. Watson et al., *Recombinant DNA* (1992), provides a summary of polymerase chain reaction and its applications.

Polyploidy

Field of study: Population genetics

Significance: *Polyploids have three or more complete sets of chromosomes in their nuclei instead of the two sets found in diploids. Polyploids are observed in both plants and animals and have a role in the evolution of species. Some specific tissues of diploid organisms are polyploid.*

Key terms

DEOXYRIBONUCLEIC ACID (DNA): the genetic material for most organisms, a long-chain macromolecule made of units called "nucleotides"

CHROMOSOMES: the structures within a cell that carry the genes (discrete units of heredity);

a chromosome consists of a long, continuous DNA molecule and proteins

ANEUPLOID: a cell with one or more missing or extra chromosomes; the opposite is "euploid," a cell with the normal chromosome number

ALLOPOLYPLOID: a type of polyploid species that contains genomes from more than one ancestral species

AUTOPOLYPLOID: a type of polyploid species that contains more than two sets of genetically identical genomes

Polyploid Plants and Animals

In the plant kingdom, polyploidy is estimated to occur in 95 percent of pteridophytes (plants, including ferns, that reproduce by spores) and perhaps as many as 80 percent of angiosperms (flowering plants that form seeds inside an ovary), although there is high variability in polyploidy between different families of angiosperms. In contrast, polyploidy is uncommon in gymnosperms (plants that have naked seeds that are not within specialized structures). Extensive polyploidy is observed in chrysanthemums, in which chromosome numbers range from 18 to 198. The basic chromosome number (haploid or gamete number of chromosomes) is nine. Polyploids from triploids (twenty-seven chromosomes) to twenty-two-ploids (198 chromosomes) are observed. The stonecrop *Sedum suaveolens*, which has the highest chromosome number of any angiosperm, is believed to be about eightyploid (720 chromosomes).

Many important agricultural crops, including wheat, corn, sugarcane, potatoes, coffee, apples, and cotton, are polyploid. Polyploidy may originate in three ways. The first form occurs when a diploid plant fails to undergo chromosome separation at mitosis in a dividing bud cell. This will produce a branch of the plant that will be tetraploid, which, if flowers and seed are produced, will eventually generate a tetraploid strain (an autoploid) of the plant. The second way that polyploidy may

Wheat is one of many important polyploid crops. (Larry Mulvehill/Rainbow)

originate is during the reduction of the number of chromosomes to the haploid number in the gametes (the process of meiosis). If meiosis is disrupted, a diploid gamete may be produced. A diploid gamete might join a normal haploid gamete to produce a triploid, or it might join another diploid gamete to produce a tetraploid. Finally, hybrids between individuals of different chromosome numbers may produce an individual with an intermediate chromosome number. Hybrids of diploids and tetraploids produce triploids.

Polyploidy occurs in a number of animals, including crustaceans, earthworms, flatworms, and insects such as weevils, sawflies, and moths. Polyploidy has been observed in a number of vertebrate organisms, including tree frogs, lizards, salamanders, and fish. It has been suggested that the genetic redundancy observed in vertebrates may be caused by ancestral polyploidy.

Polyploidy in Tissues

Most plants and animals contain particular tissues that are polyploid or polytene, while the rest of the organism is diploid. Polyploidy is observed in multinucleate cells and in cells that have undergone endomitosis, in which the chromosomes condense but the cell does not undergo nuclear or cellular division. For example, in vertebrates, liver cells are binucleate with a ploidy number of four. In addition, in humans, megakaryocytes can have polyploidy levels of up to sixty-four. A megakaryocyte is a giant bone-marrow cell with a large, irregularly lobed nucleus that is the precursor to blood platelets. A megakaryocyte does not circulate, but buds off platelets. A single megakaryocyte can produce three thousand to four thousand platelets. A platelet is an anucleate, disk-shaped cell in the blood that has a role in blood coagulation. In polytene cells, the replicated copies of the chromosomal DNA remain associated to produce giant chromosomes that have a continuously visible banding pattern. The trophoblast cells of the mammalian placenta are polytene.

Aneuploidy

Most human polyploids die as embryos or fetuses. In a few rare cases, a polyploid infant is born that lives for a few days. Aneuploidy (in which cells have one or more missing or extra chromosomes) in humans can have varying degrees of symptoms. In general, sex-chromosome aneuploidy has less severe symptoms. Aneuploidy and polyploidy have been observed in many tumors. A determination of the type of aneuploidy can be useful in identifying the type of tumor. Flow cytometry is a tool used to determine the numbers of chromosomes in cells and has been used to identify the ploidy level in tumors. An important question is whether aneuploidy plays a role in tumor progression or is simply a phenomenon caused by the cancer. Research suggests that aneuploidy actually facilitates tumor progression by increasing the rate of loss of tumor-suppressor genes. Loss of tumor-suppressor genes leads to the deregulation of cell division and the rapid cell growth of cancer.

Polyploids may be induced by the use of drugs such as colchicine, which halts cell division. Molecular techniques such as in situ hybridization (in which a DNA probe is hybridized to chromosomes) allow the identification of allopolyploids and the visualization of the ancestral genomes. This is important for an increased understanding of polyploidy genome organization and evolution.

—Susan J. Karcher

See Also: Genetic Testing; Speciation; Tumor-Suppressor Genes.

Further Reading: Neil A. Campbell, *Genetics* (1996), provides a description of alterations of chromosome number and a description of the role of polyploidy in speciation. Ricki Lewis, *Human Genetics: Concepts and Applications* (1997), gives an overview of polyploidy and euploidy in humans. Illia J. Leitch and Michael D. Bennett, "Polyploidy in Angiosperms," *Trends in Plant Science* 2 (December, 1997), describes the role of polyploidy in the evolution of higher plants. Toby J. Gibson and Jurg Spring discuss the role of polyploidy in genetic redundancy in their article "Genetic Redundancy in Vertebrates: Polyploidy and Persistence of Genes Encoding Multidomain Proteins," *Trends in Genetics* 14 (February, 1998). The possible role of aneuploidy in tumor progression is presented in "A Checkpoint on the

Road to Cancer," by Terry L. Orr-Weaver and Robert A. Weinberg, *Nature* (March 19, 1998).

Population Genetics

Field of study: Population genetics

Significance: *Population genetics is the field of genetics in which principles of Mendelian inheritance are applied to genes in a population of organisms. Researchers in the field are concerned with both theoretical and experimental investigations of changes in genetic variation caused by various forces; therefore, the field has close ties to evolutionary biology. Using population genetic tools, one can explore the evolutionary history of species, make predictions about future evolution, and predict the fate of genetic diseases in human populations.*

Key terms

ALLELE: one of the different forms of a particular gene

GENETIC DRIFT: random changes in genetic variation caused by sampling error in small populations

GENETIC VARIATION: the existence of multiple alleles (and therefore genotypes) in a population

GENOTYPE: the pair of alleles carried by an individual for a specific gene

FITNESS: a measure of a genotype's ability to survive and reproduce compared to other genotypes

HARDY-WEINBERG LAW: a mathematical model that results, under particular conditions, in a population in which the genetic variation remains constant over time, with genotypes in specific predictable proportions

MODERN SYNTHESIS: the merging of the Darwinian mechanisms for evolution with Mendelian genetics to form the modern fields of population genetics and evolutionary biology

NEUTRAL THEORY OF EVOLUTION: Motoo Kimura's suggestion that evolution at the deoxyribonucleic acid (DNA) level is caused primarily by genetic drift of alleles without much fitness difference

Basic Population Genetics: The Hardy-Weinberg Law

The branch of genetics called population genetics is based on the application of nineteenth century Austrian botanist Gregor Mendel's principles of inheritance to genes in a population. (Although, for some species, "population" can be difficult to define, the term generally refers to a geographic group of interbreeding individuals of the same species.) Mendel's principles can be used to predict the expected proportions of offspring in a cross between two individuals of known genotypes, where the genotype describes the genetic content of an individual for one or more genes. An individual carries two copies of all chromosomes (except perhaps for the sex chromosomes, as in human males) and therefore has two copies of each gene. These two copies may be identical or somewhat different. Different forms of the same gene are called "alleles." A genotype in which both alleles are the same is a "homozygote," while one in which the two alleles are different is a "heterozygote." Although a single individual can carry no more than two alleles for a particular gene, there may be many alleles of a gene present in a population.

In studying the genetics of a population, it would be very difficult to try to track the inheritance patterns of every single mating pair in that population. However, by making some simplifying assumptions about the way that mating pairs form in that population and by following Mendelian principles to determine the expected genotype proportions among offspring, it is possible to predict what will happen to the genetic content of the population over time. Working independently in 1908, the British mathematician Godfrey Hardy and the German physiologist Wilhelm Weinberg were the first to formulate a simple mathematical model describing a gene with two alleles in a population. In this model, the numbers of each allele and of each genotype are not represented as actual numbers but as proportions (known as "allele frequencies" and "genotype frequencies," respectively) so that the model can be applied to any population regardless of its size. By assuming Mendelian inheritance of alleles, Hardy and Weinberg showed that allele frequencies in a population do not change over time and that genotype frequencies will change to specific proportions determined by the allele frequencies within one generation and remain at those proportions in future generations. This result is known as the Hardy-Weinberg law, and the stable genotype proportions predicted by the law are known as Hardy-Weinberg equilibrium. It was shown in subsequent work by others that the Hardy-Weinberg law remains true in more complex models with more than two alleles and more than one gene.

The Hardy-Weinberg law presents a problem for the interpretation of observations of real populations because there are often rapid changes in genotype and allele frequencies in natural populations, while the Hardy-Weinberg prediction is that these frequencies should not change. The resolution to this problem is the recognition that the Hardy-Weinberg theoretical population is oversimplified compared to real populations in order to simplify the analysis and interpretation of the model. This oversimplification means that there are a number of implicit assumptions about the theoretical population that may or may not exist in a real population, including the absence of natural selection of genotypes, mutation of alleles, or migration in or out of the population; random formation of mating pairs; and a very large (technically infinite) population size. These may be considered conditions for Hardy-Weinberg equilibrium. Since no real population will meet these conditions, it may seem that the Hardy-Weinberg model is too unrealistic to be useful, but in fact it can be quite useful. First, the conditions of a natural population may be very close to Hardy-Weinberg conditions, and so the Hardy-Weinberg law may be approximately true for that population. Second, if genotypes in a population are not in Hardy-Weinberg equilibrium, it may be an indication that one or more of these conditions are not met, and further research into the cause of deviation from Hardy-Weinberg expectations would be warranted.

Genetic Variation and Natural Selection

Of course, population geneticists are concerned with describing and explaining the ge-netic composition of real populations, not just theoretical ones. Sampling and genetic analysis of real populations of many different types of organisms (animals, plants, fungi, protists, or bacteria) reveal that there is usually a substantial amount of genetic variation present, meaning that for a fairly large proportion of genes that are analyzed, there are multiple alleles, and therefore multiple genotypes, within populations. For example, in the common fruit fly *Drosophila melanogaster* (an organism that has been well-studied genetically since the very early 1900's), between one-third and two-thirds of the genes that have been examined by protein electrophoresis have been found to be variable. Genetic variation can be measured as allele frequencies ("allelic variation") or genotype frequencies ("genotypic variation"). A major task of population geneticists has been to describe such variation, to try to explain why it exists, and to predict if and how it will change over time.

The Hardy-Weinberg law predicts that if genetic variation exists in a population, it will remain constant over time, with genotypes in specific proportions. However, the law cannot begin to explain natural variation, since genotypes are not always found in Hardy-Weinberg proportions, and studies that involve sampling populations over time often show that genetic variation can be changing. The historical approach to explaining these observations has been to formulate more complex mathematical models based on the simple Hardy-Weinberg model that violate one or more of the implicit Hardy-Weinberg conditions.

Beginning in the 1920's and 1930's, a group of population genetic theorists, working independently, began exploring the effects of various natural forces on genetic variation in populations. At this time, Charles Darwin's ideas about natural selection as a mechanism for organic evolution had become widely accepted by the scientific community but had not yet become the pervasive influence on all branches of biology that would follow. In what has become known as the "modern synthesis," Ronald A. Fisher, J. B. S. Haldane, and Sewall Wright merged Darwin's theory of natural selection with Mendel's theory of genetic inheri-

J. B. S. Haldane, whose work was central to the development of the field of popular genetics, poses with his wife. (California Institute of Technology)

tance to create a field of population genetics that allows for genetic change. Later work by these researchers and others explored other mechanisms of change, but for many years natural selection was considered to be the primary force of genetic change in populations.

Natural selection in a simple model of a gene with two alleles in a population can be easily represented by assuming that genotypes differ in their ability to survive and produce offspring. This ability is called "fitness." In applying natural selection to a theoretical population, each genotype is assigned a fitness value between zero and one. Typically, the genotype in a population that is best able to survive and can, on average, produce more offspring than other genotypes is assigned a fitness value of one, and genotypes with lower fitness are assigned fitnesses with fractional values relative to the high-fitness genotype.

The study of this simple model of natural selection has revealed that it can alter genetic variation in different ways, depending on which genotype has the highest fitness. In the simple one-gene, two-allele model, there are three possible genotypes: two homozygotes and one heterozygote. If one homozygote has the highest fitness, it will be favored, and the genetic composition of the population will gradually shift toward more of that genotype (and its corresponding allele). This is called "directional selection." If both homozygotes have higher fitness than the heterozygote ("disruptive selection"), one or the other will be favored, depending on the starting conditions. Both of these situations will decrease genetic variation in the population, because eventually one allele will prevail. Although each of these types of selection (particularly directional) may be found for genes in natural populations, they cannot explain why genetic variation is present, and perhaps increasing, in nature.

Heterozygote advantage, in which the heterozygote has higher fitness than either homozygote, is the other possible situation in this model. In this case, because the heterozygote carries both alleles, both are expected to be favored together and therefore maintained. This is the only condition in this simple model in which genetic variation may be maintained

or increased over time. Although this seems like a plausible explanation for the observed levels of natural variation, studies in which fitnesses are measured almost never show heterozygote advantage in genes from natural populations. As a general explanation for the presence of genetic variation, this simple model of selection is unsatisfactory.

Studies of more complex theoretical models of selection (for example, those with many genes and different forms of selection) have revealed conditions that allow patterns of variation very similar to those observed in natural populations, and in some cases it seems clear that natural selection is a major factor determining patterns of genetic change. However, in many cases, selection does not seem to be the most important factor or even a factor at all. In these cases, it is necessary to consider other conditions of the Hardy-Weinberg model that, when violated, result in genetic change and patterns of genetic variation similar to those in nature.

Assortative Mating and Inbreeding

One of the implicit conditions of the Hardy-Weinberg model is that genotypes form mating pairs at random. This is not a mathematical simplification because for most genes mates are not selected based on the individual genotype. Unless the gene in question has some direct effect on mate choice, mating with respect to that gene is random. However, there are conditions in natural populations in which mating is not random. For some genes, individuals may actually choose mates based on the genotype. For example, if a gene controls fur color and mates are chosen by appropriate fur color, then the genotype of an individual with respect to that gene will determine mating success. For this gene, then, mating is not random but rather "assortative." Positive assortative mating means that individuals tend to choose similar genotypes as mates, while negative assortative mating means that individuals tend to choose dissimilar genotypes.

Variation in a population for a gene subject to assortative mating is altered from Hardy-Weinberg expectations. Although allele frequencies do not change, genotype frequencies are altered. With positive assortative mating, the result is higher proportions of homozygotes and fewer heterozygotes, while the opposite is true when assortative mating is negative. This type of nonrandom mating affects only variation for single genes, so if these patterns are present in a natural population for only certain genes, it may suggest assortative mating based on specific genotypes.

Sometimes random mating in a population is not possible because of the geographic organization of the population or general mating habits. Truly random mating would mean that any individual can mate with any other, but this is nearly impossible because of gender differences and practical limitations. In natural populations, it is often the case that mates are somewhat related, even closely related, because the population is organized into extended family groups whose members do not (or cannot, as in plants) disperse to mate with members of other groups. Mating between relatives is called "inbreeding." Because related individuals tend to have similar genotypes for many genes, the effects of inbreeding are much like those of positive assortative mating for many genes. The proportions of homozygotes for many genes tend to increase. Again, this situation has no effect on allelic variation, only genotypic variation. Clearly, the presence of nonrandom mating patterns cannot by themselves explain the majority of patterns of genetic variation in natural populations but can contribute to the action of other forces, such as natural selection.

Migration and Mutation

In the theoretical Hardy-Weinberg population, there are no sources of new genetic variation. In real populations, individual genotypes may enter or leave the population, a process called migration or sometimes "gene flow" (a more accurate term, since migration in this context means not only movement between populations but also successful reproduction to introduce genes in the new population). Also, new alleles may be introduced by mutation, the change in the deoxyribonucleic acid (DNA) sequence of an existing allele to create a new one, as a result of errors during

DNA replication or the inexact repair of DNA damage from environmental influences such as radiation or mutagenic chemicals. Both of these processes can change both genotype frequencies and allele frequencies in a population. If the tendency to migrate is associated with particular genotypes, a long period of continued migration tends to push genotype and allele frequencies toward higher proportions of one type (in general, more homozygotes) so that the overall effect is to reduce genetic variation. However, in the short term, migration may enhance genetic variation by allowing new alleles and genotypes to enter. The importance of migration depends on the particular population. Some populations may be relatively isolated from others so that migration is a relatively weak force affecting genetic variation, or there may be frequent migration among geographic populations. There are many factors involved, not the least of which is the ability of members of the particular species to move over some distance.

Mutation, because it introduces new alleles into a population, acts to increase genetic variation. Before the modern synthesis, one school of thought was that mutation might be the driving force of evolution, since one can imagine genetic change over time coming about from continual introduction of new forms of genes. In fact, it is possible to develop simple mathematical models of mutation that show resulting patterns of genetic variation that resemble those found in nature. However, to account for the rates of evolution that are commonly observed, very high rates of mutation are required. In general, mutation tends to be quite rare, making the hypothesis of evolution by mutation alone unsatisfactory.

The action of mutation in conjunction with other forces, such as selection, may account for the low-frequency persistence of clearly harmful alleles in populations. For example, one might expect that alleles that can result in genetic diseases (such as cystic fibrosis) would be quickly eliminated from human populations by natural selection. However, low rates of mutation can continually introduce these alleles into populations. In this "mutation-selection balance," mutation tends to introduce alleles while selection tends to eliminate them, with a net result of low frequencies in the population.

Genetic Drift

Real populations are not, of course, infinite in size, though some are large enough that this Hardy-Weinberg condition is a useful approximation. However, many natural populations are small, and any population with less than about one thousand individuals will vary randomly in the pattern of genetic variation from generation to generation. These random changes in allele and genotype frequencies are called "genetic drift." The situation is analogous to coin tossing. With a fair coin, the expectation is that half of the tosses will result in "heads" and half in "tails." On average, this will be true, but in practice a small sample will not show the expectation. For example, if a coin is tossed ten times, it is unlikely that the result will be exactly five heads and five tails. On the other hand, with a thousand tosses, the results will be closer to half and half. This higher deviation from the expected result in small samples is called "sampling error." In a small population, there is an expectation of the pattern of genetic variation based on the Hardy-Weinberg law, but sampling error during the union of sex cells to form offspring genotypes will result in random deviations from that expectation. The effect is that allele frequencies increase or decrease randomly, with corresponding changes in genotype frequencies. The smaller the population, the higher the sampling error and the more pronounced genetic drift will be.

Genetic drift has an effect on genetic variation that is similar to that of other factors. Over the long term, allele frequencies will drift until all alleles have been eliminated but one, eliminating variation. (For the moment, ignore the action of other forces that increase variation.) Over a period of dozens of generations, however, drift can allow variation to be maintained, especially in larger populations in which drift is minimal.

In the early days of population genetics, the possibility of genetic drift was recognized but often considered to be a minor consideration with natural selection as a dominant force. Fisher in particular dismissed its importance, engaging over a number of years in a published debate with Wright, who always felt that drift would be important in small populations. Beginning in the 1960's with the acquisition of data on DNA-level population variation, the role of drift in natural populations became more recognized.

Experimental Population Genetics and the Neutral Theory

Population genetics has always been a field in which the understanding of theory is ahead of empirical observation and experimental testing, but these have not been neglected. Although Fisher, Haldane, and Wright were mainly theorists, there were other architects of the modern synthesis who concentrated on testing theoretical predictions in natural populations. Beginning in the 1940's, for example, Theodosius Dobzhansky showed in natural and experimental populations of *Drosophila* species that frequency changes and geographic patterns of variation in chromosome variants are consistent with the effects of natural selection.

Natural selection was the dominant hypothesis for genetic changes in natural populations for the first several decades of the modern synthesis. In the 1960's, new techniques of molecular biology allowed population geneticists to examine molecular variation, first in proteins and later, with the use of restriction enzymes in the 1970's and DNA sequencing in the 1980's and 1990's, in DNA sequences. These types of studies only confirmed that there is a large amount of genetic variation in natural populations, much more than can be attributed only to natural selection. As a result, Motoo Kimura proposed the "neutral theory of evolution," the idea that most DNA sequence differences do not have fitness differences and that population changes in DNA sequences are governed mainly by genetic drift, with selection playing a minor role. This view was mostly accepted by the 1990's, although it was recognized that evolution of proteins and physical traits may be governed by selection to a greater extent.

Impact and Applications

The field of population genetics is a fundamental part of the modern field of evolutionary biology. One possible definition of evolution would be "genetic change in a population over time," and population geneticists try to describe patterns of genetic variation, document changes in variation, determine their theoretical causes, and predict future patterns. These types of research have been very valuable in studying the evolutionary histories of organisms for which there are living representatives, including humans.

In addition to the scientific value of understanding evolutionary history better, there are more immediate applications of such work. In conservation biology, data about genetic variation in a population can help to assess its ability to survive in the future. Data on genetic similarities between populations can aid in decisions about whether they can be considered as the same species or are unique enough to merit preservation.

Population genetics has had an influence on medicine, particularly in understanding why "disease genes," while clearly harmful, persist in human populations. The field has also affected the planning of vaccination protocols to maximize their effectiveness against parasites, since a vaccine-resistant strain is a result of a rare allele in the parasite population. In the 1990's it began to be recognized that effective treatments for medical conditions would need to take into account genetic variation in human populations, since different individuals might respond differently to the same treatment.

—Stephen T. Kilpatrick

See Also: Artificial Selection; Evolutionary Biology; Hardy-Weinberg Law; Inbreeding and Assortative Mating; Natural Selection; Punctuated Equilibrium; Speciation.

Further Reading: William Provine's *The Origin of Theoretical Population Genetics* (1971) is an excellent account of the early history of the field. Theodosius Dobzhansky's *Genetics and the Origin of Species* (1951) is a classic treatment of population genetics and evolution. Richard Lewontin's *The Genetic Basis of Evolutionary Change* (1974) discusses genetic variation in populations. The September, 1978, *Scientific American* (titled "Evolution") has several articles related to population genetics.

Prenatal Diagnosis

Field of study: Human genetics
Significance: *Tests performed on a pregnant woman's blood and the cells of her fetus may produce a wealth of information about genetic disorders that the fetus may have. Results can comfort or distress prospective parents and can produce new ethical dilemmas.*

Key terms

AMNIOTIC FLUID: the liquid that surrounds the developing fetus
NEURAL TUBE: the embryonic structure that becomes the brain and spinal cord
PRENATAL: the period prior to birth
PLACENTA: an organ composed of both fetal and maternal tissue through which the fetus is nourished
TRISOMY: the presence of three copies (instead of two) of a particular chromosome in a cell

Prenatal Testing

Prenatal testing is administered to a large number of women, and the tests are becoming more informative. Some of the tests are only mildly invasive to the mother, but others involve obtaining fetal cells. Some are becoming routine for all pregnant women; others are offered only when an expectant mother meets a certain set of criteria. Some physicians will not offer the testing (especially the more invasive procedures) unless the parents have agreed that they will abort the fetus if the testing reveals a major developmental problem, such as Down syndrome or Tay-Sachs disease. Others will order testing without any such guarantees, believing that test results will give the parents time to prepare themselves mentally for a special-needs baby or to locate and join a support group. The test results are also used to determine if additional medical teams should be present at the delivery to deal with a newborn who is not normal and healthy. Most often, prenatal testing is offered if the mother will be age thirty-five or older at the time of delivery, if a particular disorder is present in relatives on one or both sides of the family, or if the parents have already produced one child with an inherited disorder.

Maternal Blood Tests and Ultrasound

Screening maternal blood for the presence of alpha fetoprotein (AFP) is offered to pregnant women who are about eighteen weeks into a pregnancy. Although AFP is produced by the fetal liver, some will cross the placenta into the mother's blood. Elevated levels of AFP can indicate an open neural tube defect (such as spina bifida), although it can also indicate twins. Unusual AFP findings are usually followed up by ultrasound examination of the fetus.

Other tests of maternal blood measure the amounts of two substances that are produced by the fetal part of the placenta: hCG and UE3. When the levels of these plus AFP are compared with the average levels for fetuses at a particular age, an increased probability of a baby with an abnormal chromosome number may be found. Low levels of AFP and UE3, combined with a higher-than-average amount of hCG, increases the risk that the woman is carrying a Down syndrome (trisomy 21) fetus. For example, a nineteen-year-old woman has a baseline risk of Down syndrome of 1 in 1193. When blood-test results show low AFP and UE3, and high hCG, the risk of Down syndrome rises to 1 in 145.

During an ultrasound examination, harmless sound waves are bounced off the fetus from an emitter placed on the surface of the mother's abdomen or in her vagina. They are used to make a picture of the fetus on a television monitor. Measurements on the monitor determine the overall size, the head size, and the sex of the fetus, and whether all the arms and legs are formed and of the proper length. Successive ultrasound tests will indicate if the fetus is growing normally. Certain ultrasound findings, such as shortened long bones, may indicate an increased probability for a Down syndrome baby. Because Down syndrome is a highly variable condition, normal ultrasound findings do not guarantee that the child will be born without Down syndrome. Only a chromosome analysis can determine this for certain.

Amniocentesis, Karyotyping, and FISH

Amniocentesis is the process of collecting fetal cells from the amniotic fluid. Fetal cells collected by amniocentesis can be grown in culture; then the fluid around the cells is collected and analyzed for enzymes produced by the cells. If an enzyme is missing (as in the case of Tay-Sachs disease), the fetus may be diagnosed with the disorder before it is born. Because disorders such as Tay-Sachs disease are untreatable and fatal, a woman who has borne one Tay-Sachs child may not wish to give birth to another. Early diagnosis of a second Tay-Sachs fetus would permit her to have a therapeutic abortion.

Chromosomes in the cells obtained by amniocentesis may be stained to produce a karyotype. In a normal karyotype, the chromosomes will be present in pairs. If the fetus has Down syndrome (trisomy 21), there will be three copies of chromosome 21. Karyotypes that show three representatives of chromosome 18 (trisomy 18) or chromosome 13 (trisomy 13) will provide a firm diagnosis of these disorders. Other types of chromosome abnormalities that also appear in karyotypes are changes within a single chromosome. If a chromosome has lost a piece, it is said to contain a deletion. Large deletions will be obvious when a karyotype is read because the chromosome will appear smaller than it should. Sometimes the deletion is so small that it is not visible on a karyotype.

If chromosome analysis is needed early in pregnancy before the volume of amniotic fluid is large enough to permit amniocentesis, the mother and doctor may opt for chorionic villus sampling (CVS). The embryo produces finger-like projections (villi) into the uterine lining. Because these projections are produced by the embryo, their cells will have the same chromosome number as the rest of the embryonic cells. After growing in culture, the cells may be karyotyped in the same way as those obtained by amniocentesis. Both amniocentesis and CVS carry risks of infection and miscarriage. Normally these procedures are not offered unless the risk of having an affected child is found to be greater than the risk of complications from the procedures.

If the doctor is convinced that the fetus has a tiny chromosomal defect that is not visible on a karyotype, it will then be necessary to probe (or "FISH") the fetal chromosomes (the initials

"FISH" stand for "fluorescent in situ hybridization." A chromosome probe is a piece of deoxyribonucleic acid (DNA) that is complementary to DNA within a gene. Complementary pieces of DNA will stick together (hybridize) when they come in contact. The probe also has an attached molecule that will glow when viewed under fluorescent light. A probe for a particular gene will stick to the part of the chromosome where the gene is located and make a glowing spot. If the gene is not present because it has been lost, no spot will appear. Probes have been developed for many individual genes that cause developmental abnormalities when they are deleted from the chromosomes.

Cells obtained by amniocentesis can be probed in less time than it takes to grow and prepare them for karyotyping. Probes have been developed for the centromeres of the chromosomes that are frequently present in extra copies, such as 13, 18, 21, X, and Y. Y chromosomes that have been probed appear as red spots, X chromosomes as green spots, and number 18 chromosomes as aqua spots. A second set of probes attached to other cells from the same fetus will cause number 13 chromosomes to appear as green spots and number 21 chromosomes to appear as red spots. Cells from a girl with trisomy 21 would have two green spots and two aqua spots, but no red spot when the first set of probes is used. Some other cells from the same girl will show two red spots, but three green ones, when the second set of probes is attached.

Impact and Applications

Until the development of prenatal techniques, pregnant women had to wait until delivery day to find out the sex of their child and whether or not the baby was normal. Now much more information is available to both the woman and her doctor weeks before the baby is due. Even though tests are not available for all possible birth defects, normal blood tests, karyotypes, or FISH can be very comforting. On the other hand, abnormal test results give the parents definite information about birth defects, as opposed to the possibilities inherent in a statement of risk. The parents must decide whether to continue the pregnancy. If they do,

they must then cope with the fact that they are not going to have a normal child. In the best circumstances, the test results are explained by a genetic counselor who is equipped to help the parents deal with the strong emotions that bad news can produce. Genetic testing also has far-reaching implications. If insurance companies pay for the prenatal testing, they receive copies of the results. Information about genetic abnormalities could cause the insurance companies to deny claims arising from treatment of the newborn or to deny insurance to the individual later in life.

—*Nancy N. Shontz*

See Also: Amniocentesis; Congenital Defects; Down Syndrome; Genetic Testing; Tay-Sachs Disease.

Further Reading: Linda Heller, "Genetic Testing," *Parents* 70 (November, 1995), explores the risks of some tests. The difference between prenatal diagnosis and certainty about the outcome is discussed in Perri Klass, "The Hunt for the Perfect Baby," *Redbook* 179 (May, 1992). Edwin McConkey, *Human Genetics* (1993), contains additional information on FISH.

Prion Diseases: Kuru and Creutzfeldt-Jakob Syndrome

Field of study: Human genetics
Significance: *Kuru and Creutzfeldt-Jakob syndrome are rare, fatal diseases of the brain and spinal cord. Nerve cell death is caused by the accumulation of a protein called a "prion" that appears to be a new infectious agent that interferes with the genetic programming of nerve cells. Understanding the process of these diseases has far-reaching implications for the study of other degenerative mental disorders and for biology itself.*

Key terms

DEMENTIA: mental deterioration ranging from forgetfulness and disorientation to complete unresponsiveness

PROTEIN: a chain of amino acids (compounds composed of carbon, hydrogen, oxygen, and nitrogen)

VIRUS: a small particle consisting of a protein coat surrounding a genetic core that is capable of reproducing only when inside a cell

Causes, Symptoms, and Treatment

Kuru and Creutzfeldt-Jakob syndrome, degenerative diseases of the human central nervous system, are among a group of diseases that also affect cattle (mad cow disease) and sheep (scrapie). They have been classified in several ways, including "slow-virus" infections (because of the extremely long incubation period between contact and illness) and "spongiform encephalopathies" (because of the large holes seen in the brain after death). However, a virus that may cause such a disease has never been found, and the body does not respond to the disease as an infection. The only clue to the cause is the accumulation of a transmissible, toxic protein known as a prion; therefore, these disorders are now known simply as "prion diseases."

Creutzfeldt-Jakob syndrome is rare: Approximately 250 people die from it yearly in the United States. It usually begins in middle age with symptoms that include rapidly progressing dementia, jerking spastic movements, and visual problems. Within one year after the symptoms begin, the patient is comatose and paralyzed, and powerful seizures affect the entire body. Death occurs shortly thereafter. The initial symptom of the illness (rapid mental deterioration) is similar to other disorders; therefore, diagnosis is difficult. No typical infectious agent (bacteria or viruses) can be found in the blood or in the fluid that surrounds the brain and spinal cord. X rays and other scans are normal. There is no inflammation, fever, or antibody production. Brain wave studies are, however, abnormal, and at autopsy, the brain is found to have large holes and massive protein deposits in it.

Kuru is found among the Fore tribe of Papua New Guinea. Until the early 1960's, more than one thousand Fore died of Kuru each year. Anthropologists recording their customs described their practice of eating the brains of

The practice of feeding cattle with parts of other dead livestock may have contributed to the spread of prion diseases. (Ben Klaffke)

their dead relatives in order to gain the knowledge they contained. Clearly, some infectious agent was being transmitted during this ritual. Such cannibalism has since stopped, and Kuru has declined markedly. Kuru, like Creutzfeldt-Jakob syndrome, shows the same spongiform changes and protein deposits in the brain after death. Similarly, early symptoms include intellectual deterioration, spastic movements, and eye problems. Within a year, the patient becomes unresponsive and dies.

The outbreak of "mad cow" disease in the mid-1990's in Great Britain led to widespread fear. Thousands of cattle were killed to prevent human consumption of contaminated beef. The cows were infected by supplemental feedings that contained infected sheep meat. Animal-to-human transmission of these diseases is well documented, and research has shown that human-to-animal infection is possible as well.

Both Kuru and Creutzfeldt-Jakob syndrome, as well as the animal forms, have no known treatment or cure. Because of the long incubation period, decades may pass before symptoms appear, but once they do, the central nervous system is rapidly destroyed, and death comes quickly. It is likely that many more people die of these disorders than is known because they are so rarely diagnosed.

Properties of Prions

Most of the research on prion diseases has focused on scrapie in sheep. It became clear that the infectious particle had novel properties: It was not a virus as had been suspected, nor did the body react to it as an invader. It was discovered that this transmissible agent was an abnormal version of a common protein. This defied medical understanding. This protein, normally secreted by nerve cells and found on their outer membranes, is genetically coded for on chromosome 20 in humans. The transmissible, infectious fragment of the prion somehow disrupts the nerve cell, causing it to produce the abnormal fragment instead of the normal protein. This product accumulates to toxic levels in the tissue and fluid of the brain and spinal cord over many years, finally destroying the central nervous system.

Prion infection appears to occur from exposure to infected tissues or fluids. Transmission has occurred accidently through nerve tissue transplants and neurosurgical instruments. Prions are not affected by usual sterilization techniques; prevention requires careful handling of infected materials and exposure to steam (for at least one hour) or chlorine bleach for instrument decontamination. The agent is not spread by casual contact or air, and isolating the patient is not necessary.

Other human degenerative nervous system diseases whose causes remain unclear also show accumulations of proteins to toxic levels. Alzheimer's disease is the best-studied example, and it is possible that a process similar to that in prion diseases is at work. The discovery of prions has far-reaching implications for genetic and cellular research. Scientists have already learned a startling fact: Substances as inert as proteins and far smaller than viruses can act as agents of infections.

—Connie Rizzo

See Also: Alzheimer's Disease; Genetic Code; Protein Structure; Protein Synthesis.

Further Reading: *Cecil's Textbook of Medicine* (1996), edited by C. Bennett, is a classic medical reference text that covers all aspects of prion diseases. B. Fields and D. Knipe, *Virology* (1990), provides an excellent discussion of slow viruses, prions, and proteins. J. Ingraham and C. Ingraham, *Introduction to Microbiology* (1995), is a beautifully illustrated, clearly written text that covers the mechanisms of infectious diseases. R. Johnson, *Viral Infections of the Nervous System* (1992), contains a detailed examination of prion diseases as well as other related disorders.

Protein Structure

Field of study: Molecular genetics
Significance: *Proteins have three-dimensional structures that are major determinants of their functions, and slight changes in overall structure may significantly alter a protein's actions. The knowledge of how proteins obtain these three-dimensional structures has led to a better under-*

standing of the ways in which proteins carry out cellular processes and how these processes are controlled in the cell. Because most diseases result from improper protein function, understanding protein structure is essential for understanding proper function and for developing molecular-based disease treatments.

Key terms

AMINO ACID: the basic subunit of a protein; there are twenty commonly occurring amino acids, any of which may join together by chemical bonds to form a complex protein molecule

PROTEIN: a long chain of amino acids joined together by chemical bonds

GENE: a segment of a chromosome that codes for one protein

HYDROGEN BOND: a weak bond that helps stabilize the folding of a protein

R GROUP: a functional group that is part of an amino acid that gives each amino acid its unique characteristics

Protein Structure and Function

Proteins essentially consist of strings of individual subunits called amino acids that are chemically bonded together. Once the amino acids are bonded, the protein will begin to fold into a specific shape or conformation that is required for proper protein function. Proteins have been called the "workhorses" of the cell because they are the molecules that perform all of the activities encoded in the genes of the cell. All proteins function by binding to other molecules, frequently to other proteins. The binding of a protein to another molecule is dependent upon the protein obtaining its proper shape; if the protein does not obtain the correct shape, it will not bind to the other molecule and hence will not function properly. A protein having the correct conformation to perform its function can be likened to a craftsman having the correct tool to perform a task. Proteins fold into three-dimensional conformations that maximize the efficiency of their functions.

In 1973, Christian B. Anfinsen performed experiments that showed that the three-dimensional structure of a protein is determined by the sequence of its amino acids. He used a protein called ribonuclease, an enzyme that degrades ribonucleic acid (RNA) in the cell. The ability of ribonuclease to degrade RNA is dependent upon its ability to fold into its proper three-dimensional shape. Anfinsen showed that if the enzyme was completely unfolded by heat and chemical treatment (at which time it would not function), it formed a linear chain of amino acids. Although there were 105 possible conformations that the enzyme could take upon refolding, it would refold into the single correct functional conformation upon removal of heat and chemicals. This established that the amino acid sequences of proteins, which are specified by the genes of the cell, carry all of the information necessary for proteins to fold into their proper three-dimensional shapes.

While the final conformations of proteins are dictated by their amino acid sequences, proteins are considered to have multiple levels of structure that result in these three-dimensional conformations. These levels are referred to as primary, secondary, tertiary, and quaternary structures. The primary structure of a protein is, by definition, its amino acid sequence, and because this sequence determines higher-order structures of the protein (that is, the folding of the protein), it is necessary to have a basic understanding of amino acids.

Primary Protein Structure

Proteins are made of subunits of amino acids that are joined together by chemical bonds. There are twenty naturally occurring amino acids that are commonly found in proteins, and each of these has a common structure consisting of a nitrogen-containing amino group ($-NH_2$), a carboxyl group (-COOH), a hydrogen atom (H), and a unique functional group referred to as an R group, all bonded to a central carbon atom (known as the alpha carbon, or C_α) as shown in the following figure:

$$H_2N - \overset{\overset{\text{H}}{|}}{C_\alpha} - \overset{\overset{\text{O}}{\|}}{C} - OH$$
$$\underset{R}{|}$$

The uniqueness of each of the twenty amino acids is determined by the R group. This group may be as simple as a hydrogen atom (in the case of the amino acid glycine) or as complex as a ring-shaped structure (as found in the amino acid phenylalanine). It may be charged, either positively or negatively, or it may be uncharged.

Because they have similar structures, amino acids can join together to form peptides (strings of up to ten amino acids), polypeptides (strings of ten to one hundred amino acids), or proteins (strings of one hundred or more amino acids). The amino acids join together by forming a chemical bond called a peptide bond (in the box in the following figure) between the carbon atom of the carboxyl group of one amino acid (-COOH) and the nitrogen atom of the amino group (-NH$_2$) of the next adjacent amino acid:

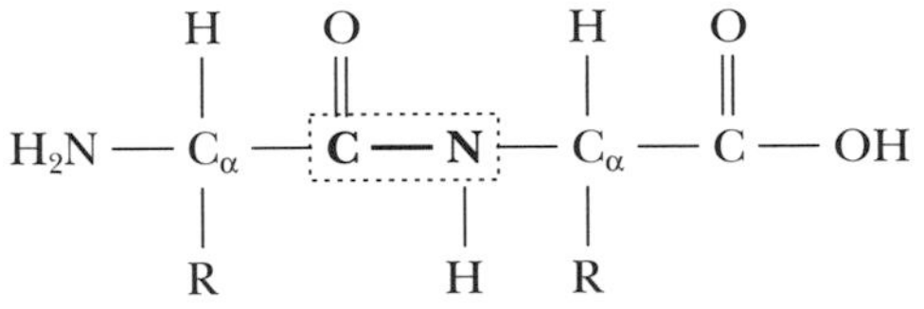

$$H_2N - C_\alpha - \boxed{C - N} - C_\alpha - C - OH$$

During the formation of the peptide bond, a molecule of water (H$_2$O) is lost (an -OH from the carboxyl group and an -H from the amino group), so this reaction is also called a dehydration synthesis. The result is a dipeptide (a peptide made of two amino acids joined by a peptide bond) that has a "backbone" of nitrogens and carbons (N-C$_\alpha$-C-N-C$_\alpha$-C) with other elements and R groups protruding from the backbone. An amino acid may be joined to the growing peptide chain by formation of a peptide bond between the carbon atom of the free carboxyl group (on the right of the preceding figure) and the nitrogen atom of the amino acid being added. Any one of the twenty commonly occurring amino acids may join to the growing peptide chain.

The atoms and R groups that protrude from the backbone are capable of interacting with each other, and these interactions lead to higher-order secondary, tertiary, and quaternary structures. Because the amino acid sequence, or primary structure, of the protein dictates which R groups are present and because R groups are unique and react in different ways within the protein to yield its overall conformation, it is the primary structure of the protein that ultimately determines the shape of the protein.

Secondary Structure

The next level of structure in the folding of a protein is secondary structure, which involves the formation of bonds between atoms attached to the backbone of the polypeptide chain; it does not involve the formation of bonds with R groups or atoms that are parts of R groups. Essentially, bonds called hydrogen bonds can form between the oxygen atoms protruding from the carboxyl groups of the protein and the hydrogen atoms protruding from the amino groups of the protein:

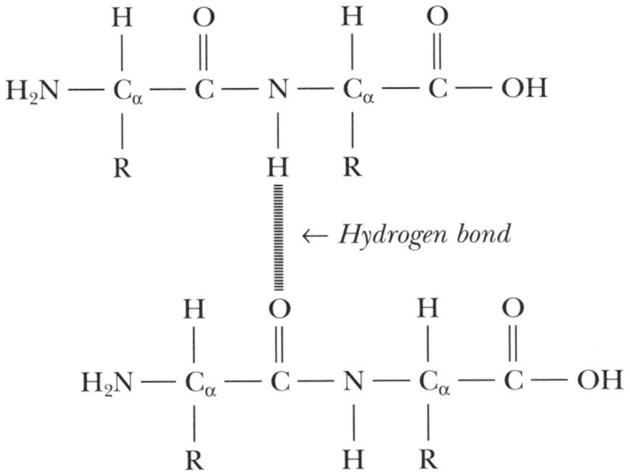

$$H_2N - C_\alpha - C - N - C_\alpha - C - OH$$

← *Hydrogen bond*

$$H_2N - C_\alpha - C - N - C_\alpha - C - OH$$

Hydrogen bonds are weak bonds that form between atoms that have a very strong attraction for electrons (such as oxygen or nitrogen) and a hydrogen atom that is bound to another atom with a very strong attraction for electrons. These hydrogen bonds between backbone molecules lead to the formation of two major types of structures: alpha helices and beta-pleated sheets. An alpha helix is a rigid structure shaped very much like a telephone cord; it spirals around as the oxygen of one amino acid of the chain forms a hydrogen bond with the hydrogen atom of an amino acid five amino acids away on the protein strand. The rigidity of the structure is caused by the large number of hydrogen bonds (individually weak but collectively strong) and the compactness of the helix that forms; many alpha helices are found in proteins that function to maintain cell structure.

Beta sheets are also formed by hydrogen bonding between backbone molecules, but rather than bonding to an amino acid five spaces away, these bonds are between amino acids in different regions (often very far apart on the linear strand) of the protein. The result is that the region takes on a shape that may be likened to the bellows of an accordion or a sheet of paper that has been folded multiple times to form pleats. Because of the large number of hydrogen bonds in them, beta sheets are also strong structures, and they form planar regions that are often found at the bottom of "pockets" inside proteins to which other molecules attach.

In addition to alpha helices or beta-pleated sheets, other regions of the protein may have no obvious secondary structure; these regions are said to have a "random coil" shape. It is the combinations of random coils, alpha helices, and beta sheets that form the secondary structure of the protein.

Tertiary Structure

The final level of protein shape (for a single protein) is called tertiary structure. Tertiary structure of a protein is caused by the numerous interactions of R groups on the amino acids of the protein itself and of the protein with its environment, which is usually aqueous (water based). Various R groups may either be attracted to and form bonds with each other, or they may be repelled from each other. For example, if an R group has an overall positive electrical charge, it will be attracted to R groups with a negative charge but repelled from other positively charged R groups. For a protein that has one hundred amino acids, if amino acid number 6 from the beginning of the chain is negatively charged, it could be attracted to a positively charged amino acid at position 74, thus bringing two ends of the protein that are linearly distant in close proximity to each other, thus folding the protein. Many other R groups in the protein will also be attracted to or repelled from each other, leading to an overall

folded shape that fits together quite well. In addition, because most proteins exist in an aqueous environment in the cell, most proteins are folded such that their amino acids with hydrophilic (attracted to water) R groups are on the outside, while their amino acids with hydrophobic (repelled from water) R groups are tucked away in the interior of the protein. This three-dimensional shape that occurs because of the interactions of the R groups of amino acids is tertiary structure and is the final structural level for proteins that function alone (not attached to another protein that is required for function).

Quaternary Structure

Many proteins are actually nonfunctional until they physically associate with another protein, forming a functional unit made up of two or more proteins. Proteins that associate in this manner are said to have quaternary structure, a fourth level of structure above that of tertiary. Quaternary structure is caused by interactions between the R groups of amino acids of two different proteins rather than within the same protein. An example of a protein with quaternary structure is hemoglobin, the protein found in red blood cells that carries oxygen. Hemoglobin actually functions as a tetramer with four protein subunits bound together.

The three-dimensional shape of a protein is dependent upon the formation of peptide bonds between amino acids (primary structure), hydrogen bonds between the backbone molecules of the protein (secondary structure), and various types of bonds between R groups (tertiary and quaternary structure). The interactions that lead to secondary, tertiary, and quaternary structure may be between amino acids that lie very close to each other on the linear protein chain, leading to slight folding, or they may be between amino acids distant from each other on the linear chain, leading to significant protein folding. Because these interactions are caused by the R groups of the specific amino acids, the folding is ultimately dictated by the amino acid sequence of the protein. Although there may be numerous possible final conformations that a protein could take, a single protein usually assumes only one of these, and this is the conformation that leads to proper protein function.

Impact and Applications

The function of a protein may be altered by changing the shape of that protein because a protein's function is dependent upon its three-dimensional shape. The implications of this in terms of normal cellular processes, disease processes, and disease treatments are numerous. Normal cellular processes are controlled by "turning on" and "turning off" proteins at the appropriate time. A protein's activity may be altered by attaching a molecule or ion to that protein that results in a change of shape. Because the shape is caused by R group interactions, binding of a charged ion such as calcium to the protein will alter these interactions and thus alter the shape and function of the protein. One molecular "on/off" switch that is used frequently within a cell involves the attachment or removal of a phosphate group (a very highly negatively charged molecule made of phosphorous and oxygens) to or from a protein. Attachment of a phosphate will significantly alter the shape of the protein by repelling negatively charged amino acids and attracting positively charged amino acids; this will either activate the protein to perform its function ("turn on") or deactivate it ("turn off").

Cancer and diseases caused by bacterial or viral infections are often the result of nonfunctional proteins that have been produced with incorrect shapes or that cannot be turned on or off by a molecular switch. The effects may be minor or major, depending upon the protein, its function, and the severity of the structural deformity. Understanding how a normal protein is shaped and how it is altered in the disease process allows for the development of drugs that may block the disease process. This may be accomplished by blocking or changing the effect of the protein of interest or by generating drugs or therapies that mimic the normal functioning of the protein. Thus, understanding protein structure is essential for understanding proper protein function and for developing molecular-based disease treatments.

—*Sarah Lea McGuire*

See Also: Genetic Code; Genetic Code, Cracking of; Molecular Genetics; Mutation and Mutagenesis; Protein Synthesis.

Further Reading: *How Scientists Think* (1996), by George B. Johnson, gives an excellent introductory-level account of Christian Anfinsen's experiments leading to the determination that primary sequence dictates protein shape. James Darnell et al., *Molecular Cell Biology* (1995), provides both summary and detailed accounts of protein structure and the chemical bonds that lead to the various levels of structure. An excellent introductory-level account of protein structure can be found in George B. Johnson and Peter H. Raven, *Biology* (1996).

Protein Synthesis

Field of study: Molecular genetics

Significance: *Cellular proteins can be grouped into two general categories: proteins with a structural function that contribute to the three-dimensional organization of a cell, and proteins with an enzymatic function that catalyze the biochemical reactions required for cell growth and function. Understanding the process by which protein is synthesized provides insight into how a cell organizes itself and how defects in this process can lead to disease.*

Key terms

RIBONUCLEIC ACID (RNA): a single-stranded molecule composed of nucleotides

SUBUNIT: a separable part of a larger complex

PEPTIDE BOND: the chemical bond between amino acids in protein

POLYPEPTIDE: a linear molecule composed of amino acids joined together by peptide bonds; all proteins are polypeptides

The Flow of Formation from Stored to Active Form

The cell can be viewed as a unit that assembles resources from the environment into biochemically functional molecules and organizes these molecules in three-dimensional space in a way that allows cellular growth and replication. In order to carry out this organizational process, a cell must have a biosynthetic means to assemble resources into useful molecules, and it must contain the information required to produce the biosynthetic and structural machinery. Deoxyribonucleic acid (DNA) serves as the stored form of this information, whereas protein is its active form. Although there are thousands of different proteins in cells, they either serve a structural role or are enzymes that catalyze the biosynthetic reactions of a cell. Following the discovery of the structure of DNA in 1953 by James Watson and Francis Crick, scientists began to study the process by which the information stored in this molecule is converted into protein.

Proteins are linear molecules composed of a unique sequence of amino acids. Twenty different amino acids are used as the protein building blocks. Although the information for the amino acid sequence of each protein is present in DNA, protein is not synthesized directly from this source. Instead, ribonucleic acid (RNA) serves as the intermediate form from which protein is synthesized. There are three roles that RNA plays during protein synthesis. Messenger RNA (mRNA) contains the information for the amino acid sequence of a pro-

tein. Transfer RNAs (tRNAs) are small RNA molecules to which individual amino acids are attached that serve as adapters that decipher the coded information present within an mRNA and bring the appropriate amino acid to the polypeptide as it is being synthesized. Ribosomal RNAs (rRNAs) act as the engine that carries out most of the steps during protein synthesis. Together with a specific set of proteins, rRNAs form ribosomes that bind the mRNA, serve as the platform for tRNAs to decode an mRNA, and catalyze the formation of peptide bonds between amino acids. Each ribosome is composed of two subunits: a small (or 40's) and a large (or 60's) subunit, each of which has its own function.

Like all RNA, mRNA is composed of just four types of nucleotides: adenine, guanine, cytidine, and uracil. Therefore, the information in an mRNA is contained in a linear sequence of nucleotides that is converted into a protein molecule composed of a linear sequence of amino acids. This process is referred to as "translation," since it converts the "language" of nucleotides that make up an mRNA into the

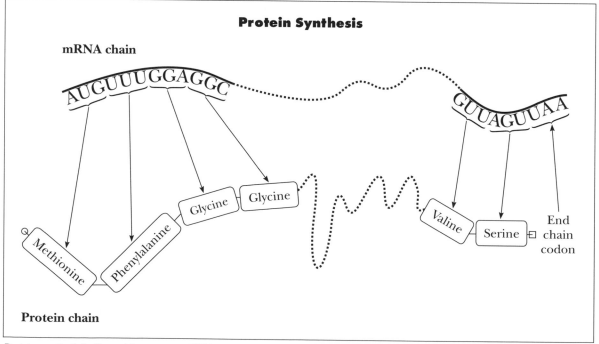

Protein Synthesis

mRNA chain

AUGUUUGGAGGC

GUUAGUUAA

Methionine

Phenylalanine

Glycine Glycine

Valine

Serine

End chain codon

Protein chain

Protein synthesis is directed by messenger RNA (mRNA). The order of the amino acids in the protein chain is controlled by the order of the bases in the mRNA chain. It takes a codon of three bases to specify one amino acid.

"language" of amino acids that make up a protein. This is achieved by a three-letter genetic code in which each amino acid in a protein is specified by a three-nucleotide sequence in the mRNA called a codon. The four possible "letters" (A for adenine, G for guanine, C for cytidine, and U for uracil) means that there are sixty-four possible three-letter "words." As there are only twenty amino acids used to make proteins, most amino acids are encoded by several different codons. For instance, there are six different codons (UCU, UCC, UCA, UCG, AGU, and AGC) that specify the amino acid serine, whereas there is only one codon (AUG) that specifies the amino acid methionine. The mRNA, therefore, is simply a linear array of codons (that is, three-nucleotide "words" that are "read" by tRNAs together with ribosomes). The region within an mRNA containing this sequence of codons is called the coding region.

Before translation can occur, mRNAs undergo processing steps at both ends to add features that will be necessary for translation. Nucleotides are structured such that they have two ends, a 5′ (five-prime) and a 3′ end, that are available to form chemical bonds with other nucleotides. Each nucleotide present in an mRNA has a 5′ to 3′ orientation that gives a directionality to the mRNA so that the RNA begins with a 5′ end and finishes in a 3′ end. The ribosome reads the coding region of an mRNA in a 5′ to 3′ direction. Following the synthesis of an mRNA from its DNA template, one guanine is added to the 5′ end of the mRNA in an inverted orientation and is the only nucleotide in the entire mRNA present in a 3′ to 5′ orientation. It is referred to as the cap. A long stretch of adenosine is added to the 3′ end of the mRNA to make what is called the poly-A tail.

Typically, mRNAs have a stretch of nucleotide sequence that lies between the cap and the coding region. This is referred to as the leader sequence and is not translated. Therefore, a signal is necessary to indicate where the coding region initiates. The codon AUG usually serves as this initiation codon; however, other AUG codons may be present in the coding region. Any one of three possible codons (UGA, UAG, or UAA) can serve as stop codons that signal the ribosome to terminate translation. Several accessory proteins assist ribosomes in binding mRNA and help carry out the required steps during translation.

The Translation Process: Initiation

Translation occurs in three phases: initiation, elongation, and termination. The function of the 40's ribosomal subunit is to bind to an mRNA and locate the correct AUG as the initiation codon. It does this by binding close to the cap at the 5′ end of the mRNA and scanning the nucleotide sequence in its 5′ to 3′ direction in search of the initiation codon. Marilyn Kozak identified a certain nucleotide sequence surrounding the initiator AUG of eukaryotic mRNAs that indicates to the ribosome that this AUG is the initiation codon. She found that the presence of an A or G three nucleotides prior to the AUG and a G in the position immediately following the AUG were critical in identifying the correct AUG as the initiation codon. This is referred to as the "sequence context" of the initiation codon. Therefore, as the 40's ribosomal subunit scans the leader sequence of an mRNA in a 5′ to 3′ direction, it searches for the first AUG in this context and may bypass other AUGs not in this context.

Nahum Sonenberg demonstrated that the scanning process by the 40's subunit can be impeded by the presence of stem-loop structures present in the leader sequence. These form from base pairing between complementary nucleotides present in the leader sequence. Two nucleotides are said to be complementary when they join together through electrostatic interactions, which is a weaker interaction than a chemical bond. For instance, the nucleotide (or base) A is complementary to U, and these two can form what is called a "base pair." Likewise, the nucleotides C and G are complementary. Several accessory proteins, called eukaryotic initiation factors (eIFs), aid the binding and scanning of 40's subunits. The first of these, eIF4F, is composed of three subunits called eIF4E, eIF4A, and eIF4G. The protein eIF4E is the subunit responsible for recognizing and binding to the cap of the mRNA. The eIF4A subunit of eIF4F, together with another factor called eIF4B, functions to remove the presence of stem-loop structures in the leader sequence through the disruption of the base pairing between nucleotides in the stem loop. The protein eIF4G is the large subunit of eIF4F, and it serves to interact with several other proteins, one of which is eIF3. It is this latter initiation factor that the 40's subunit first associates with during its initial binding to an mRNA.

Through the combined action of eIF4G and eIF3, the 40's subunit is bound to the mRNA, and through the action of eIF4A and eIF4B, the mRNA is prepared for 40's subunit scanning. As the cellular concentration of eIF4E is very low, mRNAs must compete for this protein. Those that do not compete well for eIF4E will not be translated efficiently. This represents one means by which a cell can regulate protein synthesis. One class of mRNA that competes poorly for eIF4E encodes growth-factor proteins. Growth factors are required in small amounts to stimulate cellular growth, and their overproduction can lead to the uncontrolled growth characteristic of cancer cells. Sonenberg has shown that the overproduction of eIF4E in animal cells leads to a reduction in the competition for this protein, and mRNAs such as growth-factor mRNAs that were previously poorly translated when the concentration of eIF4E was low are now translated at a higher rate when eIF4E is abundant. This in turn results in the overproduction of growth factors, which leads to uncontrolled growth.

A protein that specifically binds to the poly-A tail at the 3′ end of an mRNA is called the poly-A-binding protein (PABP). Discovered in the 1970's, the only function of this protein was thought to be to protect the mRNA from attack at its 3′ end by enzymes that degrade RNA. Daniel Gallie demonstrated another function for PABP by showing that the PABP-poly-A-tail complex was required for the function of the eIF4F-cap complex during translation initiation. The idea that a protein located at the 3′ end of an mRNA should participate in events occurring at the opposite end of an mRNA was not appealing to scientists initially. However, RNA is quite flexible and is rarely present in a straight, linear form in the cellular environment. Consequently, the poly-A tail can easily approach the cap at the 5′ end. Gallie showed that PABP interacts with eIF4G and eIF4B, two initiation factors that are closely associated with the cap, through protein-to-protein contacts. The consequence of this interaction is that the 3′ end of an mRNA is held in close physical proximity to its cap. The interaction between these proteins stabilizes their binding to the mRNA, which in turn promotes protein synthesis. Therefore, mRNAs can be thought of as adopting a circular form during translation that looks similar to a snake biting its own tail. This idea is now widely accepted by scientists.

One additional factor, called eIF2, is needed to bring the first tRNA to the 40's subunit. Along with the initiator tRNA (which decodes the AUG codon specifying the amino acid methionine), eIF2 aids the 40's subunit in identifying the AUG initiation. Once the 40's subunit has located the initiation codon, the 60's ribosomal subunit joins the 40's subunit to form the intact 80's ribosome. This marks the end of the initiation phase of translation.

The Translation Process: Elongation and Termination

During the elongation phase, tRNAs bind to the 80's ribosome as it passes over the codons of the mRNA, and the amino acids attached to the tRNAs are transferred to the growing polypeptide. Binding of the tRNAs to the ribosome is assisted by an accessory protein called eukaryotic elongation factor 1 (eEF1). A codon is decoded by the appropriate tRNA through base pairing between the three nucleotides that make up the codon in the mRNA and three complementary nucleotides within a specific region (called the anticodon) within the tRNA. The tRNA binding sites in the 80's ribosome are located in the 60's subunit. The ribosome moves over the coding region one codon at a time, or in steps of three nucleotides, in a process referred to as "translocation." When the ribosome moves to the next codon to be decoded, the tRNA containing the appropriate anticodon will bind tightly in the open site in the 60's subunit (the A site). The tRNA that bound to the previous codon is present in a second site in the 60's subunit (the P site). Once a new tRNA has bound to the A site, the ribosomal RNA itself catalyzes the formation of a peptide bond between the growing polypeptide and the new amino acid. This results in the transfer of the polypeptide attached to the tRNA present in the P site to the amino acid on the tRNA present in the A site. A second elongation factor, eEF2, catalyzes the movement of the ribosome to the next codon to be decoded. This process is repeated one codon at a time until a stop codon is reached.

The termination phase of translation begins when the ribosome reaches one of the three termination codons. These are also referred to as "nonsense" codons as the cell does not produce any tRNAs that can decode them. Accessory factors, called release factors, are also required to assist this stage of translation. They bind to the empty A site in which the stop codon is present, and this triggers the cleavage of the bond between the completed protein from the last tRNA in the P site, thereby releasing the protein. The ribosome then dissociates into its 40's and 60's subunits, the latter of which diffuses away from the mRNA. The close physical proximity of the cap and poly-A tail of an mRNA maintained by the interaction between PABP and the initiation factors (eIF4G and eIF4B) is thought to assist the recycling of the 40's subunit back to the 5′ end of the mRNA to participate in a subsequent round of translation.

Impact and Applications

The elucidation of the process and control of protein synthesis provides a ready means by which scientists can manipulate these processes in cells. In addition to infectious diseases, insufficient dietary protein represents one of the greatest challenges to world health. The majority of people living today are limited to obtaining their dietary protein solely through the consumption of plant matter. Knowledge of the process of protein synthesis may allow molecular biologists to increase the amount of protein in important crop species. Moreover, most plants contain an imbalance in the amino acids needed in the human diet that can lead to disease. For example, protein from corn is poor in the amino acid lysine, whereas the protein from soybeans is poor in methio-

nine and cysteine. Molecular biologists may be able to correct this imbalance by changing the codons present in plant genes, thus improving this source of protein for those people who rely on it for life.

—Daniel R. Gallie

See Also: Central Dogma of Molecular Biology; Genetic Code; Molecular Genetics; RNA Structure and Function; RNA World.

Further Reading: James Lake, "The Ribosome," *Scientific American* 245 (August, 1981), summarizes information about the structure and function of ribosomes. Benjamin Lewin, *Genes VI* (1997), details the translational process. Alexander Rich and Sung Hou Kim, "The Three Dimensional Structure of Transfer RNA," *Scientific American* 238 (January, 1978), present a structural description of tRNA. Francis Crick, "The Genetic Code III," *Scientific American* 215 (October, 1966), provides a good summary of the code specifying the amino acids.

Pseudohermaphrodites

Field of study: Human genetics
Significance: *Pseudohermaphrodites are individuals born with either ambiguous genitalia or external genitalia that are the opposite of their chromosomal sex. These individuals need a thorough medical evaluation and appropriate medical intervention to help ensure a healthy life.*

Key terms

AMBIGUOUS GENITALIA: external sexual organs that are not clearly male or female

GENOTYPE: the sum total of the genes present in an individual

GONADS: organs that produce reproductive cells and sex hormones

KARYOTYPE: the number and kind of chromosomes present in every cell of the body (normal female karyotype is 46,XX and normal male karyotype is 46,XY)

PHENOTYPE: the physical appearance of an individual, which depends on the interaction of one's genotype and the environment

Normal Fetal Development

Prior to nine weeks gestational age, a male and a female fetus have identical external genitalia (sexual organs) consisting of a phallus and labioscrotal folds. The phallus develops into a penis in males and a clitoris in females; labioscrotal folds become the scrotum in males and the labial folds in females. Early in development, the gonad can either develop into testes or ovaries. In a fetus with a normal male karyotype (46,XY), the primitive gonads become testes, which produce testosterone. Testosterone in turn causes enlargement of the primitive phallus into a penis. It is the presence of the Y chromosome, and in particular a small, sex-determining region of the Y chromosome termed the SRY locus, that drives the formation of the testes. The presence of the SRY locus appears to be essential for development of a normal male.

Pseudohermaphroditism

A true hermaphrodite is born with both ovarian and testicular tissue. A male pseudohermaphrodite has a 46,XY karyotype with either female genitalia or ambiguous genitalia (but only testicular tissue); a female pseudohermaphrodite has a 46,XX karyotype with either male genitalia or ambiguous genitalia (but only ovarian tissue). Ambiguous genitalia typically consist of a small, abnormally shaped, phalluslike structure, often with hypospadias (in which urine comes from the base of the penis instead of the tip) and abnormal development of the labioscrotal folds (not clearly a scrotum or labia). A vaginal opening may be present.

Most cases of pseudohermaphroditism result from abnormal exposure to increased or decreased amounts of sex hormones during embryonic development. The most common cause of female pseudohermaphroditism is exposure of a female fetus to increased levels of testosterone during the first half of pregnancy. Maternal use of anabolic steroids can cause this condition, but the most common genetic cause of increased testosterone exposure is congenital adrenal hyperplasia (CAH). CAH results from an abnormality in the enzymatic pathways of the fetus that make both cortisol (a stress hormone) and the sex steroids (such as testosterone). At several points in these pathways, there may be a nonfunctioning enzyme that results in too little production of cortisol and too much production of the sex steroids. This will result in partial masculinization of the external genitalia of a female embryo. Females with CAH are usually born with an enlarged clitoris (often mistakenly thought to be a penis) and partial fusion of the labia. Males can also have CAH, but the excess testosterone does not affect their genital development since a relatively high level of testosterone exposure is a normal part of their development.

The most common causes of male pseudohermaphroditism are abnormalities of testosterone production or abnormalities in the testosterone receptor at the cellular level. One example is a deficiency in 5-alpha-reductase, the enzyme that converts testosterone to dihydrotestosterone (DHT). When there is a deficiency of this enzyme, there will be a deficiency of DHT, which is the hormone primarily responsible for masculinization of external genitalia. A male who lacks DHT will have female-appearing external genitalia or ambiguous genitalia at birth. Often these individuals are reared as females, but at puberty they will masculinize because of greatly increased production of testosterone. These individuals may actually develop into nearly normal-appearing males. Abnormalities of the testosterone receptor can also result in a range of different conditions in affected males, from normal female appearance (a totally defective receptor) to ambiguous genitalia (partially defective receptor) in a 46,XY male. These individuals will not masculinize at puberty because no matter how much testosterone or DHT they produce, their bodies cannot respond to the hormones.

Both male and female pseudohermaphroditism can result from chromosomal abnormalities. The absence or dysfunction of the SRY locus produces an individual with normal female genitalia but a 46,XY karyotype. Individuals with a 46,XX karyotype who have the SRY locus transposed to one of their X chromosomes will have a normal male appearance.

Impact and Applications

Some forms of pseudohermaphroditism are life threatening, and so early diagnosis is imperative. Both males and females with CAH are at risk for sudden death caused by low cortisol levels and other hormone deficiencies. Early diagnosis is relatively easy in affected females since their genital abnormalities are noticeable at birth. Affected males are often not recognized until they have a life-threatening event, which usually occurs in the first two weeks of life. Treatment of CAH consists of appropriate hormone supplementation that, if instituted early in life, can help prevent life-threatening events. CAH is inherited in an autosomal recessive manner, so parents of an affected individual have a 25 percent chance of having another affected child with each pregnancy.

The sex of rearing of a child with ambiguous genitalia is usually determined by the child's type of pseudohermaphroditism. Typically, sex of rearing will be based on the chromosomal sex of the child. These children may need sex hormone supplementation or surgery to assist in developing gender-appropriate genitalia. Children with pseudohermaphroditism with normal-appearing genitalia at birth may not be recognized until puberty, when abnormal masculinization or feminization may occur. These individuals need medical evaluation and karyotype determination to guide the proper medical treatment.

—*Patricia G. Wheeler*

See Also: Gender Identity; Hermaphrodites; Testicular Feminization Syndrome.

Further Reading: More specific information on male and female pseudohermaphroditism can be found in "Disorders of the Gonads, Genital Tract, and Genitalia," by J. L. Simpson, in *Principles and Practices of Medical Genetics* (1996), edited by D. L. Rimoin et al. The surgical treatment of individuals with pseudohermaphroditism is discussed in Zoran Krstic et al., "Surgical Treatment of Intersex Disorders," *Journal of Pediatric Surgery* (September, 1995). Further information on gender identity and its relation to pseudohermaphroditism can be found in Heino Meyer-Bahlburg, "Intersexuality and the Diagnosis of Gender Identity Disorder," *Archives of Sexual Behavior* 23 (1994).

Punctuated Equilibrium

Field of study: Population genetics
Significance: *Punctuated equilibrium is a model of evolutionary change in which new species originate abruptly and then exist through a long period of stasis. This model is important as an explanation of the stepwise pattern of species change seen in the fossil record.*

Key terms

ALLOPATRIC SPECIATION: a theory that suggests that small parts of a population may become genetically isolated and develop differences that would lead to the development of a new species

HETEROCHRONY: a change in the timing or rate of development of characters relative to those same events in the ancestors

PHYLETIC GRADUALISM: the idea that evolutionary change proceeds by a progression of tiny changes, adding up to produce new species over immense periods of time

Evolutionary Patterns

Nineteenth century English naturalist Charles Darwin viewed the development of new species as occurring slowly by a shift of characters within populations, so that a gradual transition from one species to another took place. This is generally referred to as phyletic gradualism. A number of examples from the fossil record were put forward to support this view, particularly that of the horse, in which changes to the feet, jaws, and teeth seem to have progressed in one direction over a long period of time. Peter Sheldon in 1987 documented gradual change in eight lineages of trilobites over a three-million-year period in the Ordovician period of Wales. Despite these examples, it is clear that the fossil record more commonly shows a picture of populations that are stable through time but are separated by abrupt morphological breaks. This pattern was recognized by Darwin but was attributed by him to the sketchy and incomplete nature of the fossil record. So few animals become fossilized, and conditions for fossilization are so rare, that he felt only a fragmentary sampling of gradual transitions were present, giving the

Stephen Jay Gould, who, together with Niles Eldredge, developed the theory of punctuated equilibrium to explain gaps in the fossil record. (Stephen Jay Gould)

appearance of abrupt change.

One hundred years later, the incompleteness of the fossil record no longer seemed convincing as an explanation. In 1972, Niles Eldredge and Stephen Jay Gould published their theory of the evolutionary process, called by them "punctuated equilibrium." This model explains the lack of intermediates by suggesting that evolutionary change occurs only in short-lived bursts in which a new species arises abruptly from a parent species, often with relatively large morphological changes, and thereafter remains more or less stable until its extinction.

The Process of Punctuated Equilibrium

A number of explanations have been put forward to show how this process might take place. One of these, termed allopatric speciation, was first proposed by Ernst Mayr in 1963. He pointed out that a reproductive isolating mechanism is needed to provide a barrier to gene flow and that this could be provided by geographic isolation. Allopatric or geographical isolation could result when the normal range of a population of organisms is reduced or fragmented. Parts of the population become separated in peripheral isolates, and if the population is small, it may become modified rapidly by natural selection or genetic drift, particularly if it is adapting to a new environment. As the initial members of the peripheral isolate may be few in number, it might take only a few generations for the population to have changed enough to become reproductively isolated from the parent population. In the fossil record, this will be seen as a period of stasis representing the parent population, followed by a rapid morphological change as the peripheral population is isolated from it and then replaces it, either competitively or because it has become extinct or has moved to follow a shifting habitat. Because this is thought to take place rapidly in small populations, fossilization potential is low, and unequivocal examples are not common in the fossil record. However, in 1981, Peter Williamson published a well-documented example from the Tertiary period of Lake Turkana in Kenya, which showed episodes of stasis and rapid change in populations of freshwater mollusks. The increases in evolutionary rate were apparently driven by severe environmental change that caused parts of the lake to dry up.

Punctuated changes may also have taken place because of heterochrony, which is a change in the rate of development or timing of appearance of ancestral characters. Paedomorphosis, for example, would result in the retention of juvenile characters in the adult, while its opposite, peramorphosis, would result in an adult morphologically more advanced than its ancestor. Rates of development could be affected by a mutation, perhaps resulting in the descendent growing for much longer than the ancestral form, thus producing a giant version. These changes would be essentially instantaneous and thus would show as abrupt changes of species in the fossil record.

Impact and Applications

The publication of the idea of punctuated equilibrium ignited a storm of controversy that was still raging twenty years later. It demonstrated that speciation can be very rapid, but more important, it showed the prevalence of stasis over long periods of time, something that had not been anticipated. Species had been viewed as flexible and responsive to the environment, but fossil species showed no change over long periods despite a changing environment. Biologists have thus had to review their ideas about the concept of species and the processes that operate on them. Species are now seen as real entities that have characteristics that are more than the sum of their component populations. Thus the tendency of a group to speciate rapidly or slowly may be intrinsic to the group as a whole and not dependent on the individuals that compose it. This debate has helped show that the fossil record can be important in detecting phenomena that are too large in scale for biologists to observe.

—*David K. Elliott*

See Also: Evolutionary Biology; Natural Selection; Speciation.

Further Reading: *The Dynamics of Evolution* (1992), edited by Albert Somit and Steven Peterson, provides an overview of the punctuated equilibrium debate, including updates by Stephen Jay Gould and Niles Eldredge. An example of punctuated equilibrium in the fossil record is discussed in Peter Williamson, "Paleontological Documentation of Speciation in Cenozoic Mollusks from Turkana Basin," *Nature* 293 (1981). General discussions of the concept of punctuated equilibrium are provided in *Invertebrate Paleontology and Evolution* (1993), by Ewen Clarkson, and *Bringing Fossils to Life* (1997), by Donald Prothero.

Quantitative Inheritance

Field of study: Population genetics

Significance: *Quantitative inheritance is the genetic mechanism that influences metric traits. Quantitatively inherited traits are those generally associated with adaptation, reproduction, yield, form, and function. They are thus of great importance to evolution, conservation biology, psychology, and especially to improvement programs for agricultural organisms.*

Key terms

METRIC TRAITS: traits that vary in a continuous manner through the effects of multiple genes with small individual effects and a continuously varying environmental effect

GENETIC VARIATION: a measure of the availability of genetic differences within a population

GENOTYPE: the genetic makeup of an organism at all loci that affect a quantitative trait

PHENOTYPE: the observed value for a trait that results from the combined effects of the genotype and the environment to which the organism has been exposed

HERITABILITY: the proportion of differences among animals for a trait that are a result of genetic differences

The Genetics Underlying Metric Traits

An understanding of the genetics affecting metric traits came with the unification of the Mendelian and biometrical schools of genetics early in the 1900's. The statistical relationships involved in inheritance of metric traits such as height of humans was well known in the late 1800's. Soon after that, Gregor Mendel's breakthrough on particulate inheritance, obtained from work utilizing traits such as colors and shapes of peas, was rediscovered. However, some traits did not follow the patterns of inheritance predicted based on Mendel's work. As an example, Francis Galton crossed pea plants having uniformly large seeds with those having uniformly small seeds. The seed size of the progeny was intermediate. However, when the progeny were mated among themselves, seed size formed a distribution from small to large with many intermediate sizes.

How could genetic factors with a particulate nature explain inheritance of traits with a continuous distribution? The solution was described by Swedish plant breeder Herman Nilsson-Ehle from the results of experiments with wheat performed early in the twentieth century. Nilsson-Ehle crossed red wheat with white. The resulting progeny were light red in color. When matings were made within the progeny, the resulting kernels of wheat ranged in color from white to red. He was able to categorize the wheat into five colors: red, intermediate red, light red, pink, and white. Intermediate colors occurred with greater frequency than extreme colors. Nilsson-Ehle deduced that particulate genetic factors (now known as alleles) were involved, with red wheat inheriting four red alleles, intermediate red inheriting three red alleles, light red inheriting two red alleles, pink inheriting one red allele, and white inheriting no red alleles. These results were consistent with Mendel's findings, except that two sets of factors (now known as loci) were controlling this trait rather than the single locus observed for the traits considered by Mendel. Further, these results could be generalized to account for inheritance patterns controlled by more than two loci. These extensions were mathematically described by British statistician and geneticist Ronald A. Fisher.

Under many circumstances, the environment also modifies observed traits. A combination of many loci with individually small effects provides an underlying bell-shaped distribution of genetic effects. Environmental effects are truly continuous and occur randomly with respect to the genetic effects. The environmental effects blur the boundaries of the genetic categories and can make it impossible to identify the effects of individual loci for many quantitative traits. The distribution of phenotypes, reflecting combined genetic and environmental effects, is bell shaped and smooth.

Genetic and environmental effects jointly influence the value of most metric traits. The relative magnitude of these two effects is measured by the heritability. Although all are statistically equivalent, the heritability has several practical definitions. One important definition states that the heritability is equal to the proportion of observed differences among organisms for a trait that are caused by genetic differences. For example, if one-quarter of the differences among cows for the amount of milk they produce are caused by differences among their genotypes, the heritability of milk production is 25 percent. The remaining 75 percent of differences among the animals are attributed to environmental effects. A second definition is that heritability is equal to the proportion of differences among sets of parents that are passed on to their progeny. For example, if the average height of a pair of parents is 8 inches more than the mean of their population and the heritability of height is 50 percent, their progeny would be expected to average 4 inches taller than their peers in the population.

Fundamental Relationships of Quantitative Genetics

Two relationships are fundamental to the understanding and application of quantitative genetics. First, there is a tendency for likeness among related individuals. Although similarities of human stature and facial appearance within families are familiar to most people, similar relationships hold for genetically influenced metric traits for all organisms. Correlation among relatives exists for such diverse traits as blood pressure, plant height, grain yield, and egg production. These correlations are caused by relatives sharing a portion of genes in common. The more closely the individuals are related, the greater the proportion of genes that are shared. Identical twins share all their genes, and full brothers and sisters or parent and offspring are expected to share one-half their genes. This relationship is commonly utilized in the improvement of agricultural organisms. Individuals are chosen to be parents based on the performance of their relatives. For example, bulls of dairy breeds are chosen to become widely used as sires based on the milk-producing ability of their sisters and daughters.

The second fundamental relationship is that in organisms that do not normally self-fertilize, vigor is depressed in progeny that result from the mating of closely related individuals within a species. This penalty of close mating is known as inbreeding depression. It may be the basis of the social taboos regarding incestuous relationships in humans and for the dispersal systems for some other species of mammals such as wolves. Physiological barriers have evolved to prevent fertilization between close relatives in many species of plants. Some mechanisms function as an anatomical inhibitor to prevent union of pollen and ova from the same plant; in maize, for example, the male and female flower are widely separated on the plant. Indeed, in some species such as asparagus and holly trees, the sexes are separated in different individuals; thus all seeds must consequently result from cross-pollination. In other systems, cross-pollination is required for fertile seeds to result. The pollen must originate from a variety of plant different from the source of the ova. These phenomena are known as self-incompatibility and are present in species such as broccoli, radishes, some clovers, and many fruit trees.

The corollary to inbreeding depression is hybrid vigor, a phenomenon of improved fitness that is often evident in progeny resulting from the mating of individuals less related than the average in a population. Hybrid vigor has been utilized in breeding programs to achieve remarkable productivity of hybrid seed corn as well as crossbred poultry and livestock. Hybrid vigor results in increased reproduction and efficiency of nutrient utilization. The mule, which results from mating a male donkey to a female horse, is a well-known example of a hybrid that has remarkable strength and hardiness compared to the parent species.

Quantitative Traits of Humans

Like other organisms, many traits of humans are quantitatively inherited. Psychological characteristics, intelligence quotient (IQ), and birth weight have been studied extensively. The heritability of IQ has been reported to be high. Other personality characteristics such as incidence of depression, introversion, and enthusiasm have been reported to be heritable. Musical ability is another characteristic under some degree of genetic control. These results have been consistent across replicated studies and are thus expected to be reliable; however, some caution must be exercised when considering the reliability of results from individual studies. Most studies of heritability in humans have involved likeness of twins reared together and apart. The difficulty in obtaining such data results in a relatively small number of observations contributing to the individual studies, at least relative to similar experiments that might be conducted in animals. An unfortunate consequence of studies of quantitative inheritance in humans was the eugenics movement.

Birth weight of humans is of interest because it is both under genetic control and subject to influence by well-known environmental factors, such as smoking by the mother. Birth weight is subject to stabilizing selection, in which individuals with intermediate values have the highest rates of survival. This results in genetic pressure to maintain the average birth weight at a relatively constant value. Birth weight provides an example of this phenomenon, which is likely to be common in nature.

Quantitative Characters in Agricultural Improvement

The ability to meet the demand for food by the growing world population is dependent upon continuous increases in productivity of agronomic crops and livestock. The world no longer has reserves of high-quality farmland that can be brought into production, nor can the oceans provide a sustainable increase in the harvest of fish. Many countries that struggle to meet the food demands of their populations are too poor to increase agricultural yields through increased inputs of fertilizer and chemicals. Increased food production will, therefore, largely depend on genetic improvement of the organisms, both plants and animals, produced by farmers worldwide.

Most characteristics of economic value in agriculturally important organisms are quantitatively inherited. Traits such as grain yield, baking quality, milk and meat production, and efficiency of nutrient utilization are under the influence of many genes as well as the production environment. The task of breeders is not only to identify organisms with superior genetic characteristics but also to identify those breeds and varieties well adapted to the specific environmental conditions in which they will be produced. The type of dairy cattle that most efficiently produces milk under the normal production circumstances of the United States, which includes high health status, unlimited access to high-quality grain rations, and protection from extremes of heat and cold, may not be ideal under conditions in New Zealand in which the cow is required to compete with herdmates for high-quality pasture forage. Neither of these animals may be ideal under tropical conditions in which the animal may be exposed to extremely high temperatures, disease, and parasites.

Remarkable progress has been made in many important food organisms. Grain yield has responded to improvement programs. Development of hybrid corn increased yield severalfold over the last few decades of the twentieth century. Development of improved varieties of small grains resulted in an increased ability of many developing countries to be self-sufficient in food production. Grain breeder Norman Borlaug won the Nobel Peace Prize in 1970 for his role in developing grain varieties that contributed to the green revolution.

Can breeders continue to make improvements in the genetic potential for crops, livestock, and fish to yield food adequate for the growing human population? Tools of biotechnology are expected to increase the rate at which breeders can make genetic change. Ultimately, the answer depends upon the genetic variation available in the global populations of food-producing organisms and their relatives. That genetic variation is the raw material upon which breeders rely for continued improvement of yields. The potential for genetic improvement of some species has been relatively untapped. Domestication of fish for use in aquaculture and utilization of potential crop species such as amaranth are possible food reserves. Wheat, corn, and rice provide a large proportion of the calories supporting the world population. The yields of these three crop species have already benefited from many generations of selective breeding. For continued genetic improvement, it is critical that variation not be lost through the extinction of indigenous strains and wild relatives of important food-producing organisms.

Quantitative Characters Under Evolution

Metric traits have been shown to be of importance in adaptation. Natural selection acts on characters that are of importance to the fitness of an organism and can result in modification that may, for example, increase its ability to compete for food or avoid detection by predators. Henry B. D. Kettlewell demonstrated that genetic variation for a qualitative character (color) of peppered moths influenced their ability to escape detection by predatory birds. Kettlewell showed that a light-colored variant was favored in nonindustrialized areas in England, while a dark-colored variant had an advantage in smoky industrialized areas through their ability to blend into trees that were stained with soot.

Extensive studies of finches in the Galápagos Islands by Peter Grant have shown an interaction among climatic environment, feeding behavior, and beak shape. Charles Darwin's finches are a group of closely related species of which some are particularly similar and vary most notably in body size and size and shape of beak, all quantitatively inherited traits. These characteristics impact the type of seeds that provide their primary forage. It has been observed that during years of abundant food, all finches have a high reproductive rate, hybridization occurs, and divergence between the species is diminished. During years of drought, birds most specialized to feed on their particular type of seeds tend to be most successful, and intermediate types have poorer reproductive success and survival. In addition, it has been shown through experiments using decoys that birds focus on choosing mates with a beak shape similar to their own. These data, collected over a series of many years, provide direct evidence that natural selection on quantitatively inherited traits has an evolutionary impact on creating divergence between relatively young species. Natural selection for birds adapted to al-

Grain breeder Norman Borlaug received a 1970 Nobel Peace Prize for his development of improved crop varieties. (The Nobel Foundation)

ternative niches has been observed to eliminate the intermediate forms, thus reinforcing the divergence between the species.

—*William R. Lamberson*

See Also: Artificial Selection; Heredity and Environment; High-Yield Crops; Natural Selection; Population Genetics.

Further Reading: Colin Tudge, *The Engineer in the Garden* (1993), provides a historical overview of genetics and explores the potential ramifications of past, present, and future genetic advances. I. Michael Lerner, *Heredity, Evolution, and Society* (1968), provides a clear discussion of polygenic inheritance, particularly with respect to humans. Anticipating discovery of the genetic basis for phenotypic variation, Charles Darwin describes the remarkable variability of domesticated plants and animals in *Variation of Plants and Animals Under Domestication* (1875). Jonathan Weiner, *The Beak of the Finch* (1994), describes the observations of Peter Grant on forces actively shaping the evolution of Darwin's finches on the Galápagos Islands.

Race

Field of study: Human genetics

Significance: *Humans typically have been catego-rized into a small number of races based on com-mon traits, ancestry, and geography. Knowledge of human genomic diversity has increased aware-ness of ambiguities associated with traditional racial groups. The sociopolitical consequences of using genetics to devalue certain races are pro-found.*

Key terms

CLASSIFICATION: a hierarchical grouping of or-ganisms based on shared characteristics and evolutionary relationships

EUGENICS: a movement concerned with the im-provement of human genetic traits, pre-dominantly by the regulation of mating

HUMAN GENOME DIVERSITY PROJECT: an exten-sion of the Human Genome Project in which deoxyribonucleic acid (DNA) of native peo-ple around the world is collected for study

POPULATION: a group of geographically local-ized, interbreeding individuals

RACE: a collection of geographically localized populations with well-defined genetic traits

History of Racial Classification

Efforts to classify humans into a number of distinct types date back at least to the ancient Greeks. Applying scientific principles to sepa-rate people into races has been a goal for more than two centuries. In 1758, the founder of biological classification, Swedish botanist Carolus Linnaeus, arranged humans into four principal races: *Americanus, Europeus, Asiaticus,* and *Afer.* Although geographic location was his primary organizing factor, Linnaeus also de-scribed the races according to subjective traits such as temperament. Despite his use of ar-chaic criteria, Linnaeus did not give superior status to any of the races. This is evidenced by his listing the *Americanus* race first and *Europeus* second.

Johann Friedrich Blumenbach, a German naturalist and admirer of Linnaeus, developed a classification with lasting influence. Many of his contemporaries believed that different groups of humans arose separately in several regions of the world. Blumenbach, on the other hand, strongly believed in one form of human and believed that physical variations among races were chiefly caused by differences in environment. Therefore, his scheme sought to show a gradual change in bodily appearance, all deviating from an original type. Blumen-bach maintained that the original forms, which he named "Caucasian," were those primarily of European ancestry. His final classification, pub-lished in 1795 in *On the Natural Variety of Man-kind,* consisted of five races: Caucasian, Malay, Ethiopian, American, and Mongolian. Two races directly radiated from the Caucasians: the Malay and the American. The Malay (Pacific islanders) then generated the Ethiopian (Afri-cans), while the American (from the New World) gave rise to the Mongolian (East Asians). The fifth race, the Malay, was added to Linnaeus's classification to show a step-by-step change from the original body type.

Since the time of Linnaeus and Blumen-bach, many variations of their categories have been formulated, chiefly by biologists and an-thropologists. Classification "lumpers" com-bine people into only a few races (for example, black, white, and Asian). "Splitters" separate the traditional groups into many different races. One classification scheme divided all Europeans into Alpine, Nordic, and Mediterra-nean races. Others have split Europeans into ten different races. No one scheme of racial classification is accepted throughout the scien-tific community.

Genetic Diversity Among Races

The genetic components of a population are produced by three primary factors: natu-ral selection, nonadaptive genetic change, and mating between neighboring populations. The first two factors may result in differences be-tween populations, and reproductive isolation, either voluntary or because of geographic iso-lation, perpetuates the distinctions. Natural se-lection refers to the persistence of genetic traits favorable in a specific environment. For exam-ple, a widely held assumption concerns skin

In the mid-eighteenth century, the pioneering biologist and botanist Carolus Linnaeus classified humans into four separate races. (Archive Photos)

population with elevated melanin production genes. Therefore, genes coding for melanin production are favorable and persist in these environments.

The second factor contributing to the genetic makeup of a population is nonadaptive genetic change. This process involves random genetic mutations. Mutations are changes resulting in modified forms of the same gene. For example, certain genes are responsible for eye color. Individuals contain alternate forms of these genes, or alleles, which result in observed differences in eye color. Alleles resulting from nonadaptive genetic change may remain in the population because of their neutral nature. In other words, they are not harmful or beneficial. Because these traits are impartial to environmental influences, they may endure from generation to generation. Different populations will spontaneously produce, persist, and delete them. Genetic difference between populations caused by these random mutations and isolation is called genetic drift.

color, primarily a result of the pigment melanin. Melanin offers some shielding from ultraviolet solar rays. According to this theory, people living in regions with concentrated ultraviolet exposure have increased melanin synthesis and, therefore, dark skin color conferring additional protection against skin cancer. Individuals with genes for increased melanin have enhanced survival rates and reproductive opportunities. The reproductive opportunities produce offspring that inherit those same genes for increased melanin. This process results in a higher percentage of the

The third factor, mating between individuals from neighboring groups, tends to merge traits from several populations. This genetic mixing often results in offspring with blended characteristics and only moderate variations between adjacent groups.

Several studies have compared the overall genetic complement of various human populations. On average, any two people of the same or a different race diverge genetically by a mere 0.2 percent. It is estimated that only 0.012 percent contributes to traditional racial variations. Hence, most of the genetic dissimilarities between a person of African descent and a person of European descent are also different between two individuals with the same ancestry. The genes do not differ. It is the proportion of individuals expressing a specific allele of a gene that varies from population to population.

Upon closer examination, it was found that Africa is unequaled with respect to cumulative genetic diversity. If overall genetic distinctness is evaluated, numerous races are found in Africa, Khoisan Africans of southern Africa being the most distinct. According to one theory, the remainder of the human species (including Asians, Europeans, and aboriginal Australians) corresponds to only one other race.

Conflicts Concerning Definitions of Race

Linnaeus developed the scientific system of classification still in use. This approach involves separating all organisms first into broad groups based on general characteristics. These large groups are broken down further into smaller and smaller groups, each subdivision containing individuals with more similarities. For example, humans are found within the large kingdom containing all types of animals. Animals are separated based on the formation of a backbone. Of those animals containing a backbone, humans are placed into a set with all mammals and then further cataloged with other primates. Each succeeding classification unit contains individuals more alike, since the characteristics used to define each subdivision are more specific. Eventually, all organisms are placed into a species category. Humans belong in the species *Homo sapiens*. By definition, a sexually reproducing species contains all individuals that can mate and produce fertile offspring. Race is analogous to a more specific unit, the subspecies, a fundamentally distinct subgroup within one species.

For a racial or subspecies classification scheme to be objective and biologically meaningful, researchers must decide carefully which heritable characteristics (passed to future generations genetically) will define, or separate, the races. Several principles are considered. First, the discriminating traits must be discrete. In other words, differences among races must be distinguishable, not continually changing by small degrees between populations. Second, everyone placed within a specific race must possess the selected trait's defining variant. Features used to describe a race must agree. This means that all of the selected characteristics are found consistently in each member. For example, if blue eyes and brown hair are chosen as defining characteristics, everyone designated as belonging to that race must share both of those characteristics. Individuals placed in other races should not exhibit this particular combination. The purpose of using these characteristics is to distinguish groups. Consequently, if traits are shared by members of two or more races, their defining value is poor. Third, individuals of the same race must have descended from a common ancestor, unique to those people. Many shared characteristics present in individuals of a race may be traced to that ancestor by heredity. Based on the preceding defining criteria (selection of discrete traits, agreement of traits, and common ancestry), pure representatives of each racial category should be detectable.

Many researchers maintain that traditional races do not conform to accepted scientific principles of subspecies classification. For example, the traits used to define traditional human races are rarely discrete. Skin color, a prominent characteristic employed, is not a well-defined trait. Approximately five genes influence skin color significantly, but fifty or so likely contribute. Pigmentation in humans results from a complex series of biochemical pathways regulated by amounts of enzymes (molecules that control chemical reactions) and enzyme inhibitors, along with environmental factors. Like most complex traits involving many genes, human skin color varies on a continuous gradation. From lightest to darkest, all intermediate pigmentations are represented. Color may vary widely even within the same family. The boundary between black and white is an arbitrary, human-made border, not one imposed by nature.

In addition, traditional defining racial characteristics, such as skin color and facial characteristics, are not found in all members of a race; they are not in agreement. For example, many

Melanesians, indigenous to Pacific islands, have pigmentation as dark as any human but are not classified as "black." Another example concerns unclassifiable populations. For example, many individuals native to India have Caucasoid facial features and very dark skin, yet live in Asia. When traditional racial characteristics are examined closely, many groups are left with no conventional race. No "pure" genetic representatives of any traditional race exist.

Common ancestry, or evolutionary relationships, must also be considered. Genetic studies have shown that Africans do not belong to a single "black" heritage. In fact, several lineages are found in Africa. An even greater variance is found in African Americans. Besides a diverse African ancestry, it is estimated that, on average, 20 to 30 percent of African American heritage is European or Native American. Yet all black Americans are consolidated into one race.

The true diversity found in humans is not patterned according to accepted standards of the subspecies. Only at extreme geographical distances are notable differences found. However, "in-between" populations have always been in existence because of mating, and therefore gene flow, between neighboring groups. Consequently, human populations in close proximity have more genetic similarities than distant populations. It is the population itself that best illustrates the pattern of human diversity. Well-defined genetic borders between human populations are not observed, and racial boundaries in classification schemes are often formed arbitrarily.

Theories of Human and Racial Evolution

Advances in deoxyribonucleic acid (DNA) technology have greatly aided researchers in their quest to reconstruct the history of *Homo sapiens* and its various subgroups. Analysis of human DNA has been performed on both nuclear and mitochondrial DNA. Mitochondria are organelles responsible for generating cellular energy. Each mitochondrion contains a single, circular DNA molecule accounting for approximately 0.048 percent of the entire genetic complement. In 1987, geneticist Rebecca L. Cann compared mitochondrial DNA from

many populations: African, Asian, Caucasian, Australian, and New Guinean. Agreeing with other mitochondrial and nuclear DNA studies, the results indicated that Africans were the most genetically variable by a significant extent. The results suggested to Cann that Africa was the root of all humankind and that humans first arose there 100,000 to 200,000 years ago. Several lines of research, including DNA analysis of humanoid fossils, provide evidence for this theory.

Many scientists are using genetic markers to decipher the migrations that fashioned past and present human populations. For example, DNA comparisons revealed three Native American lineages. Some scientists believe one migration crossed the Bering Strait, most likely from Mongolia. Only after further migration throughout the Americas were the three American Indian lineages formed. Another theory states that three separate Asian migrations occurred, each bringing a different lineage. Another example is the South African Lemba community. DNA analysis gives credence to their claim as one of the lost tribes of Israel. Considering the cumulative evidence, many scientists regard a more correct depiction of human populations to be a roughly inverted version of Blumenbach's. Asians arose from Africans, and Europeans are Asian and North African hybrids. However, interpretations of DNA analyses are, almost inevitably, controversial. Multiple theories abound and are revised as additional research is performed.

Sociopolitical Implications

Race is often portrayed as a natural, biological division, the result of geographic isolation and adaptation to local environment. However, confusion between biological and cultural classification obscures perceptions of race. When individuals describe themselves as "black," "white," or "Hispanic," for example, they are usually describing cultural heredity as well as biological similarities. The relative importance of perceived cultural affiliations or genetics varies depending on the circumstances. Examples illustrating the ambiguities are abundant. Nearly all people with African American ancestry are labeled black, even if they have a white parent. In addition, dark skin color designates one as belonging to the black race, including Africans and aboriginal Australians, who have no common genetic lineage. State laws, some on the books until the late 1960's, required a "Negro" designation for anyone with one-eighth black heritage (one black great-grand-parent).

Unlike biological boundaries, cultural boundaries are sharp, repeatedly motivating discrimination, genocide, and war. The frequent use of biology to devalue certain races and excuse bigotry has profound implications for individuals and society. The eugenics movement, advocating the genetic improvement of the human species, translated into laws against interracial marriage, sterilization programs, and mass murder. Harmful effects include accusations of deficiencies in intelligence or moral character based on traditional racial classification.

The frequent use of biology to devalue certain races and excuse bigotry has profound implications for individuals and society. Blumenbach selected Caucasians (who inhabit regions near the Caucasus Mountains, a Russian and Georgian mountain range) as the original form of humans because in his opinion they were the most beautiful. All other races deviated from this ideal and were, therefore, less beautiful. Despite Blumenbach's efforts not to demean other groups based on intelligence or moral character, the act of ranking in any form left an ill-fated legacy.

Many scientists are attempting to reconcile the negativities associated with racial studies. A genetic investigation tracing African Americans to their African origins is planned by researcher Georgia Dunston. The Human Genome Diversity Project, a global undertaking, has requested that researchers collect and store DNA from indigenous populations around the world. These samples will be available to all qualified scientists. Results of the studies may include gene therapy treatments and greater success with organ transplantation. A more thorough understanding of the genetic diversity and unity present in the species *Homo sapiens* will be possible.

—*Stacie R. Chismark*

See Also: Biological Determinism; Eugenics; Human Genome Project; Miscegenation and Antimiscegenation Laws; Population Genetics.

Further Reading: *Discover* released a special issue, "The Science of Race" (November, 1994), dedicated to an overview of biological perspectives concerning race. Stephen Jay Gould presents a historical commentary on racial categorization and a refutation of theories espousing a single measure of genetically fixed intelligence in *The Mismeasure of Man* (1996). Luigi L. Cavalli-Sforza et al., *The History and Geography of Human Genes* (1994), often referred to as a "genetic atlas," contains fifty years of research comparing heritable traits, such as blood groups, from more than one thousand human populations.

Restriction Enzymes

Field of study: Genetic engineering and biotechnology
Significance: *Restriction enzymes are bacterial enzymes capable of cutting DNA molecules at specific sites. Discovery of these enzymes was a pivotal event in the development of genetic engineering technology, and they are routinely and widely used in molecular biology.*

Key terms

DEOXYRIBONUCLEIC ACID (DNA): the genetic material of the cell found in the nucleus and composed of "nucleotides"

NUCLEOTIDES: the building blocks of nucleic acids composed of a sugar, a phosphate group, and nitrogen-containing "bases"

DOUBLE-STRANDED MOLECULE: a DNA molecule that has two strands or chains of DNA nucleotides connected in the middle by hydrogen bonds between complementary bases; the alternative is a single-stranded molecule, which is a single chain of DNA nucleotides

ENZYME: a molecule, usually a protein, that is used by cells to facilitate and speed up a chemical reaction

METHYLATION: the process of adding a methyl chemical group (one carbon atom and three hydrogen atoms) to a particular molecule, such as a DNA nucleotide

The Discovery and Role of Restriction Enzymes in Bacteria

Nucleases are a broad class of enzymes that destroy nucleic acids by breaking the sugar-phosphate backbone of the molecule. Until 1970, the only known nucleases were those that destroyed nucleic acids nonspecifically—that is, in a random fashion. For this reason, these enzymes were not considered useful tools for working with nucleic acids. In 1970, molecu-

lar biologist Hamilton Smith discovered a type of nuclease that could fragment deoxyribonucleic acid (DNA) molecules in a specific and therefore predictable pattern. This nuclease, HindII, was the first restriction enzyme. Smith was working with the bacterium *Haemophilus influenzae* (*H. influenzae*) when he discovered this enzyme, which was capable of destroying DNA from other bacterial species but not the DNA of *H. influenzae* itself. The term "restriction" refers to the basic mechanism of these enzymes in selectively fragmenting only certain types of DNA, usually DNA molecules from other strains or species of bacteria or even bacterial viruses.

As more restriction enzymes from a wide variety of bacterial species were discovered in the 1970's, it became increasingly clear to molecular biologists that these enzymes could be useful tools for creating and manipulating DNA fragments in ways that were never possible before. What was not clear, however, was the role of these enzymes in the bacterial cells that produced them. An important piece of evidence came from comparisons of DNA molecules that could be fragmented by a particular restriction enzyme with DNA molecules that remained intact in the presence of the enzyme. There was one important difference between these two groups of molecules: The DNA molecules that were protected from the effects of the restriction enzyme had extra molecules at various places on the DNA strand. These extra molecules were shown to be methyl, or CH_3 groups, giving this phenomenon the name "DNA methylation."

Hamilton O. Smith received a 1978 Nobel Prize for his discovery of restriction enzymes. (The Nobel Foundation)

Closer examination of the methylation sites revealed that they were the same sites that were normally recognized and cleaved by the restriction enzyme if the methylation was absent. The conclusion was that the methylation somehow protected the DNA from attack, and this could account for Smith's observation that *H. influenzae* DNA was not destroyed by its own restriction enzyme; presumably the enzyme recognized a specific methylation pattern on the DNA molecule and left it alone. Foreign DNA (from another species, for example) would not have the correct methylation pattern, or it might not be methylated at all, and could therefore be fragmented by the restriction enzyme. Hence, restriction enzymes are now regarded as part of a simple yet effective bacterial defense mechanism to guard against foreign DNA, which can enter bacterial cells with relative ease.

Mechanism of Action

To begin the process of cleaving a DNA molecule, a restriction enzyme must first recognize the appropriate place on the molecule. The recognition site for most restriction enzymes involves a short (usually four to six nucleotides) "palindromic" sequence (a palindrome is a word or phrase that reads the same backward and forward, such as "Otto" or "madam"). In terms of DNA, a palindromic sequence is one that reads the same on each strand of DNA but in opposite directions because of the physical nature of the DNA molecule. EcoRI (derived from the bacterium *Escherichia coli*) is an example of an enzyme that has a recognition site composed of nucleotides arranged in a palindromic sequence:

——GAATTC——
——CTTAAG——

If the top sequence is read from left to right or the bottom sequence is read from right to left, it is always GAATTC.

An additional consideration in the mechanism of restriction enzyme activity is the type of cut that is made. When a restriction enzyme cuts DNA, it is actually breaking the "backbone" of the molecule, consisting of a chain of sugar and phosphate molecules. This breakage occurs at a precise spot on each strand of the double-stranded DNA molecule. The newly created ends of the DNA fragments are then informally referred to as "sticky ends" or "blunt ends." These terms refer to whether single-stranded regions of DNA are generated by the cutting activity of the restriction enzyme. For example, the enzyme EcoRI is a "sticky end" cutter; when the cuts are made at the recognition site, the result is:

—GAATTC— → —G   AATTC—
—CTTAAG—   —CTTAA   G—

The break in the DNA backbone is made just after the G in each strand; this helps weaken the connections between the nucleotides in the middle of the site, and the DNA molecule splits into two fragments. The single-stranded regions, where the bases TTAA are not paired with their complements (AATT) on the other strand, are called "overhangs"; however, the bases in one overhang are still capable of pairing with the bases in the other overhang as they did before the DNA strands were cut. The ends of these fragments will readily stick to each other if brought close together, hence the name "sticky ends."

Enzymes that create blunt ends make a flush cut and do not leave any overhangs, as demonstrated by the cutting site of the enzyme AluI:

——AGCT—— → ——AG   CT——
——TCGA——   ——TC   GA——

Because of the lack of overhanging single-strand regions, these two DNA fragments will not readily rejoin. In practice, either type of restriction enzyme may be used, but enzymes that produce sticky ends are generally favored over blunt-end-cutting enzymes because of the ease with which the resulting fragments can be rejoined.

Impact and Applications

It is no exaggeration to say that the entire field of genetic engineering would have been impossible without the discovery and widespread use of restriction enzymes. On the most basic level, restriction enzymes allow scientists to create recombinant DNA molecules (hybrid molecules containing DNA from different sources, such as humans and bacteria). No matter what the source, DNA molecules can be cut with restriction enzymes to produce fragments that can then be rejoined in new combinations with DNA fragments from other molecules. This technology has led to advances such as the production of human insulin by bacterial cells such as *Escherichia coli*.

The DNA of most organisms is relatively large and complex; it is usually so large, in fact, that it becomes difficult to manipulate and study the DNA of some organisms, such as humans. Restriction enzymes provide a convenient way to specifically cut very large DNA molecules into smaller fragments that can then be used more easily in a variety of molecular genetics procedures, such as cloning, DNA sequencing, and the polymerase chain reaction (PCR).

Another area of genetic engineering that is possible because of restriction enzymes is the production of restriction maps. A restriction map is a diagram of a DNA molecule showing where particular restriction enzymes cut the molecule and the molecular sizes of fragments that are generated. The restriction sites can then be used as markers for further study of the DNA molecule and to help geneticists locate important genetic regions. Use of restriction enzymes has also revealed other interesting and useful markers of the human genome called restriction fragment length polymorphisms (RFLP). RFLP refers to changes in the size of restriction fragments caused by mutations in the recognition site for a particular restriction enzyme. More specifically, the recognition site is mutated so that the restriction enzyme no longer cuts there; the result is one long fragment where, before the mutation, there would have been two shorter fragments. These changes in fragment length can then be used as markers for the region of DNA in question. Because they result from mutations in the DNA sequence, they are inherited from one generation to the next. Thus these mutations have been a valuable tool for molecular biologists trying to produce a map of human DNA

and for those scientists involved in "fingerprinting" individuals by means of their DNA.

—*Randall K. Harris*

See Also: Bacterial Genetics and Bacteria Structure; Cloning; DNA Isolation; Human Genome Project; Shotgun Cloning.

Further Reading: *Recombinant DNA* (1992), by James D. Watson et al., is an excellent resource for the general reader wishing to understand the basics of genetic engineering. Karl Drlica, *Understanding DNA and Gene Cloning: A Guide for the Curious* (1996), also provides basic information about restriction enzymes and their use in cloning. *Genes VI* (1997), by Benjamin Lewin, provides a detailed yet highly readable explanation of restriction and methylation systems in bacteria.

RNA Structure and Function

Field of study: Molecular genetics

Significance: *Ribonucleic acid (RNA), a molecule that plays many roles in the effective usage of genetic information, exists in several forms, each with its own unique function. RNA functions in the process of protein synthesis, during which information from DNA is used to direct the construction of a protein, and possesses enzymatic and regulatory capabilities. Understanding the structure and function of RNA is important to a fundamental knowledge of genetics; in addition, many developing medical therapies will undoubtedly utilize special RNAs to combat genetic diseases.*

Key terms

RIBONUCLEIC ACID (RNA): a single-stranded genetic molecule that consists of ribonucleotide subunits joined together by phosphodiester chemical bonds

MESSENGER RNA (mRNA): the form of RNA that carries genetic instructions, copied from deoxyribonucleic acid (DNA), to the ribosome to be decoded during translation

RIBOSOMAL RNA (rRNA): a type of RNA that forms a major part of the structure of the ribosome

TRANSFER RNA (tRNA): the form of RNA that acts to decode genetic information present in mRNA, carries a particular amino acid, and is vital to translation

RETROVIRUS: a special type of virus that carries its genetic information as RNA

RIBOZYME: an RNA molecule that can function catalytically as an enzyme

TRANSCRIPTION: the synthesis of an RNA molecule directed by RNA polymerase using a DNA template

TRANSLATION: the synthesis of a protein molecule directed by the ribosome using information provided by an mRNA

The Chemical Nature of RNA

Ribonucleic acid (RNA) is a complex biological molecule that is classified along with deoxyribonucleic acid (DNA) as a nucleic acid. Chemically, RNA is a polymer (long chain) consisting of subunits called ribonucleotides linked together by phosphodiester bonds. Each ribonucleotide consists of three parts: the sugar ribose (a five-carbon simple sugar), a negatively charged phosphate group, and a nitrogen-containing base. There are four types of ribonucleotides, and the difference between them lies solely in which of four possible bases they contain. The four bases are adenine (A), guanine (G), cytosine (C), and uracil (U).

The structures of DNA and RNA are very similar, but there are three important differences. The sugar found in the nucleotide subunits of DNA is deoxyribose, which is related to but differs slightly from the ribose found in the ribonucleotides of RNA. In addition, while DNA nucleotides also contain four possible bases, there is no uracil in DNA; instead, DNA nucleotides may contain a different base called thymine (T). Finally, while DNA exists as a double-stranded helix in nature, RNA is almost always single stranded. Like DNA, a single RNA strand has a five-prime (5′) to three-prime (3′) polarity.

The Folding of RNA Molecules

The significance of many types of RNA lies in the order of their nucleotides, which often represents information copied from DNA. This nucleotide order is called the primary structure of the molecule. An important aspect of many biological molecules, however, is the way their primary structures fold to create a three-di-

mensional shape. A single strand of DNA, for example, associates with another strand in a particular way to form the famous double helix, which represents its actual three-dimensional shape in nature. Similarly, protein molecules, especially enzymes, must be folded into a very specific three-dimensional shape if they are to perform their functions; loss of this shape will cause their inactivation.

Since RNA is single stranded, it was recognized shortly after the discovery of some of its major roles that its capacity for folding is great and that this folding might play an important part in the functioning of the molecule. The nucleotides in an RNA molecule can form hydrogen-bonded base pairs, according to the same rules that govern DNA base pairing. Cytosine binds to guanine, and uracil binds to adenine. What this means is that in a particular single-stranded RNA molecule, complementary portions of the molecule are able to fold back and form base pairs with each other. These are often local interactions, and a common structural element that is formed is called a "hairpin loop" or "stem loop." A hairpin loop is formed when two complementary regions are separated by a short stretch of bases so that when they fold back and pair, some bases are left unpaired, forming the loop. The net sum of these local interactions is referred to as the RNA's secondary structure and is usually important to an understanding of how the RNA works. All transfer RNAs (tRNAs), for example, are folded into a secondary structure that contains three stem loops and a fourth stem arranged onto a "cloverleaf" shape.

Finally, local structural elements may interact with other elements in long-range interactions, causing more complicated folding of the molecule in space. The full three-dimensional structure of a tRNA molecule from yeast was finally confirmed in 1978 by several groups independently, using the method of X-ray diffraction. In this process, crystals of a molecule are bombarded with X rays, which causes them to scatter; an expert can tell by the pattern of scattering how the different atoms in the molecule are oriented in space. The cloverleaf arrangement of a tRNA undergoes further folding so that the entire molecule takes on a roughly *L*-shaped appearance. An understanding of the three-dimensional shape of an RNA molecule is crucial to understanding its function. By the late 1990's, the three-dimensional structures of many tRNAs had been worked out, but it had proven difficult to do X-ray diffraction analyses on most other RNAs because of technical problems. More advanced computer programs and alternate structure-determining techniques are enabling research in this field to proceed.

Synthesis and Stability of RNA

RNA molecules of all types are continually being synthesized and degraded in a cell; even the longest-lasting ones exist for only a day or two. How is the synthesis of RNA accomplished? Shortly after the structure of DNA was established, it became clear that RNA was synthesized using a DNA molecule as a template, and the mechanism was worked out shortly thereafter. The entire process by which an RNA molecule is constructed using the information contained in DNA is called "transcription." An enzyme called RNA polymerase is responsible for assembling the ribonucleotides of a new RNA chain according to information provided by a specific DNA segment (gene). Only one strand of the relevant DNA is used as a template, and the ribonucleotides are initially arranged according to the base-pairing rules. A DNA sequence called the "promoter" is a site that attracts RNA polymerase and allows the process of RNA synthesis to begin. At the appropriate starting site, RNA polymerase begins to assemble and connect the nucleotides according to the DNA instructions. This process continues until another sequence, called a "terminator," is reached. At this point, the RNA polymerase stops transcription, and a new RNA molecule is released.

Much attention is rightfully focused on transcription, since it controls the rate of synthesis of any particular RNA. It has become increasingly clear, however, that the amount of RNA in the cell at a given time is also strongly dependent on the RNA's stability (the rate at which it is degraded). Every cell contains several enzymes called ribonucleases (RNases) whose job it is to cut up RNA molecules into their nucleo-tide subunits, rendering them unable to function. Some RNAs last only 30 seconds, while some may last up to a day or two. The process by which the RNases are signaled to begin degrading a particular RNA is being studied, but the details are somewhat unclear. Certain secondary structural elements have been identified that seem to contribute to the process. What is important to remember is that both the rates of synthesis (transcription) and degradation ultimately determine the amount of functional RNA in a cell at any given time.

Three Classes of RNA

While all RNAs are produced by transcription, several classes of RNA are created, and each has a particular function. By the late 1960's, three major classes of RNAs had been identified, and their respective roles in the process of protein synthesis had been elucidated. In general, protein synthesis refers to the assembly of a protein using information encoded in DNA, with RNA acting as an intermediary to carry information and assist in protein building. In 1956, Francis Crick, one of the scientists who had discovered the double-helical structure of DNA, referred to this information flow as the "central dogma," a term that continues to be used.

Messenger RNA (mRNA) is the molecule that carries a copy of the DNA instructions for building a particular protein. It usually represents the information provided by a single gene and carries this information to the ribosome, the site of protein synthesis. This information must be decoded so that it will specify the order of amino acids in a protein. Nucleotides are read in groups of three (codons). In addition to the information required to order amino acids, the mRNA contains signals that tell the protein-building machinery where to start and stop reading the genetic information.

Ribosomal RNA (rRNA) exists in three distinct sizes and is part of the structure of the ribosome. The three ribosomal RNAs interact with many proteins to complete the ribosome, the organelle (specific part of the cell) that directs the events of protein synthesis. One of the functions of the rRNA is to interact with mRNA at a particular location and orient it properly so that reading of its genetic code can begin at the correct location. Another rRNA acts to facilitate the transfer of the growing polypeptide chain from one tRNA to another (peptidyl transferase activity).

Transfer RNA (tRNA) serves the vital role of decoding the genetic information. There are at least twenty and usually fifty to sixty different tRNAs in a given cell. On one side, they contain an "anticodon" loop, which can base-pair to the mRNA codon according to its sequence and the base-pairing rules. On the other side, they contain an amino acid binding site, to which is attached the appropriate amino acid for its anticodon. In this way, tRNA allows the recognition of any particular mRNA codon and matches it up with the appropriate amino acid. The process continues until an entire new protein molecule has been constructed.

How does a particular tRNA know to which specific amino acid it should attach? This important aspect of protein synthesis is facilitated by a group of enzymes called "tRNA amino acyl synthetases." These enzymes recognize any tRNA as a particular type and allow the attachment of the correct amino acid to its amino acid binding site. The integrity of this process is crucial to translation; if only one tRNA is attached to an incorrect amino acid, the proteins being built will likely be nonfunctional.

Split Genes and mRNA Processing in Higher Organisms

In bacterial genes, there is a colinearity between the segment of a DNA molecule that is transcribed and the resulting mRNA. In other words, the mRNA sequence is complementary to its template and is the same length, as would be expected. In the late 1970's, several groups of scientists made a seemingly bizarre discovery regarding mRNAs in eukaryotic organisms (organisms whose cells contain a nucleus, including all living things that are not bacteria): The sequences of mRNAs isolated from eukaryotes were not collinear with the DNA from which they were transcribed. The coding regions of the corresponding DNA were interrupted by seemingly random sequences that served no immediately obvious function. These "introns," as they came to be known, were apparently transcribed along with the coding regions (exons) but were somehow removed before the mRNA was translated. This completely unexpected observation led to further investigations that revealed that mRNA is extensively processed, or modified, after its transcription in eukaryotes.

After a eukaryotic mRNA is transcribed, it contains all of the intervening sequences and is referred to as immature, or a "pre-mRNA." Before it can become mature and functional, three major processing events must occur: splicing and the addition of a "cap" and "tail." The process of splicing is a complex one that occurs in the nucleus with the aid of the "spliceosome," a large complex of RNAs and proteins that identify intervening sequences and cut them out of the pre-mRNA. In addition, the spliceosome must rejoin the sequences from which the intron-encoded nucleotides were removed so that a complete, functional mRNA results. The process must be extremely specific, since a mistake that caused the removal of only one extra nucleotide could change the protein product of translation so radically that it might fail to function. While splicing is occurring, two other vital events are being performed to make the immature mRNA ready for action. A so-called cap, which consists of a modified G nucleotide, is added to the beginning (5' end) of the pre-mRNA by an unconventional linkage. The cap appears to function by interacting with the ribosome, helping to orient the mature mRNA so that translation begins at the proper location. A tail, which consists of many A nucleotides (often two hundred or more), is also attached to the 3' end of the pre-mRNA. This so-called poly-A tail, which virtually all eukaryotic mRNAs contain, seems to be involved in determining the relative stability of an mRNA. These important steps must be performed after transcription in eukaryotes to enable the creation of a mature, functional messenger RNA molecule that is now ready to leave the nucleus and be translated.

Other Important Classes of RNA and Specialized Functions

The traditional roles of RNA in protein synthesis were originally considered the only roles

Thomas Cech, who shared the 1989 Nobel Prize in Chemistry with his colleague Sidney Altman for their discoveries relating to RNA. (The Nobel Foundation)

RNA was capable of performing. RNA in general, while considered an important molecule, was thought of as a "helper" in translation. This all began to change in 1982, when the molecular biologists Thomas Cech and Sidney Altman, working independently and with different systems, reported the existence of RNA molecules that had catalytic activity. This means that RNA molecules can function as enzymes; until this time, it was believed that all enzymes were protein molecules. The importance of these findings cannot be overstated, and Cech and Altman ultimately shared the 1989 Nobel Prize in Chemistry for the discovery of these RNA enzymes, or "ribozymes." Both of these initial ribozymes catalyzed reactions that involved the cleavage of other RNA molecules—that is, they acted as nucleases. Subsequently, many ribozymes have been found in various organisms, from bacteria to humans. Some of them are able to catalyze different types of reactions, and there are new ones reported every year. Thus ribozymes are not a mere curiosity but play an integral role in the molecular machinery of many organisms. Their discovery also gave rise to the idea that at one point in evolutionary history, molecular systems composed solely of RNA performing many roles existed in an "RNA world."

At around the same time as these momentous discoveries, still other classes of RNAs were being discovered, each with its own specialized functions. In 1981, Jun-Ichi Tomizawa discovered RNAI, the first example of what would become another major class of RNAs, the "antisense RNAs." The RNAs in this group are complementary to a target molecule (usually an mRNA) and exert their function by binding to that target via complementary base pairing. These antisense RNAs usually play a regulatory role, often acting to prevent translation of the relevant mRNA to modulate the expression of the protein for which it codes. Most of these antisense RNAs are encoded by the same gene as their target, but a group called the "transencoded antisense RNAs" actually have their own genes, which are separate and distinct from their target molecule's gene. This is especially significant because the complementarity between antisense RNA and target is often not perfect, resulting in interesting interactions with unique structural features. The prototype of this class of RNAs, micF RNA, was discovered in 1983 by Masayori Inouye and subsequently characterized by Nicholas Delihas. An understanding of the binding of this special type of antisense RNA to its target will provide insights into RNA-RNA interactions that may be vital for use in genetic therapy.

Another major class of RNAs, the small nuclear RNAs (snRNAs), was also discovered in the early 1980's. Molecular biologist Joan Steitz was working on the autoimmune disease systemic lupus when she began to characterize the snRNAs. There are six different snRNAs, now called U1-U6 RNAs. These RNAs exist in the nucleus of eukaryotic cells and play a vital role in mRNA splicing. They associate with proteins in the spliceosome, forming so-called ribonucleoprotein complexes (snRNPs), and play a prominent role in detecting proper splice sites and directing the protein enzymes to cut and paste at the proper locations.

It has been known since the late 1950's that many viruses contain RNA, and not DNA, as their genetic material. This is another fascinating role for RNA in the world of biology. The viruses that cause influenza, polio, and a host of other diseases are RNA viruses. Of particular note are a class of RNA viruses known as "retroviruses" because they have a particularly interesting life cycle. These retroviruses, which include human immunodeficiency virus (HIV), the virus that causes acquired immunodeficiency syndrome (AIDS) in humans, use a special enzyme called "reverse transcriptase" to make a DNA copy of their RNA instructions when they enter a cell. That DNA copy is inserted into the DNA of the host cell, where it is referred to as a "provirus," and never leaves. Clearly, understanding the structures and functions of the RNAs associated with these viruses will be important in attempting to create effective treatments for the diseases associated with them.

An additional role of RNA was noted during the elucidation of the mechanism of DNA replication. It was found that a small piece of RNA, called a "primer," must be laid down by the enzyme primase before DNA polymerase adds DNA nucleotides to this initial RNA sequence, which is subsequently removed. Also, it is worth mentioning that the universal energy-storing molecule of all cells, adenosine triphosphate (ATP), is in fact a version of the RNA nucleotide containing adenine (A).

Impact and Applications

The discovery of the many functions of RNA, especially its catalytic ability, has radically changed the understanding of the functioning of genetic and biological systems and has revolutionized the views of the scientific community regarding the origin of life. The key to understanding how RNA can perform all of its diverse functions lies in elucidating its many structures, since structure and function have an inseparable relationship. Much progress has been made in establishing the structures of hundreds of RNA molecules; several methods, including advanced computer programs, are making it easier to predict and analyze RNA structure. Three-dimensional modeling is much more difficult, and while the three-dimensional structures of several RNAs have been worked out, much work remains to be done in this realm.

In terms of basic research and genetic engineering, the discovery of antisense RNAs and ribozymes has facilitated many procedures, providing insight at the molecular level of genetic processes that would have been difficult to obtain without this knowledge and the tools it has made available. Additionally, plants, bacteria, and animals have been genetically engineered to alter the expression of some of their genes, in many cases making use of the new RNA technology. An example is the genetically engineered tomato, which does not ripen until it is treated when it reaches its point of sale. This tomato was created by inserting a gene that produces an antisense RNA; when it is expressed, it inactivates the mRNA that codes for the enzyme that allows production of the substance that naturally causes ripening to occur.

The prospects for human genetic therapy have also been dramatically increased by the recognition of the many capabilities of RNA and the realization of how to manipulate and utilize it. Success has been limited, but the usage of retroviruses to introduce ribozymes, antisense RNAs, or a combination of both into genetically defective cells offers great promise for the future in fighting a wide variety of genetic diseases, from AIDS and cancer to cystic fibrosis and sickle-cell anemia. One thing is clear: RNA will play an important role in increasing the understanding of genetics and in the revolution of gene therapy. RNA is one of the most structurally interesting and functionally diverse of all the biological molecules.

—Matthew M. Schmidt

See Also: Central Dogma of Molecular Biology; DNA Structure and Function; Gene Therapy; Genetic Code; Molecular Genetics; Protein Structure; Protein Synthesis; RNA World.

Further Reading: *RNA Structure and Function* (1998), edited by Robert W. Simons and Marianne Grunberg-Manago, is an advanced text that takes a detailed look at the various structures of RNA, their relationships to function, and the techniques for determining RNA structure. *Catalytic RNA* (1996), edited by Fritz Eckstein and David M. J. Lilly, offers a comprehensive overview of ribozyme diversity and function. *Antisense RNA and DNA* (1992), edited by James A. H. Murray, and *Gene Regulation: Biology of Antisense RNA and DNA* (1992), edited by Robert P. Erickson and Jonathan G. Izant, both provide a comprehensive overview of natural antisense RNA function and prospects for its uses in gene therapy. *Molecular Biology of the Gene* (1988), by James D. Watson et al., provides an insightful discussion of the various structures of RNA and their relationships to function.

RNA World

Field of study: Molecular genetics
Significance: *The "RNA World" is a theoretical time in the early evolution of life, during which RNA molecules played important genetic and enzymatic roles that were later taken over by molecules of DNA and proteins. Ideas about RNA's ancient functions have led to new concepts of the origin of*

life and have important implications in the use of gene therapy to treat diseases.

Key terms

DEOXYRIBONUCLEIC ACID (DNA): the genetic material for most organisms, a double-stranded molecule composed of units called "nucleotides"

RIBONUCLEIC ACID (RNA): a single-stranded genetic molecule that consists of ribonucleotides

ENZYME: a molecule, usually a protein, that facilitates chemical reactions without itself being altered by its participation

PROTEIN: a biological molecule that consists of amino acid subunits joined together by peptide chemical bonds

RIBOZYME: an RNA molecule that can function catalytically as an enzyme

The Central Dogma and the Modern Genetic World

Soon after the discovery of the double-helical structure of deoxyribonucleic acid (DNA) in 1953 by James Watson and Francis Crick, Crick proposed an idea regarding information flow in cells that he called the "central dogma of molecular biology." Crick correctly predicted that in all cells, information flows from DNA to ribonucleic acid (RNA) to protein. DNA was known to be the genetic material, the "library" of genetic information, and it had been clear for some time that the enzymes that actually did the work of facilitating chemical reactions were invariably protein molecules. The discovery of three classes of RNA during the 1960's seemed to provide the link between the DNA instructions and the protein products.

In the modern genetic world, cells contain three classes of RNA that act as helpers in the synthesis of proteins from information stored in DNA, a process called "translation." A messenger RNA (mRNA) is "transcribed" from a segment of DNA (a gene) that contains information about how to build a particular protein and carries that information to the cellular site of protein synthesis, the ribosome. Ribosomal RNAs (rRNAs) interacting with many proteins make up the ribosome, whose major job is to coordinate and facilitate the protein-building procedure. Transfer RNAs (tRNAs) act as decoding molecules, reading the mRNA information and correlating it with a specific amino acid. As the ribosome integrates the functions of all three types of RNA, proteins are built one amino acid at a time. These proteins can then function as enzymes, ultimately determining the capabilities and properties of the cell in which they act.

While universally accepted, the central dogma led many scientists to question how this complex, integrated system came about. It seemed to be a classic "chicken and egg" dilemma: Proteins could not be built without instructions from DNA, but DNA could not replicate and maintain itself without help from protein enzymes. The two seemed mutually dependent upon each other in an inextricable way. An understanding of the origins of the modern genetic system seemed far away.

The Discovery of Ribozymes

In 1983, a discovery was made that seemed so radical it was initially rejected by most of the scientific community. Molecular biologists Thomas Cech and Sidney Altman, working independently and in different systems, announced the discovery of RNA molecules that possessed catalytic activity. This meant that RNA itself can function as an enzyme, obliterating the idea that only proteins could function catalytically.

Cech had been working with the protozoan *Tetrahymena*. In most organisms except bacteria, the coding portions of DNA genes (exons) are interrupted by noncoding sequences (introns), which are transcribed into mRNA but which must be removed before translation. Protein enzymes called "nucleases" are usually responsible for cutting out the introns and joining together the exons in a process called "splicing." The molecule Cech was working with was a ribosomal RNA that contained introns but could apparently remove them and rejoin the coding regions without any help. It was a self-splicing RNA molecule, which clearly indicated its enzymatic capability. Altman was working with the enzyme RNase P in bacteria, which is responsible for cutting mature tRNA molecules out of an immature

Sidney Altman, who shared a 1989 Nobel Prize with colleague Thomas Cech for their discovery of ribozymes. (The Nobel Foundation)

RNA segment. RNase P thus also acts as a nuclease. It was known for some time that RNase P contains both a protein and an RNA constituent, but Altman was ultimately able to show that it was the RNA rather than the protein that actually catalyzed the reaction.

The importance of these findings cannot be overstated, and Cech and Altman ultimately shared the 1989 Nobel Prize in Chemistry for the discovery of these RNA enzymes, or "ribozymes." Subsequently, many ribozymes have been found in various organisms, from bacteria to humans. Some of them are able to catalyze different types of reactions, and new ones are constantly being reported. Ribozymes have thus proven to be more than a mere curiosity, playing an integral role in the molecular machinery of many organisms.

At around the same time as these important discoveries, still other functions of RNA were being identified. While perhaps not as dramatic as the ribozymes, antisense RNAs, small nuclear RNAs, and a variety of others further proved the versatility of RNA. While understanding the roles of ribozymes and other unconventional RNAs is important to the understanding of genetic functioning in present-day organisms, these discoveries were more intriguing to many scientists interested in the origin and evolution of life. In a sense, the existence of ribozymes was a violation of the central dogma, which implied that information was ultimately utilized solely in the form of proteins. While the central dogma was not in danger of becoming obsolete, a clue had been found that might possibly allow a resolution, at least in theory, to questions about whether the DNA or the protein came first. The exciting answer: perhaps neither.

The RNA World Theory and the Origin of Life

Given that RNA is able to store genetic information (as it certainly does when it functions as mRNA) and the new discovery that it could function as an enzyme, there was no longer any need to invoke the presence of either DNA or protein as necessities in the first living system. The first living molecule would have to be able to replicate itself without any help, and just

such an "RNA replicase" has been proposed as the molecule that eventually led to life as it is now known. Like the self-splicing intron of *Tetrahymena*, this theoretical ribozyme could have worked on itself, catalyzing its own replication. This RNA would therefore have functioned as both the genetic material and the replication enzyme, allowing it to make copies of itself without the need for DNA or proteins. Biologist Walter Gilbert coined the term "RNA World" for this interesting theoretical period dominated by RNA. Modern catalytic RNAs can be thought of as molecular fossils that remain from this period and provide clues about its nature.

How might this initial RNA have come into being in the first place? Biologist Alexander Oparin predicted in the late 1930's that if simple gases thought to be present in the earth's early atmosphere were subjected to the right conditions (energy in the form of lightning, for example), more complex organic molecules would be formed. His theory was first tested in 1953 and was resoundingly confirmed: a mixture of methane, ammonia, water vapor, and hydrogen gas was energized with high-voltage electricity, and the products were impressive: several amino acids and aldehydes, among other organic molecules. Subsequent experiments have been able to produce ribonucleotide bases. It seems reasonable, then, that nucleotides could have been present on the early earth and that their random linkage could lead to the formation of an RNA chain.

After awhile, RNA molecules would have found a way to synthesize proteins, which are able to act as more efficient and diverse enzymes than ribozymes by their very nature. Why are proteins better enzymes than ribozymes? Since RNA contains only four bases that are fundamentally similar in their chemical properties, the range of different configurations and functional capabilities is somewhat limited as opposed to proteins. Proteins are constructed of twenty different amino acids whose functional groups differ widely in terms of their chemical makeup and potential reactivity. It is logical to suppose, therefore, that proteins eventually took over most of the roles of RNA enzymes because they were simply better suited to doing so. Several of the original or efficient ribozymes would have been retained, and those are the ones that can be observed today.

How could a world composed strictly of RNAs, however, be able to begin protein synthesis? While it seems like a tall order, scientists have envisioned an early version of the ribosome that was composed exclusively of RNA. Biologist Harry Noller reported in the early 1990's that the activity of the modern ribosome that is responsible for catalyzing the formation of peptide bonds between amino acids is in fact carried out by the ribosomal RNA. This so-called peptidyl-transferase activity had always been attributed to one of the ribosomal proteins, and ribosomal RNA had been envisioned as playing a primarily structural role. Noller's discovery that the large ribosomal RNA is actually a ribozyme allows scientists to picture a ribosome working in roughly the same way that modern ones do, without containing any proteins. As proteins began to be synthesized from the information in the template RNAs, they slowly began to assume some of the RNA roles and probably incorporated themselves into the ribosome to allow it to function more efficiently.

The transition to the modern world would not be complete without the introduction of DNA as the major form of the genetic material. RNA, while well suited to diverse roles, is actually a much less suitable genetic material than DNA for a complex organism (even one only as complex as a bacterium). The reason for this is that the slight chemical differences between the sugars contained in the nucleotides of RNA and DNA cause the RNA to be more reactive and much less chemically stable; this is good for a ribozyme but clearly bad if the genetic material is to last for any reasonable amount of time. Once DNA initially came into existence, therefore, it is likely that the relatively complex organisms of the time quickly adopted it as their genetic material; shortly thereafter, it became double stranded, which facilitated its replication immensely. This left RNA, the originator of it all, relegated to the status it enjoys today; molecular fossils exist that uncover its former glory, but it functions mainly as a helper in protein synthesis.

This still leaves the question of how DNA evolved from RNA. At least two protein enzymes were probably necessary to allow this process to begin. The first, ribonucleoside diphosphate reductase, converts RNA nucleotides to DNA nucleotides by reducing the hydroxyl group located on the 2′ carbon of ribose. Perhaps more important, the enzyme reverse transcriptase would have been necessary to transcribe RNA genomes into corresponding DNA versions. Examples of both of these enzymes exist in the modern world.

Some concluding observations are in order to summarize the evidence that RNA and not DNA was very likely the first living molecule. No enzymatic activity has ever been attributed to DNA; in fact, the 2′ hydroxyl group that RNA possesses and DNA lacks is vital to RNA's ability to function as a ribozyme. Furthermore, ribose is synthesized much more easily than deoxyribose under laboratory conditions. All modern cells synthesize DNA nucleotides from RNA precursors, and many other players in the cellular machinery are RNA-related. Important examples include adenosine triphosphate (ATP), the universal cellular energy carrier, and a host of coenzymes such as nicotinamide adenine dinucleotide (NAD), derived from B vitamins and vital in energy metabolism.

Impact and Applications

The discovery of ribozymes and the other interesting classes of RNAs has dramatically altered the understanding of genetic processes at the molecular level and has provided compelling evidence in support of exciting new theories regarding the origin of life and cellular evolution. The RNA World theory, first advanced as a radical and unsupported hypothesis in the early 1970's, has gained almost universal acceptance by scientists. It is the solution to the evolutionary paradox that has plagued scientists since the discovery and understanding of the central dogma: Which came first, DNA or proteins? Since they are inextricably dependent upon each other in the modern world, the idea of the RNA World proposes that rather than one giving rise to the other, they are both descended from RNA, that most ancient of genetic and catalytic molecules.

Apart from this exciting theoretical break-through, the discoveries that led to its inception are beginning to have a more practical impact in the fields of industrial genetic engineering and medical gene therapy. The unique ability of ribozymes to find particular sequences and initiate cutting and pasting at desired locations makes them powerful tools. Impressive uses have already been found for these tools in theoretical molecular biology and in the genetic engineering of plants and bacteria. Most important to humans, however, are the implications for curing or treating genetically related disease using this powerful new RNA-based technology.

Gene therapy, in general, is based on the idea that any faulty, disease-causing gene can theoretically be replaced by a genetically engineered working replacement. While theoretically a somewhat simple idea, in practice it is technically very challenging. Retroviruses may be used to insert DNA into particular target cells, but the results are often not as expected; the new genes are difficult to control or may have adverse side effects. Molecular biologist Bruce Sullenger has pioneered a new approach to gene therapy, which seeks to correct the genetic defect at the RNA level. A ribozyme can be engineered to seek out and replace damaged sequences before they are translated into defective proteins. Sullenger has shown that this so-called transsplicing technique can work in nonhuman systems and, in 1996, began trials to test his procedure in humans.

Many human diseases could potentially be corrected using gene therapy technology of this kind, from inherited defects such as sickle-cell anemia to degenerative genetic problems such as cancer. Even pathogen-induced conditions such as acquired immunodeficiency syndrome (AIDS), caused by the human immunodeficiency virus (HIV), could be amenable to this approach. It is ironic and gratifying that an understanding of the ancient RNA World holds promise for helping scientists to solve some of the major problems in the modern world of DNA-based life.

—Matthew M. Schmidt

See Also: DNA Replication; Gene Therapy; Protein Synthesis; RNA Structure and Function; Sickle-Cell Anemia.

Further Reading: *The RNA World* (1993), edited by Raymond F. Gesteland and John F. Atkins, takes an advanced, detailed look at the theories behind the RNA World, the evidence for its existence, and the modern "fossils" that may be left from this historic biological period. A discussion of the use of ribozymes in gene therapy is presented in Stephen Hart, "RNA's Revising Machinery," *Bioscience* 46 (May, 1996). "The Beginnings of Life on Earth," by Christian de Duve, *American Scientist* 83 (September-October, 1995), discusses several scenarios regarding exactly how RNA may have been involved in life's origins. John Horgan, "The World According to RNA," *Scientific American* 189 (January, 1996), summarizes the accumulated evidence that RNA molecules once served both as genetic and catalytic agents. *Molecular Biology of the Gene* (1988), by James D. Watson et al., provides an extremely comprehensive discussion of all aspects of the RNA World.

Sheep Cloning

Field of study: Population genetics

Significance: *The cloning of the sheep Dolly, the first cloning of a mammal from adult DNA, caused a worldwide sensation. Critics worried that the achievement carried with it the potential to reshape the genetic makeup of entire species, including, perhaps, the human race.*

Key terms

DEOXYRIBONUCLEIC ACID (DNA): the genetic material found in the cells of most organisms, a double-stranded substance composed of subunits called "nucleotides"

CHROMOSOME: a DNA molecule containing genetic information, or genes, in each cell

PROTEIN: a macromolecule made of amino acids used by cells for structure and function

GENETIC RECOMBINATION: the swapping of genes between a pair of chromosomes

The Birth of Dolly

In February of 1997, Scottish scientists announced that they had succeeded in creating a cloned copy of an adult mammal. The clone, a sheep, named Dolly, was unique to the animal kingdom, since hers was the first successful cloning of a mammal using adult deoxyribonucleic acid (DNA). All life on earth is possible because of the simple DNA molecule, which contains a code that directs protein synthesis in cells. The code is written using four nucleotides, or bases, called adenine (A), guanine (G), cytosine (C), and thymine (T). Genes, or discrete units of information contained within DNA (or chromosomes), are like words that each have a different meaning to a cell. It takes approximately 100,000 genes to make the 50,000 or so proteins that human cells need to function.

In sexually reproducing organisms, such as mammals and humans, each parent contributes one-half of their chromosomes to their offspring. During the formation of gametes (sperm and egg), a reshuffling of genes occurs via genetic recombination so that each offspring of the same two parents will get a different set of genes from each parent. This is why full brothers and sisters do not look exactly like each other or either parent. A clone is an organism or group of organisms with the exact same set of genes. Clones of livestock have been produced since the 1980's, but these clones were made by separating and culturing the cells of an eight-celled embryo. These clones were genetically identical to each other, like identical twins, but no more related to either parent than normal. A global science project related to cloning called the Human Genome Project was started in 1990 with the goal of identifying all 100,000 or so genes required to build a human and mapping them to specific sites on specific chromosomes.

The Cloning Process

Geneticists had traditionally agreed that cloning of an adult organism was not feasible because of the way DNA works in controlling development as compared to controlling a differentiated (adult) cell. Essentially, specific groups of genes control embryonic development. Differentiated cells, or cells that have become a functional part of tissue and organs, are under the direction of different sets of genes; the developmental genes have been "turned off" in these cells. It was believed that once these developmental genes were turned off, they were not likely ever to be turned on again by a cell. The sheep cloning experiment disproved this hypothesis.

To clone Dolly, the Scottish scientists collected tissue from an adult ewe udder and cultured it. An embryo from another ewe was treated to remove the nucleus, thus removing the DNA of that cell. The enucleated embryo was then fused with the adult cells. After trials with approximately 280 embryos, the fusion was finally successful and the new embryo, now containing the DNA from an adult, was implanted into a surrogate ewe. Dolly was born and became the first mammal known to be created from the DNA of an adult and thus the first clone of an adult mammal. Humans will most definitely benefit from such animal cloning. Farmers and pharmaceutical companies will be able to make identical copies of adult

Dolly, the world's first adult mammal clone. (Reuters/Jeff Mitchell/Archive Photos)

animals with desired characteristics such as high milk yield and the ability to produce insulin.

Much of the impact of adult sheep cloning relates to the ethical dilemmas raised by the potential for cloning humans. Some scientists have suggested that human clones could be used to assist parents that have lost a young child to re-create a genetically identical child, provide couples that are unable to produce their own offspring with new options, or grow replacement organs in vitro. A subtle but valid question about human cloning is just how similar a human clone will be to the adult donor from which it originates. Research on brain development during the 1980's and 1990's indicated that most of the development and "wiring" of the human cerebrum (80 percent of the brain, consisting of neuronal networks) occurs after birth and is directly related to environmental input. Therefore, an essential fact concerning human behavior that is tied directly to the cloning issue is that early environment plays a major role in what each person becomes. If a clone is made from adult human DNA, the environmental input (stimuli and conditions) to which it is exposed after birth could never be identical to the conditions to which the adult donor had been exposed; in theory, therefore, the result would be a completely new human who would be physically identical to but behaviorally different from the donor human.

The Human Genome Project and human cloning may also lead toward a level of eugenics that could never have been dreamed of before. Eugenics, popular in the early twentieth century and carried to terrible extremes in Nazi Germany, refers to attempts to assume a direct, underlying genetic cause for a person's appearance and behavior. Will a market develop so that individuals of proven talent and skills can sell their DNA for such uses? Will single women begin having children without paternal input? Will employers and health insurers "scan" applicants' genes for potential genetic problems and deny employment or raise insurance rates? Will wealthy people clone themselves and prevent the poorer or less desirable segments of society from being cloned? Such questions serve as reminders that novels such as English author Aldous Huxley's *Brave New World* (1932), in which citizens are programmed for different levels of work in society, may be more than mere science fiction.

Impact and Applications

The potential of cloning will be determined by how humankind utilizes the technology. Hypothetical scenarios such as those previously mentioned will most likely be sorted out by the free market, capitalism, and community ethical standards. Thus, leaving specifics to time, one may reflect on the potential cloning of humans in a larger context. As members of the animal kingdom, humans are most likely influenced by the same forces that control speciation in other living organisms, with the difference that humans have a much more profound effect on the environment. In the face of overpopulation, pollution, and resource-depletion problems, many humans have begun to realize that they may not be immune to forces such as natural selection that eliminate other species on a daily basis. It is possible that the Human Genome Project and mammalian cloning are technologies that will provide new avenues by which humans may continue to flourish in the face of these environmental pressures. On the other hand, these same technologies could lead to a substantial decrease in the genetic diversity of *Homo sapiens* (the human species) and, along with resource-depletion problems, leave nature with no choice but to add them to the millions of species that have come and gone throughout Earth's history.

—*W. W. Gearheart*

See Also: Cloning; Developmental Genetics; Eugenics; Genetic Code; Natural Selection.

Further Reading: A good review of sheep cloning is given in Christopher Wills, "A Sheep in Sheep's Clothing: Genetic Clonic Not Necessarily Unethical," *Discover* 18 (January, 1998). Fred Ross et al., *Diversity of Life* (1996), presents an understandable overview of evolution, basic genetics, and biotechnology. Research and other aspects of sheep cloning are covered in "The Start of Something Big? Dolly Has Become a New Icon for Science," *Scientific American* 276 (May, 1997).

Shotgun Cloning

Field of study: Genetic engineering and biotechnology

Significance: *Shotgun cloning is the random insertion of a large number of different DNA fragments into a cloning vector. A large number of different recombinant DNA molecules are generated, which are then introduced into a host, often a bacteria, and amplified. Because a large number of different recombinant DNAs are generated, there is a high likelihood of obtaining a clone of the specific DNA of interest.*

Key terms

DEOXYRIBONUCLEIC ACID (DNA): the genetic material for most organisms; a long-chain macromolecule composed of units called "nucleotides"

CLONING VECTOR: a plasmid or virus into which foreign DNA can be inserted to amplify the number of copies of the foreign DNA

MARKER: a gene that encodes an easily detected product that is used to indicate that foreign DNA is in an organism

RESTRICTION ENDONUCLEASE: a protein (an enzyme) that recognizes a specific nucleotide sequence in a piece of DNA and causes cleavage of the DNA

RECOMBINANT DNA: a novel DNA molecule formed by the joining of DNAs from different sources

Recombinant DNA Cloning and Shotgun Cloning

Before the development of recombinant deoxyribonucleic acid (DNA) cloning, it was very difficult to study DNA sequences. Cloning a DNA fragment allows a researcher to obtain large amounts of that specific DNA sequence to analyze without interference from the presence of other DNA sequences. There are many uses for a cloned DNA fragment. For example, a DNA fragment can be sequenced to determine the order of the nucleotides that compose it. This information is used to determine where a gene is located within the DNA and what the likely amino acid sequence of the protein coded by that gene is. Cloned pieces of DNA are also useful in DNA hybridization studies. Because DNA is made of two strands that are complementary to each other, a cloned piece of DNA can be used in hybridizations to determine if copies of that DNA are present in a sample. The cloned DNA can also be used to produce the protein product coded by that DNA, which may have practical uses.

Shotgun cloning of DNA begins with the isolation of DNA from the organism to be cloned, followed by the isolation of the cloning vector DNA. In separate test tubes, the DNA to be cloned and the cloning vector DNA are cut with an appropriate restriction endonuclease and then mixed together. Many restriction endonucleases create single-stranded ends that are complementary, so the end of any DNA molecule cut with that endonuclease can join to the end of any other DNA cut with the same endonuclease. The DNA molecules join randomly and are then sealed using DNA ligase, which seals breaks in DNA molecules. This creates recombinant DNA molecules—different DNA fragments in the cloning vector DNA molecules. The recombinant DNA molecules are then introduced into a host cell where the cloning vector can replicate. The presence of the cloning vector in host cells is determined by selecting for a marker gene in the cloning vector. It is also possible to select or screen for the presence of foreign DNA in the cloning vector. For example, the restriction endonuclease site in the vector used to clone foreign DNA may be in a gene that codes for a second antibiotic resistance. If foreign DNA has been inserted into that site, that antibiotic resistance gene will not function, and the host cell will be sensitive to that antibiotic. A host cell containing a cloning vector with a piece of foreign DNA in it will not be able to grow on a medium containing the second antibiotic. In the final step of shotgun cloning, the inserted DNA is analyzed. There are a variety of ways to select or screen for DNA of interest.

Alternatives to Shotgun Cloning

In shotgun cloning, many different DNA fragments from an organism are cloned, and then the specific DNA clone of interest is identified. It is also possible to select the DNA to be cloned first. If the DNA of interest is known to be in a restriction endonuclease fragment of a certain size, the DNA can be size-selected before cloning. This can be done by using gel electrophoresis, in which an electric current carries DNA fragments through the pores or openings of an agarose gel. DNA migrates through the gel based on the size of the DNA fragment, with the smaller fragments traveling through the pores of the gel more rapidly than the larger fragments. DNA of a particular size can be isolated from the gel and then used for cloning. Finally, to clone a piece of DNA known to code for a protein, scientists can use an enzyme called reverse transcriptase to make a DNA copy (called a complementary DNA or cDNA) of a messenger RNA. The cDNA is then cloned.

—*Susan J. Karcher*

See Also: Biotechnology; Cloning; Cloning Vectors; Genetic Engineering; Recombinant DNA.

Further Reading: James D. Watson et al., *Recombinant DNA* (1992), provides a description of shotgun cloning as well as an overview of many other methods of cloning. K. J. Denniston and L. W. Enquist, *Recombinant DNA* (1981), is a collection of reprints of many of the scientific papers describing the advancements that were needed to achieve cloning of DNA and also includes comments about the contributions these papers made. Initial concerns about the safety of recombinant DNA cloning are summarized in the following sources: James D. Watson and J. Tooze, *The DNA Story: A Sourcebook for Documents on the Recombinant DNA Debate* (1981); R. A. Zilinskas and B. K. Zimmerman, *The Gene-Splicing Wars: Reflections on the Recombinant DNA Controversy* (1986); and Clifford Grobstein, "The Recombinant DNA Debate," *Scientific American* 237 (July, 1977).

Sickle-Cell Anemia

Field of study: Human genetics
Significance: *Sickle-cell anemia is a treatable hereditary blood disease estimated to occur in 250,000 births worldwide each year. Its occurrence primarily among people of African, Caribbean,*

and Mediterranean descent has led to concerns, particularly in the United States, that it might be used as a surrogate for discrimination against particular racial groups. It is one of the most well-documented examples of an evolutionary process known as heterozygote advantage, an important means by which genetic variability is preserved.

Key terms

GENES: localized regions of chromosomes whose precise deoxyribonucleic acid (DNA) sequence directs the assembly of amino acids into proteins

HEMOGLOBIN: a molecule made up of two alpha and two beta amino acid chains whose precise chemical and structural properties normally allow it to bind with oxygen in the lungs and transport it to other parts of the body

HETEROZYGOUS: having two different forms of the same gene, each inherited from a different parent

HOMOZYGOUS: having the same form of a gene from both parents

MUTATION: rare, heritable changes in the content of a gene from one form to another

Genetics and Early Research

Sickle-cell anemia, also known as sickle-cell disease, is a hereditary blood disease found primarily among people of African, Caribbean, and Mediterranean descent. Studies of the incidence of the disease in families led to recognition that the illness is manifested only in individuals who receive the sickle-cell gene from both parents. In most circumstances, individuals who inherit the sickle-cell gene from only one parent display no symptoms of the disease; however, they are carriers of the sickle-cell gene and may pass it on to their children.

In 1910, James B. Herrick, a Chicago physician, first described the characteristically "sickle" or bent appearance of the red blood cells after which the disease is named in blood taken from an anemic patient. In the mid-1930's, Linus Pauling, working with graduate student Charles Coryell, demonstrated that hemoglobin undergoes a dramatic structural change as it combines and releases oxygen. Upon learning that red blood cells from sickle-cell anemia patients only assume their characteristic form in the oxygen-deprived venous blood system, Pauling proposed in 1949 that sickle-cell anemia was the result of a change in the normal amino acid sequence of hemoglobin that interferes with its binding properties.

Three years later, while working with another graduate student, Harvey Itano, Pauling isolated normal hemoglobin and sickle-cell hemoglobin from an individual with anemia using a technique known as electrophoresis. They conducted this investigation by loading hemoglobin onto a paper medium and subjecting it to an electrical current, the presumption being that if the two molecules differed in overall electrical charge, one would migrate along the path of the current faster than the other. In this way, Pauling and Itano established that normal and sickle-cell hemoglobins differ in their respective electrical charges, and people who are heterozygous for the sickle-cell gene have hemoglobin of both types.

In the mid-1950's, Vernon Ingram approached the problem using a more sophisticated version of Pauling's procedure. Ingram first treated hemoglobin of the two types with an enzyme (trypsin) that broke the complex hemoglobin molecules into polypeptides and then used electrophoretic techniques on the resulting polypeptides to determine precisely where in their respective amino acid sequences the two hemoglobins differed from one another. Ingram was able to show that normal and sickle-cell hemoglobin differ by only a single amino acid out of a total of over three hundred: Where the normal hemoglobin gene codes for glutamic acid in the sixth position of the beta acid chain, the sickle-cell gene substitutes an-

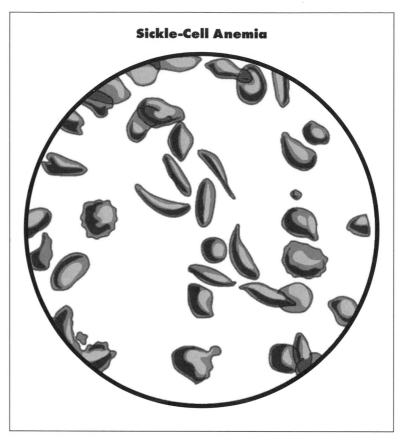

Sickle-Cell Anemia

The red blood cells are sickle-shaped rather than round, which causes blockage of capillaries.

other amino acid (usually lysine). Ingram's work provided proof of Pauling's earlier proposal, making sickle-cell anemia the very first example of a genetic disease being traced to its precise origin at the molecular level.

Physiological Basis, Symptoms, and Treatments

This substitution of lysine for glutamic acid in the beta chain of the hemoglobin molecule has a profound effect on its biological properties under conditions of oxygen deprivation. Hemoglobin coded for by the sickle-cell gene causes the beta chains of the hemoglobin to stick to one another as long, rigid rods and consequently deforms the normally smooth, donut-shaped appearance of the red blood cell to a characteristic sickle shape that prevents it from squeezing through tiny blood capillaries.

Symptoms of the disease appear about six

months after birth, when the last of fetal hemoglobin, a type of hemoglobin that increases the oxygen supply of blood, leaves the infant's body. The severity of the illness varies widely among individuals. Some develop severe anemia as deformed red blood cells are removed more rapidly from the bloodstream (an average "life" of seventeen versus forty days). They may also experience periodic bouts of severe pain ("pain crises"), strokes, and blindness, all thought to be the direct result of sickled cells clogging blood vessels and thereby depriving tissues of oxygen. Heterozygous carriers of the gene normally display no symptoms, although some have been known to become ill under extreme circumstances, such as high altitudes.

A great deal of progress has been made in the diagnosis and treatment of sickle-cell anemia. This includes both a variety of pain management therapies and the use of antibiotics such as penicillin to prevent infections. Although there is no cure, several promising experimental therapies for this disease are under investigation, including the use of bone marrow transplants (transplants of the mast cells that give rise to red blood cells from people not having the disease) and hydroxyurea, a chemical thought to stimulate the production of fetal hemoglobin.

Attention to and funding for research on sickle-cell anemia has increased since World War II, although misinformation about the disease persists. Many have raised concerns that carriers of the disease are discriminated against, both by potential employers and insurance companies. Several organizations were established in the United States in the early 1970's to promote education, treatment, and research for the disease, including Howard University's Center for Sickle-Cell Disease, founded by Ronald B. Scott in 1972. By the mid-1990's, forty U.S. states, the District of Columbia, Puerto Rico, and the Virgin Islands were screening newborns for the sickle-cell trait.

Evolutionary Significance

In most cases, hereditary diseases with such negative consequences as those associated with sickle-cell anemia are kept at low frequencies in populations by natural selection; that is, individuals who carry genes for hereditary diseases are less likely to survive and reproduce than those who carry the normal form of the gene. The continued presence of defective genes in a population therefore reflects the action of chance mutations. Yet the sickle-cell gene is much more common than one would expect if its frequency in a population was caused by mutation alone.

In some areas that are associated with a high incidence of malaria, such as the equatorial belt of Africa, some tribes have been found to have frequencies of the sickle-cell gene as high as 40 percent. This curious correlation between high frequencies of the sickle-cell gene and areas where malaria is common led Anthony C. Alison to suggest, in 1953, that the sickle-cell gene provides an advantage in such environments. Malaria is a deadly, mosquito-borne disease caused by a microscopic parasite, *Plasmodium vivax*, which uses human red blood cells as hosts for part of its life cycle. People who have normal hemoglobin are vulnerable to the disease, and people who are homozygous for the sickle-cell gene in malaria-infested regions die quite early in life because of anemia and other complications. However, when the red cells of people who are heterozygous for the sickle-cell gene are invaded by the malarial parasite, the red cells adhere to blood vessel walls, become deoxygenated, and assume the sickled shape, prompting both their destruction and that of their parasitic invader. This provides the heterozygous carrier with a natural resistance to malaria and explains the relatively high frequency of the sickle-cell gene in such environments. Sickle-cell anemia thus represents a particularly well-documented example of a selective process known as heterozygote advantage, in which individuals heterozygous for a given gene have a greater probability of surviving or reproducing than either homozygote. This is an important phenomenon from an evolutionary standpoint because it provides a mechanism by which genetic diversity in a population may be preserved.

—*David Wijss Rudge*

See Also: Genetic Counseling; Genetic Screening; Hereditary Diseases; Natural Selection; Race.

Further Reading: In "How Do Genes Act?," *Scientific American* 204 (January, 1958), Vernon M. Ingram summarizes his research on the amino acid structure of hemoglobin. Merry France-Dawson and Martin Richards offer brief personal narratives of some of the psychosocial affects of the disease in *The Troubled Helix: Social and Psychological Implications of the New Human Genetics* (1996). Anthony Allison, "Sickle Cells and Evolution," *Scientific American* 202 (August, 1956), discusses the phenomenon of sickle-cell disease as an evolutionary phenomenon.

Sociobiology

Field of study: Population genetics

Significance: *Sociobiology attempts to explain social interactions among members of animal species from an evolutionary perspective. The application of the principles of sociobiology to human social behavior initiated severe criticism and accusations of racism and sexism.*

Key terms

ALTRUISM: an action performed by one individual that benefits another individual of the same species at some cost to the actor

KIN SELECTION: a special type of altruistic behavior in which the benefactor is related to the actor

EUSOCIALITY: an extreme form of altruism and kin selection in which most members of the society do not reproduce but rather feed and protect their relatives

RECIPROCAL ALTRUISM: a type of altruism in which the benefactor may be expected to return the favor

SOCIETY: a group of individuals of the same species in which members interact in relatively complex ways

The History of Sociobiology

Sociobiology is best known from the works of Edward O. Wilson, especially his 1975 book *Sociobiology: The New Synthesis.* This work both synthesized the concepts of the field and initiated the controversy over the application of sociobiological ideas to humans. However, the concepts and methods of sociobiology did not start with Wilson; they can be traced to Charles Darwin and others who studied the influence of genetics and evolution on behavior. Sociobiologists attempt to explain the genetics and evolution of social activity of all types, ranging from flocking in birds and herd formation in mammals to more complex social systems such as eusociality. "The new synthesis" attempted to apply genetics, population biology, and evolutionary theory to the study of social systems.

When sociobiological concepts were applied to human sociality, many scientists, especially social scientists, feared a return to scientific theories of racial and gender superiority. They rebelled vigorously against such ideas. Wilson was vilified by many of these scientists, and some observers assert that the name "sociobiology" generated such negative responses that scientists who studied in the field began using other names for it. At least one scientific journal dropped the word "sociobiology" from its title, perhaps in response to its negative connotations. However, the study of sociobiological phenomena existed in the social branches of animal behavior and ethology long before the term was coined. Despite the criticism, research has continued under the name sociobiology as well as other names such as "behavioral ecology."

Sociobiology and the Understanding of Altruism

Sociobiologists have contributed to the understanding of a number of aspects of social behavior, such as altruism. Illogical in the face of evolutionary theory, apparently altruistic acts can be observed in humans and other animal groups. Darwinian evolution holds that the organism that leaves the largest number of mature offspring will have the greatest influence on the characteristics of the next generation. Under this assumption, altruism should disappear from the population as each individual seeks to maximize its own offspring production. If an individual assists another, it uses energy, time, and material it might have used for its own survival and reproduction and simultaneously contributes energy, time, and material to the survival and reproductive effort of the recipient. As a result, more members of the next generation should be like the assisted organism than like the altruistic one. Should this continue generation after generation, altruism would decrease in the population and selfishness would increase. Yet biologists have catalogued a number of altruistic behaviors.

When a prairie dog "barks," thus warning others of the presence of a hawk, the prairie dog draws the hawk's attention. Should it not just slip into its burrow, out of the hawk's reach? When a reproductively mature acorn woodpecker stays with its parents to help raise the next generation, the woodpecker is bypassing its own reproduction for one or more years. Should it not leave home and attempt to set up its own nest and hatch its own young? Eusocial species, such as honeybees and naked mole rats, actually have many members who never reproduce; they work their entire lives to support and protect a single queen, several reproductive males, and their offspring. It would seem that all these altruistic situations should produce a decrease in the number of members of the next generation carrying altruistic genes in favor of more members with "selfish" genes.

Sociobiologists have reinterpreted some of these apparently altruistic acts as camouflaged selfishness. The barking prairie dog, for example, may be notifying the hawk that it sees the predator, that it is close to its burrow and cannot be caught; therefore, the hawk would be better off hunting someone else. Perhaps the young acorn woodpecker learns enough from the years of helping to make its fewer reproductive years more successful than its total reproductive success without the training period.

It is difficult, however, to explain the worker honeybee this way. The worker bee never gets

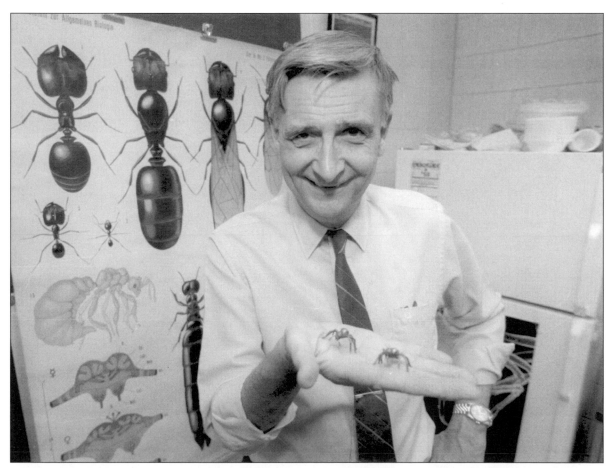

Edward O. Wilson's studies of insect behavior influenced his controversial theories of sociobiology. (AP/Wide World Photos)

Sociobiologists attempt to explain such phenomena as the apparently unselfish behavior of the prairie dog. (Ben Klaffke)

an opportunity to reproduce. Sociobiologists explain this and other phenomena by invoking kin selection. Since the worker bees are closely related (sisters or daughters) to the queen, to reproductive males, and to other workers they help feed and protect, they share a large number of genes with them. If they help raise enough brothers and sisters (especially males and queens) to more than make up for the offspring they do not produce themselves, they will actually increase the proportion of individuals similar to themselves more than if they "selfishly" reproduced.

The prairie dog's behavior might be explained this way as well. The organisms the prairie dog is warning are primarily relatives. By warning them, the prairie dog helps preserve copies of its own genes in its relatives. If the cost of the behavior (an occasional barking prairie dog being captured by a hawk because the warning call drew the hawk's attention) is more than compensated for by the number of relatives saved from the hawk by the warning, kin selection will preserve the behavior. The helper acorn woodpecker's behavior may be explained in similar ways, not as an altruistic act but as a selfish act to favor copies of the helper's genes in its relatives. Another explanation of altruism set forth by sociobiologists is reciprocity or reciprocal altruism: If the prairie dog is sometimes warned by others and returns the favor by calling out a warning when it sees a predator, the prairie dog town will be safer for all prairie dogs.

Opposition to the Application of Sociobiology to Humans

Wilson's new synthesis attempted to incorporate biology, genetics, population biology, and evolution into the study and explanation of social behavior. When the analyses turned to human sociality, critics feared that they would lead back to the sexist, racist, and determinist viewpoints of the early twentieth century. The argument over the relative importance of heredity or environment (nature or nurture) in determining individual success had been more or less decided in favor of the environment, at least by social scientists. Poor people were not poor because they were inherently inferior but

because the environment they lived in did not give them an equal chance. Black, Hispanic, and other minority people were not inordinately represented among the poor because they were genetically inferior but because their environment kept them from using their genetic capabilities.

Sociobiologists entered the fray squarely on the side of an appreciable contribution from genetic and evolutionary factors. Few, if any, said that the environment was unimportant in the molding of racial, gender, and individual characteristics; rather, sociobiologists claimed that the genetic and evolutionary history of human individuals and groups played an important role in determining their capabilities, just as they do in other animals. Few, if any, claimed that this meant that one race, gender, or group was superior to another. However, many (if not all) sociobiologists were accused of promoting racist, sexist, and determinist ideas with their application of sociobiological concepts to humans.

Extremists on both sides of the question have confused the issues. Such extremists range from opponents of sociobiological ideas who minimize genetic or evolutionary influence on the human cultural condition to sociobiologists who minimize the role of environmental influences. In at least some minds, extremists in the sociobiological camp have done as much damage to sociobiology as its most ardent opponents. Sociobiology (by that or another name) will continue to contribute to the understanding of the social systems of animals and humans. The biological, genetic, and evolutionary bases of human social systems must be studied. The knowledge obtained may prove to be as enlightening as has sociobiology's contribution to the understanding of social systems in other animals.

—*Carl W. Hoagstrom*

See Also: Altruism; Behavior; Eugenics; Evolutionary Biology; Natural Selection.

Further Reading: *On Human Nature* (1978), by Edward O. Wilson, is the third book in Wilson's trilogy on sociobiology; references are given to the other two books, *The Insect Societies* (1971) and *Sociobiology: The New Synthesis* (1975). Wilson has also written an interesting article, "The Biological Basis of Morality," in *The Atlantic Monthly* (April, 1998). Chapter 5 of Richard Brewer's *The Science of Ecology* (1994) outlines sociobiology nicely and treats a number of sociobiological subjects from an ecological perspective. *Behavioural Ecology: An Evolutionary Approach* (1991), edited by John R. Krebs and Nicholas B. Davies, covers sociobiological phenomena extensively and quite well.

Speciation

Field of study: Population genetics

Significance: *The process of speciation, the biological formation of new species, has produced the wide variety of living things on earth. Extinction has the opposite effect: It reduces the diversity of life. Knowledge of the genetic basis of evolution and speciation has helped scientists produce new varieties of food crops, hybrid animals, and useful pharmaceuticals.*

Key terms

SPECIES: a class of organisms with common attributes; individuals are usually able to produce fertile offspring only when mating with members of their own species

POPULATION: a group of organisms of the same species in the same place at the same time and thus potentially able to mate; populations are the basic unit of speciation

ALLOPATRIC SPECIATION: the genetic divergence of populations caused by separation from each other by a geographic barrier such as a mountain range or an ocean

SYMPATRIC SPECIATION: the genetic divergence of populations that are not separated geographically

REPRODUCTIVE ISOLATING MECHANISM: a characteristic that prevents an individual of one species from interbreeding (hybridizing) with a member of another species

Species Concepts

Before the time of Charles Darwin, physical appearance was the only criterion for classifying an organism. This "typological species concept" was associated with the idea that species never change (fixity of species). This way

of defining a species causes problems when males and females of the same species look different (as with peacocks and peahens) or when there are several different color patterns among members of a species (as with many insects). Variability within species, whether it is a visible part of their anatomy, an invisible component of their biochemistry, or another characteristic such as behavior, is an important element in understanding how species evolve.

The "biological species concept" uses reproduction to define a species. It states that a species is composed of individuals that can mate and produce fertile offspring in nature. This concept cannot be used to classify organisms such as bacteria, which usually do not mate. It also cannot be used to classify dead specimens or fossils. This definition emphasizes the uniqueness of each individual (variability) in sexually reproducing species. For example, in the human species (*Homo sapiens*), there are variations in body build, hair color and texture, ability to digest milk sugar (lactose), and many other anatomical, biochemical, and behavioral characteristics. All of these variations are based on genetic mutations, or changes in genes.

According to evolutionary scientist Ernst Mayr, to a "population thinker," variation is reality and type is an abstraction or average; to a "typological thinker," variation is an illusion and type is the reality. Typological thinking is similar to typecasting or stereotyping, and it cannot explain the actual variability seen in species, just as stereotyping does not recognize the variability seen in people. Additional definitions, such as the "evolutionary species concept," include the continuity of a species' genes through time or other factors not addressed by the biological species concept.

Isolation and Divergence of Populations

Species are composed of unique individuals that are nevertheless similar enough to be able to mate and produce fertile offspring. However, individuals of a species are infrequently in close enough proximity to be able to choose a mate from all opposite-sex members of the same species. Groups of individuals of the same species that are at least potential mates because of proximity are called "populations."

The basic mechanism of speciation in most sexually reproducing organisms is believed to be "allopatric," in which geographic isolation (separation) of the species into two or more populations is followed by accumulation of differences (divergence) between the populations that eventually prevent them from interbreeding. These differences are caused by natural selection of characteristics advantageous to populations in different environments. If both populations were in identical environments after geographic isolation, they would not diverge or evolve into new species.

Another way that speciation can occur is through "sympatric" means, in which populations are not separated geographically, but reproduction between them cannot occur for some other reason (reproductive isolation). For example, one population may evolve a mutation that makes the fertilized egg (zygote) resulting from interbreeding with the other population incapable of surviving. Another possibility is a mutation that changes where or when individuals are active so that members of the different populations never encounter each other.

Darwin thought that divergence and thus speciation occurred gradually by the slow accumulation of many small adaptations "selected" by the environment. More recently, it has been recognized that a very small population, or even a "founder" individual, may be the genetic basis of a new species that evolves more rapidly. This process, called "genetic drift," is essentially random. For example, which member of an insect species is blown to an island by a storm is not determined by genetic differences from other members of the species but by a random event (in this case, the weather). This individual (or small number of individuals) is highly unlikely to contain all of the genetic diversity of the entire species. Thus the new population begins with genetic differences that may be enhanced by its new environment. Speciation proceeds according to the allopatric model, but faster. However, extinction of the new population may also occur.

Plants are able to form new species by hybridization (crossbreeding) more often than are animals. When plants hybridize, postmating incompatibility between the chromosomes of the parents and the offspring may immediately create a new, fertile species rather than a sterile hybrid, as in animals such as the mule. A frequent method of speciation in plants is "polyploidy," in which two or more complete sets of chromosomes end up in the offspring. (Usually, one complete set is made up of half of each parent's chromosomes.)

Many species reproduce asexually (without the exchange of genes between individuals that defines sexual reproduction). These include bacteria and some plants, fish, salamanders, insects, rotifers, worms, and other animals. In spite of the fact that reproductive isolation has no meaning in these organisms, they are species whose chromosomes and genes differ from those of their close relatives.

Impact and Applications

Environmentalists and scientists recognize that the biodiversity created by speciation is essential to the functioning of the earth's life-support systems for humans as well as other species. Some practical benefits of biodiversity include medicines, natural air and water purification, air conditioning, and food.

The impact of understanding the genetic basis of evolving species cannot be underestimated. Artificial selection (in which humans decide which individuals of a species survive and reproduce) of plants has produced better food crops (for example, modern corn from teosinte) and alleviated hunger in developing nations by creating new varieties of existing species (for example, rice). Hybridization of animals has resulted in mules and beefaloes for the farm (both of which are sterile hybrids rather than species.) Artificial selection of domesticated animals has produced the many breeds of horses, dogs, and cats (each of which is still technically one species.) Genetic engineering promises to create crops that resist pests, withstand frost or drought, and contain more nutrients. Finally, understanding the genetics of the evolving human species has broad implications for curing disease and avoiding birth defects.

—Barbara J. Abraham

See Also: Artificial Selection; Evolutionary Biology; Natural Selection; Population Genetics.

Further Reading: Charles Darwin, *On the Origin of Species* (1859), provides the first explanation of how new species originate through the process of natural selection. *One Long Argument: Charles Darwin and the Genesis of Modern Evolutionary Thought* (1991), by Ernst Mayr, includes a chapter ("How Species Originate") that points out that Darwin's explanation of speciation was limited by his lack of understanding of the origin of genetic variation (mutation and recombination). Two well-illustrated, general references on speciation and evolution include *Origin of Species* (1981), by the staff of the British Museum of Natural History, which contains a section on further reading, and the Time-Life Understanding Science and Nature volume *Evolution of Life* (1992), which contains a glossary. Richard Attenborough's *Life on Earth: A Natural History* (1979) is based on a spectacular British Broadcasting Corporation (BBC) television series also available on video. Both do an excellent job of explaining how natural history and biodiversity arise through evolution and speciation.

Sterilization Laws

Field of study: Human genetics

Significance: *Forced sterilization for eugenic reasons became legal throughout much of the United States and many parts of the world during the first half of the twentieth century. Though sterilization is an ineffective mechanism for changing the genetic makeup of a population, sterilization laws remain in effect in many states in the United States and other countries throughout the world.*

Key terms

NEGATIVE EUGENICS: the effort to improve the human species by discouraging or eliminating reproduction among those deemed to be socially or physically unfit

POSITIVE EUGENICS: the effort to encourage more prolific breeding among "gifted" individuals

STERILIZATION: an operation to make reproduction impossible; in tubal ligation, doctors sever the Fallopian tubes so that a woman cannot conceive a child

The Eugenics Movement and Sterilization Laws

The founder of the eugenics movement is considered to be Sir Francis Galton, who carried out extensive genetic studies of human traits. He thought that the human race would be improved by encouraging humans with desirable traits (such as intelligence, good character, and musical ability) to have more children than those people with less desirable traits (positive eugenics). With the development of Mendelian genetics shortly after the beginning of the twentieth century, research on improving the genetic quality of plants and animals was in full swing. Success with plants and domestic animals made it inevitable that interest would develop in applying those principles to the improvement of human beings. As some human traits became known to be under the control of single genes, some geneticists began to claim that all sorts of traits (including many behavioral traits and even social characteristics and preferences) were under the control of a single gene with little regard for the possible impact of environmental factors.

The Eugenics Record Office at Cold Springs Harbor, New York, was set up by Charles Davenport to gather and collate information on human traits. The eugenics movement became a powerful political force that led to the creation and implementation of laws restricting immigration and regulating reproduction. Some geneticists and politicians reasoned that since mental retardation and other "undesirable" behavioral and physical traits were affected by genes, society had an obligation and a moral right to restrict the reproduction of individuals with "bad genes" (negative genetics).

The state of Indiana passed the first sterilization law in 1907, which permitted the involuntary sterilization of inmates in state institutions. Inmates included not only "imbeciles," "idiots," and others with varying degrees of mental retardation (described as "feeble-minded") but also people who were committed for behavioral problems such as criminality, swearing, and slovenliness. By 1911, similar laws had been passed in six states, and, by the end of the 1920's, twenty-four states had similar sterilization laws. Although not necessarily strictly enforced, twenty-two states currently have sterilization laws on the books.

The U.S. Supreme Court, in its 1927 *Buck v. Bill* decision, supported the eugenic principle that states could use involuntary sterilization to eliminate genetic defects from the population. The vote of the Court was eight to one. The court's reasoning went as follows:

> We have seen more than once that the public welfare may call upon the best citizens for their lives. It would be strange if it could not call upon those who already sap the strength of the state for these lesser sacrifices, often not felt to be such by those concerned, in order to prevent our being swamped with incompetence. It is better for all the world, if instead of waiting to execute degenerate offspring for crime, or to let them starve for their imbecility, society can prevent those who are manifestly unfit from continuing their kind. The principle that sustains compulsory vaccination is broad enough to cover cutting the Fallopian tubes.

It is a sad state of affairs to realize that the sterilization laws of the United States and Canada served as models for the eugenics movement in Nazi Germany in its program to ensure so-called racial purity and superiority.

Impact and Applications

Two problems associated with eugenics are the subjective nature of deciding which traits are desirable and determining who should decide. These concerns aside, the question of whether there is a sound scientific basis for the desire to manipulate the human gene pool remains. Does the sterilization of individuals who are mentally retarded or who have some other mental or physical defect improve the human genetic composition? Involuntary sterilization of affected individuals would quickly reduce the incidence of dominant genetic traits. Individuals who were homozygous for recessive traits would also be eliminated. However, most harmful recessive genes are carried by individuals who appear normal and, there-

fore, would not be "obvious" for sterilization purposes. These "normal" people would continue to pass the "bad" gene on to the next generation, and a certain number of affected people would again be born. It would take an extraordinary number of generations to significantly reduce the frequency of harmful genes.

Although the number of involuntary sterilizations in the United States is now minimal, the impact sterilization laws had on the population through 1960 was far-reaching, as nearly sixty thousand people were sterilized. Other countries also had laws that allowed forced sterilizations, with many programs continuing into the 1970's. The province of Alberta, Canada, sterilized three thousand people before its law was repealed. Another sixty thousand were sterilized in Sweden. The story of sterilization and "euthanasia" in Germany needs no retelling. With the ability to decipher the human genome and implement improved genetic testing procedures, a danger exists that new programs of eugenics and involuntary sterilization might once again emerge.

—*Donald J. Nash*

See Also: Eugenics; Eugenics: Nazi Germany; Genetic Testing: Ethical and Economic Issues; Miscegenation and Antimiscegenation Laws.

Further Reading: *In the Name of Eugenics* (1995), by Daniel K. Kevles, is a comprehensive introduction to the history of the eugenics movement and the development of sterilization laws. A discussion of important research findings on genes and human behavior is provided in Dean Hamer and Peter Copeland, *Living with Our Genes* (1998). The fascinating developments and applications of genetic testing are described in *Does It Run in the Family?* (1997), by Doris Teichler Zallen, which covers the medical, psychological, and social implications of genetic testing.

Synthetic Antibodies

Field of study: Immunogenetics
Significance: *Synthetic antibodies are artificially produced replacements for natural human antibodies. They are used to treat a variety of illnesses*

and promise to be an important part of medical technology in the future.

Key terms

ANTIBODY: a protein molecule that binds to a substance in order to remove, destroy, or deactivate it

ANTIGEN: the substance to which an antibody binds

B CELLS: white blood cells that produce antibodies

MONOCLONAL ANTIBODIES: identical antibodies produced by identical B cells

The Development of Antibody Therapy

Natural antibodies are protein molecules produced by white blood cells known as B cells in response to the presence of foreign substances. A specific antibody binds to a specific substance, known as an antigen, in a way that renders it harmless or allows it to be removed from the body or destroyed. The production of natural antibodies can be activated by exposing the patient to harmless versions of an antigen, a process known as "active immunization." Active immunization was the first form of antibody therapy to be developed and is used to prevent diseases such as measles and polio.

The oldest method of producing therapeutic antibodies outside the patient's own body is known as "passive immunization." This process involves exposing an animal to an antigen so that it develops antibodies to it. The antibodies are separated from the animal's blood and administered to the patient. Passive immunization is used to treat diseases such as rabies and diphtheria. A disadvantage of antibodies derived from animal blood is the possibility that the patient may develop an allergic reaction. Because the animal's antibodies are foreign substances, the patient's own antibodies may treat them as antigens, leading to fever, rash, itching, joint pain, swollen tissues, and other symptoms. Antibodies derived from human blood are much less likely to cause allergic reactions than antibodies from the blood of other animals. This led researchers to seek a way to develop synthetic human antibodies.

A major breakthrough in the search for synthetic antibodies was made in 1975 by Cesar Milstein and Georges Köhler. They developed a technique that allowed them to produce a specific antibody outside the body of a living animal. This method involves exposing an animal to an antigen, causing it to produce antibodies. Instead of obtaining the antibodies from the animal's blood, B cells are obtained from the animal's spleen. These cells are then combined with abnormal B cells known as "myeloma" cells. Unlike normal B cells, myeloma cells can reproduce identical copies of themselves an unlimited number of times. The normal B cells and the myeloma cells fuse to form cells known as "hybridoma" cells. Hybridoma cells are able to reproduce an unlimited number of times and are able to produce the same antibodies as the B cells. Those hybridoma cells that produce the desired antibody are separated from the others and allowed to reproduce. The antibodies produced this way are known as "monoclonal antibodies."

Because human B cells do not normally form stable hybridoma cells with myeloma cells, B cells from mice are usually used. Because mouse antibodies are not identical to human antibodies, they may be treated as antigens by the patient's own antibodies, leading to allergic reactions. During the 1980's and 1990's, researchers began to develop methods of producing synthetic antibodies that were similar or identical to human antibodies. An antibody consists of a variable region, which binds to the antigen, and a constant region. The risk of allergic reactions can be reduced by combining variable regions derived from mouse hybridoma cells with constant regions from human cells. The risk can be further reduced by identifying the exact sites on the mouse variable region that are necessary for binding and integrating these sites into human variable regions. This method produces synthetic antibodies that are very similar to human antibodies.

Other methods exist to produce synthetic antibodies that are identical to human antibodies. A species of virus known as the Epstein-Barr virus can be used to change human B cells in such a way that they will fuse with myeloma cells to form stable hybridoma cells that produce human antibodies. Another method involves using genetic engineering to produce mice

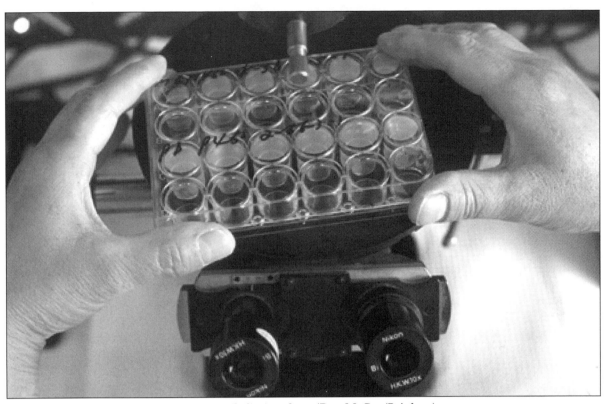

Synthetic antibodies in culture. (Dan McCoy/Rainbow)

with B cells that produce human antibodies rather than mouse antibodies. One of the most promising techniques involves creating a "library" of synthetic human antibodies. This is done by using a substance known as polymerase to produce multiple copies of the genetic material within B cells. This genetic material contains the information that results in the production of proteins that come together to form antibodies. By causing these proteins to be produced and allowing them to combine at random, researchers are able to produce millions of different antibodies. The antibodies are then tested to detect those that bind to selected antigens.

Impact and Applications

Some synthetic antibodies are used to help prevent the rejection of transplanted organs. An antibody that binds to the heart drug digoxin can be used to treat overdoses of that drug. Antibodies attached to radioactive isotopes are used in certain diagnostic procedures. Synthetic antibodies have also been used in patients undergoing a heart procedure known as a percutaneous transluminal coronary angioplasty (PTCA). The use of a particular synthetic antibody has been shown to reduce the risk of having one of the blood vessels that supply blood to the heart shut down during or after a PTCA. Researchers also hope to develop synthetic antibodies to treat acquired immunodeficiency syndrome (AIDS) and septic shock, a syndrome caused by toxic substances released by certain bacteria.

The most active area of research involving synthetic antibodies in the 1990's was in the treatment of cancer. On November 26, 1997, the U.S. Food and Drug Administration approved a synthetic antibody for use in non-Hodgkin's lymphoma, a cancer of the white blood cells. This was the first synthetic antibody approved for use in cancer.

—Rose Secrest

See Also: Antibodies; Hybridomas and Monoclonal Antibodies; Immunogenetics.

Further Reading: A detailed discussion of the history and future of antibody therapy can be found in Andy Coghlan, "A Second Chance for Antibodies," *New Scientist* 129 (February 9, 1991). A dramatic account of the promise of synthetic antibodies in the treatment of cancer is presented in *Cell Wars: The Immune System's Newest Weapons Against Cancer* (1988), by Marshall Goldberg. Methods of synthetic antibody production are described in detail in *Designing Antibodies* (1993), by Ruth D. May.

Synthetic Genes

Field of study: Genetic engineering and biotechnology

Significance: *Synthetic genes have been shown to function in biological organisms. Scientists hope that it will prove possible to restore normal function in diseased humans, animals, and plants by replacing defective natural genes with appropriately modified synthetic genes.*

Key terms

DEOXYRIBONUCLEIC ACID (DNA): a nucleic acid characteristic of chromosomes

GENE: a sequence of nucleotides that determines a trait or phenotype in an organism

RESTRICTION ENZYME: an enzyme that cleaves, or cuts, DNA at specific sites with sequences recognized by the enzyme

REVERSE TRANSCRIPTION: the synthesis of DNA from ribonucleic acid (RNA)

A Brief History

In 1871, Swiss physician Johann Friedrich Miescher reported that the chief constituent of the cell nucleus was nucleoprotein, or nuclein. Later it was established that the nuclei of bacteria contained little or no protein, so the hereditary material was named nucleic acid. At the end of the nineteenth century, German biochemist Albrecht Kossel identified the four nitrogenous bases: the purines adenine (A) and guanine (G) and the pyrimidines cytosine (C) and uracil (U). In the 1920's, Phoebus A. Levene and others indicated the existence of two kinds of nucleic acid: ribonucleic acid (RNA) and deoxyribonucleic acid

(DNA); the latter contains thymine (T) instead of uracil.

The chemical identity of genes began to unfold in 1928, when Frederick Griffith discovered the phenomenon of genetic transformation. Oswald Avery, Colin MacLeod, and Maclyn McCarty (in 1944) and Alfred Hershey and Martha Chase (in 1952) demonstrated that DNA was the hereditary material. Following the elucidation of the structure of DNA in 1953 by James Watson and Francis Crick, pioneering efforts by several scientists led to the eventual synthesis of a gene. The successful enzymatic synthesis of DNA in vitro (in the test tube) in 1956, by Arthur Kornberg and colleagues, and that of RNA by Marianne Grunberg-Manago and Severo Ochoa also contributed to the development of synthetic genes. In 1961, Marshall Nirenberg and Heinrich Matthaei synthesized polyphenylalanine chains using a synthetic messenger RNA (mRNA). In 1965, Robert W. Holley and colleagues determined the complete sequence of alanine transfer RNA (tRNA) isolated from yeast. The interpretation of the genetic code by several groups of scientists throughout the 1960's was also clearly important. In 1970, Gobind Khorana, along with twelve associates, synthesized the first gene: the gene for an alanine tRNA in yeast. There were no automatic DNA synthesizers available then. In 1976, Khorana's group synthesized the tyrosine suppressor tRNA gene of *Escherichia coli* (*E. coli*). The lac operator gene (twenty-one nucleotides long) was also synthesized, introduced into *E. coli*, and demonstrated to be functional. It took ten years to synthesize the first gene; by the mid-1990's, gene machines could synthesize a gene in hours.

Gene Synthesis

Protein engineering is possible by making targeted changes in a DNA sequence to produce a different product (protein) with a desired result, such as stress tolerance. The process of targeting a specific change in the nucleotide sequence (site-directed mutagenesis) allows one to study how gene structure relates to protein function. Rapid sequencing and gene synthesis techniques are also available. Rapid sequencing facilitates determina-

Gobind Khorana, who led the team that created the first synthetic gene, poses with a model of a DNA molecule. (AP/Wide World Photos)

tion of the order of nucleotides that make up a gene. Restriction enzymes are used to cut a DNA molecule into a set of smaller fragments. The sequence of nucleotides in these fragments is then deciphered with radioactive labeling and electrophoresis. Frederick Sanger of Cambridge University in England and Walter Gilbert of Harvard University did the pioneering work on DNA sequencing.

Once the sequence of a gene is known, it can be synthesized from organic chemicals using gene machines. A gene machine is simply a chemical synthesizer made up of tubes, valves, and pumps that stitches nucleotides together in the right order under the direction of a computer. One needs only to specify the nucleotide sequence to the computerized system. An intelligent person with a minimum of training can produce synthetic genes. A gene may be isolated from an organism using restriction enzymes (any of the several enzymes found in bacteria that serve to chop up the DNA of invading viruses), or it may be made on a gene machine. For example, the chymosin gene (an enzyme used in cheese making) in calves can be synthesized from its known nucleotide sequence instead of isolating it from calf DNA using restriction enzymes. Alternatively, chymosin mRNA can be obtained from calf stomach cells, which can be transformed into DNA through reverse transcription.

The sequencing of an organism's genome (an entire single set of genes or DNA) should unravel the complete set of instructions for manufacturing that organism. The computerized gene synthesizers will be expected to enhance the already remarkable pace of recombinant DNA research and raise genetic engineering to a higher industrial scale. New or modified genes may be manufactured to obtain a desired product. Gene synthesis, coupled with automated rapid sequencing and protein analysis, has yielded remarkable dividends in medicine and agriculture. Genetic engineers are designing new proteins from scratch from new genes to learn more about protein function and architecture. With synthetic genes, the process of mutagenesis can be explored in a greater depth. It should be possible to produce various alterations at will in the nucleotide sequence of a gene (point mutations) and observe their effects on protein function, allowing potential diseases or defects to be predicted and possible solutions to be devised.

—*Manjit S. Kang*

See Also: Biotechnology; DNA Structure and Function; Genetics, Historical Development of.

Further Reading: Susan Aldridge, *The Thread of Life: The Story of Genes and Genetic Engineering* (1996), provides a guide to DNA and genetic engineering. Robert J. Henry, *Practical Applications of Plant Molecular Biology* (1997), gives protocols for important plant molecular biology techniques. George P. Rédei, *Genetics* (1982), gives a complete historical account of genetics up to 1982.

Tay-Sachs Disease

Field of study: Human genetics

Significance: *Tay-Sachs disease (TSD) is an especially tragic disease that is inherited as an autosomal recessive disorder. Children are normal at birth, but symptoms are usually noticed by six months of age and progressively worsen; the child usually dies by four years of age. There is no cure for this severe disorder of the nervous system, but an understanding of the genetic nature of the disorder has led to effective population screening, prenatal diagnosis, and genetic counseling.*

Key terms

ASHKENAZI JEWS: Jews of Central and Eastern European origin who have a relatively high incidence of Tay-Sachs disease

GENETIC SCREENING: the testing of individuals for a disease-causing gene

HEXOSAMINIDASE A (HEX A): a lysosomal enzyme, the absence of which leads to Tay-Sachs disease

LYSOSOME: an organelle or structure in the cytoplasm of a cell that contains enzymes involved in the breakdown of metabolic products

PRENATAL DIAGNOSIS: the identification of a gene or disease in an embryo or fetus

Symptoms of Tay-Sachs Disease

Tay-Sachs disease (TSD) is an inherited birth defect that is named after Warren Tay, an English ophthalmologist, and Bernard Sachs, an American neurologist. TSD is one of the lysosomal storage disorders, as are Herler's syndrome, Hunter's syndrome, Gaucher disease, and Fabry disease. Lysosomes are organelles found in the cytoplasm of cells and contain many enzymes that digest the cell's food and waste. TSD is caused by the lack of the enzyme hexosaminidase A (Hex A), which facilitates the breakdown of fatty substances, gangliosides in the brain, and nerve cells. When Hex A is sufficiently lacking, as in TSD, gangliosides accumulate in the body and eventually lead to the destruction of the nervous system.

Children with TSD appear normal at birth and up to six months of age. During this time, they may show an exaggerated startle response to sound. Shortly after six months, more obvious symptoms appear. The child may show poor head control and an involuntary back-and-forth movement of the eyeball. Also distinctive of TSD is the "cherry red spot" on the retina of the eye, first described by Tay, that usually appears after one year of age as atrophy of the optic nerve head occurs. The symptoms are progressive, and the child loses all the motor and mental skills developed to that point. Convulsions, increased motor tone, and blindness develop as the disease progresses. The buildup of storage material in the brain causes the head to enlarge, and brain weight may be 50 percent greater than normal at the time of death. There is no cure for TSD, and death usually occurs between two and four years of age, with the most common cause of death being pneumonia.

There are several forms of Tay-Sachs disease in addition to the classical, or infant, form already described. There is a juvenile form in which similar symptoms appear between two to five years of age, with death occurring around age fifteen. A chronic form of TSD has symptoms beginning at age five that are far milder than those of the infant and juvenile forms. Late onset Tay-Sachs disease (LOTS) is a rare form in which there is some residual Hex A activity so that symptoms appear later in life and the disease progresses much more slowly.

Genetics of Tay-Sachs Disease

All forms of TSD are inherited as autosomal recessive disorders. One of the interesting features of TSD, as is true of some other genetic disorders, is its variation in incidence in different ethnic groups. The Ashkenazi Jewish population, ancestors of most of the Jewish people in the United States, is a group of Jews of Eastern European descent. This group has a high incidence of TSD, about 1 in 3,600. Approximately one in thirty Ashkenazi Jews is a heterozygote (a person who carries one copy of the gene but does not show symptoms), compared to a figure of perhaps one in three hundred for the rest of the world's population. It is possible to screen the population and identify

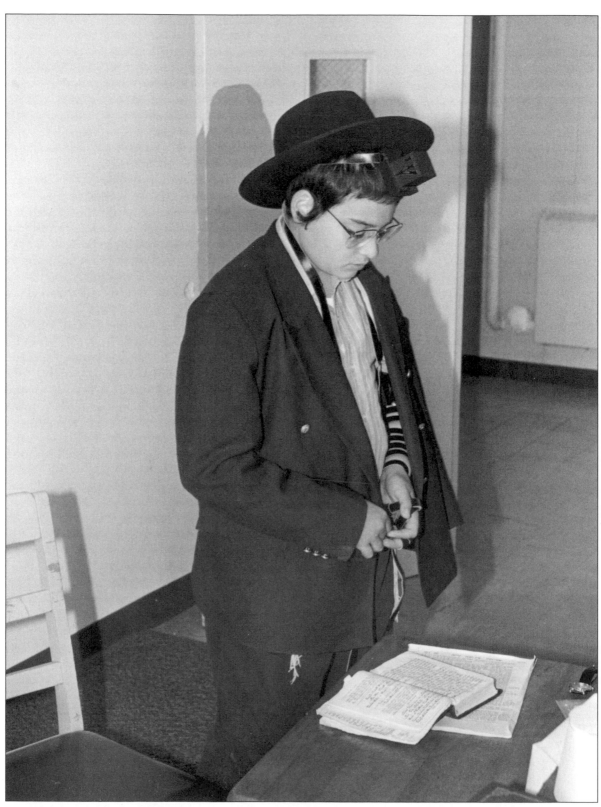

Tay-Sachs disease occurs almost ten times more often among Jews than among the general population. (James L. Shaffer)

those individuals who are heterozygotes by means of a blood plasma assay that detects differences in Hex A activity. If two people are carriers of the gene, they have a one-quarter chance of having a child with TSD. If one or both individuals are not carriers, they can be reassured that their child will not have TSD. If both people are carriers, once pregnancy ensues, prenatal diagnosis can determine whether the developing fetus will develop TSD. If a positive diagnosis of TSD is made for the fetus, the couple may be counseled regarding therapeutic pregnancy termination.

Impact and Applications

Although much has been learned about the genetics of the Tay-Sachs gene and the protein that causes the gene, there is still no cure for the disease. It is still not known how to alter or stop the disease processes in the brain and nervous system. Nevertheless, TSD provides an excellent example of how to approach genetic control and reduction in the incidence of a genetic disease. The effective screening of populations at risk for TSD and prenatal detection of fetuses with TSD have served to dramatically reduce the overall incidence of this terrible disease.

—*Donald J. Nash*

See Also: Genetic Testing; Inborn Errors of Metabolism; Isolates and Genetic Disease; Prenatal Diagnosis.

Further Reading: *The Family Genetic Sourcebook* (1990), by Benjamin A. Pierce, provides short descriptions about many genetic disorders as well as information about genetic testing and genetic counseling. John Rennie, "Grading the Gene Tests," *Scientific American* (June, 1994), evaluates the effectiveness of different genetic tests. A guide to testing for genetic disorders, including Tay-Sachs disease, is given in *Does It Run in the Family?* (1997), by Doris Teichler Zallen.

Telomeres

Field of study: Molecular genetics
Significance: *Telomeres, the end sections of chromosomes of higher organisms, become shorter as or-*ganisms age. They are thought to act biologically to slow chromosome shortening, which can lead to cell death caused by the loss of genes and may be related to aging and diseases such as cancer.*

Key terms

DEOXYRIBONUCLEIC ACID (DNA): the genetic material in the chromosomes of most organisms, a double-stranded polymer made of units called "nucleotides"

EUKARYOTE: a unicellular or multicellular organism with cells that contain a membrane-bound nucleus, multiple chromosomes, and membrane-bound subcellular organelles

ENZYME: a molecule, most often a protein, that accelerates a biochemical reaction without being altered by the reaction

PROKARYOTE: a unicellular organism with a single chromosome and lacking a nucleus or any other membrane-bound organelles

PROTEIN: a complex, chainlike substance composed of many conjoined amino acids

RIBONUCLEIC ACID (RNA): a second kind of genetic material made from short sections of DNA (genes) and composed of units called "ribonucleotides"

Eukaryote Chromosomes and Telomeres

The deoxyribonucleic acid (DNA) of bacteria and other related simple organisms (prokaryotes) consists of one duplex (double-stranded) DNA molecule. The molecule, a prokaryotic chromosome shaped like a double circle with no ends, contains all the genes that produce ribonucleic acid (RNA) and all the proteins needed to make and reproduce the cell. Structurally and functionally, the prokaryotic chromosome contains one copy of most genes as well as DNA regions that control expression of these genes. A large portion of prokaryotic gene expression depends upon a cell's moment-to-moment needs. An entire prokaryotic chromosome usually encodes about one thousand genes. These genes are the organism's genome, or its complete genetic makeup.

The genomes of higher organisms (eukaryotes) are much more complex and may include 100,000 or more genes. The number of chromosomes that are found in different types of eukaryotes can range from several to sixty or more. Each of these huge DNA molecules is a linear duplex, rather than a circular molecule of the type seen in prokaryotes. In addition, many individual segments of the DNA of eukaryotes exist in multiple copies. For example, about 10 percent of the DNA of a eukaryote consists of "very highly repetitive segments" (VRSs), units that are less than ten deoxyribonucleotides long that are repeated up to several million times per cell. DNA segments that are several hundred deoxyribonucleotide units long are about twice as plentiful as VRSs. They are repeated one thousand times or more per cell. The rest of the eukaryote DNA (from 65 to 70 percent of the total) consists of larger segments repeated once or a few times, the genes, and the DNA regions that control the expression of the genes.

Much of the repetitive DNA, called "satellite" DNA, does not seem to be involved in coding for proteins or RNAs involved in making proteins (hence, it is "nongenetic"). Telomeres are part of this DNA. Telomeres consist of several thousand-deoxyribonucleotide-unit-long pieces of DNA, found at both chromosome ends. They are believed to act to stabilize the gene-containing portions of the eukaryote chromosome in which they occur. Researchers have concluded this for two reasons. First, the enzymes that make two chromosomes every time a cell reproduces are unable to operate at the chromosome ends. Hence, the repeated reproduction of a eukaryote cell and its DNA will lead to the creation of shorter and shorter chromosomes, a process that can cause cell death when essential genes are lost. Second, as organisms age, the telomeres of their cells become shorter and shorter.

Telomeres end with DNA duplex segments that contain the four deoxyribonucleotide types (A, T, G, and C). These types form pairs in each duplex strand, with A weakly bonded to T and G weakly bonded to C (A-T or G-C bonds).

Telomerase Enzymes

Chromosomes are made in a complex process known as replication. Replication requires the synthesis of many single-stranded DNA regions, each bounded by duplex regions, and the joining of the resultant new duplex segments into a new chromosome. The short duplex part of such a region, where synthesis begins, is called a "primer." At the time when replication is finished and a new cell has been manufactured, the entire chromosome has been duplicated, and each "daughter duplex" enters one of the two cells.

Making the end of a linear chromosome is a problem, however, because the essential primer cannot be produced by any of the usual nucleic-acid-making enzymes. This problem is solved by enzymes known as telomerases, which add telomeres to eukaryote chromosomes. Any telomerase contains a nucleic acid component (RNA) about 150 ribonucleotides long. This is equivalent to 1.5 copies of the appropriate repeat in the DNA telomere to be made. The enzyme uses this piece of RNA to make the desired DNA strand of the telomere. How the telomerase in any given species identifies the correct length of telomere repeat for a specific chromosome is not clearly understood, nor is the exact mechanism by which the DNA strand is made.

Telomerase activity can be lost in certain strains of simple eukaryotes, such as protozoa. When this happens to a given cell line, each cell division leads to the additional shortening of its telomeres. This process continues for a fixed number of cell divisions; it then ends with the death of the telomerase-deficient cell line.

A related observation has been made in humans. It has been shown that when human fibroblasts are grown in tissue culture, telomere length is longest when cells are obtained from young individuals. They are shorter in cells taken from the middle-aged, and very short in cells taken from the aged. Similar observations have been made with the fibroblasts from other higher eukaryotes as well as with other human cell types. In contrast, the process of telemere shortening does not happen when germ-cell lines—which in the whole organism produce sperm and ova—are grown in tissue culture. This suggests a basis for differences in longevity of the germ cells and the somatic cells that make up other human tissues.

Impact and Applications

The discovery and study of telomeres and telomerases adds a new aspect to consideration of many processes associated with DNA synthesis, the numbers of times a cell can reproduce, and the aging process. One interesting point that must be made is that the existence of circular duplex DNAs in bacteria makes it possible for them to undergo many more cycles of reproduction than the somatic cells of the eukaryotes that contain huge, linear chromosomes. The linear eukaryote chromosome may have evolved because such DNA molecules were too big to survive as circular molecules because of the rigidity and fragility of DNA duplexes. In addition, the observation of telomere shortening in simple and complex eukaryotes raises the fascinating possibility that the lifespans of organisms may be related to the conservation of telomeres associated with the replication of these structures by telomerases.

Moreover, the presently elusive understanding of the additional enzymes involved in the process presents valid opportunities to unravel other issues associated with life and death. Furthermore, the observation that germ cells do not shorten in tissue culture, and thus may be relatively immortal, is exciting. This is because cancer cells, which are also immortal, may get some of this immortality from their ability to conserve and resynthesize telomeres. It is also possible that discovery of inhibitors and activators of the telomerases and other telomere-generating enzymes—which would either slow or speed telomere production, respectively—may add substantially to the number of drugs available to treat numerous DNA-related diseases. For example, inhibitors of DNA-making enzymes known as reverse transcriptases affect the actions of telomerases. Researchers believe that increased understanding of telomeres, telomerases, and other related phenomena will yield even more exciting results.

—Sanford S. Singer

See Also: DNA Replication; Protein Synthesis.

Further Reading: "A Mutant with a Defect in Telomere Elongation Leads to Senescence in Yeast," *Cell* (April, 1989), by V. Lundblad and Jack Szostak, shows that shortened telomeres harm or kill cells. "Telomeres, Telomerase, and Cancer," *Scientific American* (February, 1996), by Elizabeth H. Blackburn, gives good background on telomeres, telomerases, and their potential importance in carcinogenesis. *The Telomere* (1995), edited by David Kipling, is an interesting technical book describing telomeres, telomerases, relationships to cancer, and other aspects of potential telomere action. *Telomeres* (1995), edited by Elizabeth H. Blackburn and Carol W. Greider, contains articles covering the discovery, synthesis, and potential effects of telomeres on normal life, aging, neoplasms, and other pathology. Its concepts are presented in a technical but clear fashion.

Testicular Feminization Syndrome

Field of study: Human genetics

Significance: *The sex of a baby is usually determined at conception, but other genetic events can alter the outcome from what the chromosome makeup predicts. One such condition is testicular feminization syndrome, which causes a child with male chromosomes to be born with feminized genitals. Information gained from the study of this and similar conditions is being used to challenge the validity of sex-determination tests for athletes.*

Key terms

ANDROGEN RECEPTORS: molecules in the cytoplasm of cells that join with circulating male hormones

ANDROGENS: hormones that promote male body characteristics

DIFFERENTIATION: the process of changing from an unspecialized condition to a final specialized one

PHENOTYPE: the physical appearance of an individual

SEX DETERMINATION: events that cause an embryo to become male or female

Development of Testicular Feminization Syndrome

Introductory biology courses teach that a fertilized egg that receives two X chromosomes at conception will be a girl, whereas a fertilized egg that receives an X and a Y chromosome will become a boy. Research has produced evidence that there is more to the development of a person's gender. Gender development in mammals begins at conception with the establishment of chromosomal sex (the presence of XX or XY chromosomes). Even twelve weeks into development, male and female embryos have the same external appearance. Internal structures for both sexes are also similar. However, the machinery has been set in motion to cause the external genitals to become male or female, with corresponding internal structures of the appropriate sex. The baby is usually born with the proper phenotype to match its chromosomal sex. However, development of the sex organs is controlled by several genes. This leaves a great deal of room for developmental errors to occur.

The primary gene involved in sex determination is carried on the Y chromosome. It is responsible for converting the early unisex gonads into testes. Once formed, the testes then produce the balance of androgen and estrogen that pushes development in the direction of the male phenotype. In the absence of this gene, the undetermined gonads become ovaries, and the female phenotype emerges. Therefore, the main cause of sex determination is not XX or XY chromosomes, but rather the presence or absence of the gene that promotes testis differentiation.

In order for the male hormones to have an influence on the development of the internal and external reproductive structures, the cells of those structures must receive a signal that they are part of a male animal. The androgens produced by the testes are capable of entering a cell through the cell membrane. Inside the cell, the androgens attach to specific protein receptor molecules (androgen receptors). The process of attaching causes the receptors to move from the cytoplasm into the nucleus of the cell. Once in the nucleus, the receptor-plus-steroid complexes bind to deoxyribonucleic acid (DNA) near genes that are designed to respond to the presence of these hormones. The binding event is part of the process that turns on specific genes—in this case, the genes that di-

rect the process of building male genitals from the unisex embryonic structures as well as those that suppress the embryonic female uterus and tubes present in the embryo's abdomen.

In cases of testicular feminization, androgen receptors are missing from male cells. This is the result of a recessive allele located on the X chromosome. Because normal males have only one X, the presence of a recessive allele on that X will result in no manufacture of the androgen receptor in that individual. The developing embryo is producing androgen in the testes; without the receptor molecules, however, the cells of the genitals are unable to sense the androgen and respond to it. For this reason, the disorder is sometimes known by an alternate name: androgen-insensitivity syndrome. The cells of the genitals are still capable of responding to estrogen from the testes. As a result, the genitals become feminized: labia and clitoris instead of a scrotum and penis, and a short, blind vagina. To the obstetrician and parents, the baby appears to be a perfect little girl. An internal examination would show the presence of testes rather than ovaries and the lack of a uterus and tubes, but there would normally be no reason for such an examination to be performed.

Impact and Applications

Several events may lead to the diagnosis of this condition. The attempted descent of the testes into a nonexistent scrotum will cause pain that may be mistaken for the pain of a hernia; the presence of testes in the apparent girl will be discovered when the child undergoes repair surgery. In other cases, the child may seek medical help in the mid-teen years because she does not menstruate. Exploratory surgery would then reveal the presence of testes and the absence of a uterus. As a general rule, the testes are left in the abdomen until after puberty because they are needed as a source of estrogen to promote the secondary sex characteristics, such as breast development. Without this estrogen, the girl would remain childlike in body form. After puberty, the testes are usually removed because they have a tendency to become cancerous.

As a result of its phenotypic sex, an infant

with testicular feminization is normally raised as a girl whose only problem is an inability to bear children. If the girl has athletic ability, however, other problems may arise. Since 1966, female Olympic athletes have had to submit to a test for the presence of the correct chromosomal sex. In the past, this has meant microscopic examination of cheek cells to count X chromosomes. In 1992, this technique was replaced by a test for the Y chromosome. Individuals who fail the "sex test," including those with testicular feminization syndrome, cannot compete against other women. Proponents argue that androgens aid muscle development, and the extra testosterone produced by the testes of a normal male would provide an unfair physical advantage. However, because people with testicular feminization syndrome are lacking androgen receptors, their muscle development would be unaffected by the extra androgen produced by the testes, and thus they would not be any stronger than well-conditioned women.

—*Nancy N. Shontz*

See Also: Gender Identity; Hermaphrodites; Pseudohermaphrodites.

Further Reading: Elaine Mange and Arthur Mange, *Basic Human Genetics* (1994), provides a more detailed discussion of testicular feminization syndrome. The syndrome's relationship to athletes appears in Michael Lemonick, "Genetic Tests Under Fire," *Time* 139 (February 24, 1992) and "Experts Slam Olympic Gene Test," *Science* 255 (February 28, 1992).

Thalidomide and Other Teratogens

Field of study: Human genetics
Significance: *Teratogenesis is the prenatal toxicity of the embryo or fetus that arises from chemical, radiational, or environmental poisoning and leads to abnormal physical formation and development. Thalidomide, a sedative whose ingestion by pregnant women led to the birth of abnormal babies in the late 1950's and early 1960's, represents the most significant of all chemical teratogens.*

Key terms

CONGENITAL DEFECT: a defect or disorder that usually occurs before or during the time of birth

PEROMELIA: the congenital absence or malformation of the extremities caused by abnormal formation and development of the limb bud from about the fourth to the eighth week after conception; the ingestion of thalidomide by pregnant women was found to create this disorder in fetuses

Teratogenesis and Its Causes

Teratogenesis is any prenatal toxicity characterized by structural, functional, or developmental abnormalities found in an embryo or fetus. The term, derived from the Greek words *teras* (monster) and *genesis* (birth), also refers to intrauterine growth and retardation that may lead to the death of the affected subject. The phenomenon is usually attributed to exposure of the mother to poison during the early stages of pregnancy. Such poisoning may be induced by environmental chemicals, excessive radiation counts, viral infections, or drug administration. Approximately 3 percent of the developmental abnormalities are attributed to drugs. Drugs that are taken by the male partner may be teratogenic only if they damage the chromosomes of the spermatozoa.

From many centuries, the impression that malformed babies were conceived as a result of the intercourse between humans and devils or animals dominated society. Seventeenth century English physiologist William Harvey attributed teratogenesis to embryonic development. In the nineteenth century, the French brothers Etienne and Isidore Geoffroy Saint-Hilaire outlined a systematic study on the science of teratology. In the United States, the importance of teratogens was first widely covered during the 1940's, when scientists discovered that pregnant women who were affected by German measles (rubella) often gave birth to babies that had one or more birth defects. In the 1940's and 1950's, the consumption of diethylstilbestrol (DES) before the ninth week of gestation to prevent miscarriage was found to produce transplacental carcinogenesis. Animal studies have also shown that defective offspring

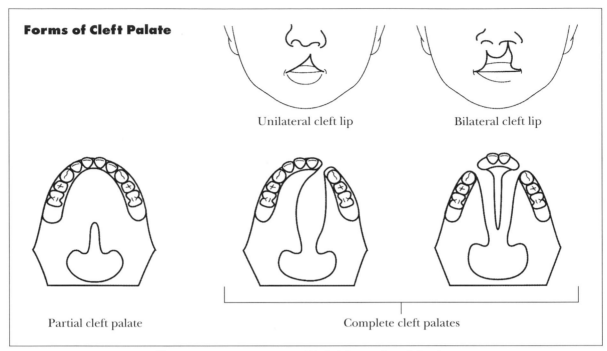

Forms of Cleft Palate

Unilateral cleft lip

Bilateral cleft lip

Partial cleft palate

Complete cleft palates

Teratogens can cause a range of birth defects such as cleft palate.

result from the use of hallucinogens such as lysergic acid diethylamide (LSD).

A broader definition of teratogenesis may include other minor birth defects that are more likely to be genetically linked, such as clubfoot, cleft lip, and cleft palate. These defects can often be treated in a much more effective way than those caused by toxic substances. Clubfoot, for example, which can be detected by the unusual twisted position of one or both feet, may be treated with surgery and physical therapy within the first month after birth. Brachydactyly (short digits) in rabbits has been linked to a recessive gene that causes a local breakdown of the circulation in the developing bud of the embryo, which is followed by necrosis (tissue death) and healing. In more extreme cases of agenesis, such as limb absence, a fold of amnion (embryonic membrane) was found to cause strangulation of the limb. Agenesis has been observed with organs such as kidneys, bladders, testicles, ovaries, thyroids, and lungs. Other genetic teratogenic malformations include anencephaly (absence of brain at birth), microcephaly (small-size head), hydrocephaly (large-size head caused by

accumulation of large amounts of fluids), spina bifida (failure of the spine to close over the spinal cord), cleft palate (lack of fusion in the ventral laminae), and hermaphrodism (presence of both male and female sexual organs).

Thalidomide and Its Impact

Thalidomide resembles glutethimide in its sedative action. Laboratory studies of the late 1950's and early 1960's had shown thalidomide to be a safe sedative for pregnant women. As early as 1958, the West German government made the medicine available without prescription. Other Western European countries followed, with the medicine available only upon physician's prescription. It took several years for the human population to provide the evidence that laboratory animals could not. German physician Widukind Lenz established the role of thalidomide in a series of congenital defects. He proved that administration of the drug during the first twelve weeks of the mother's pregnancy led to the development of phocomelia, a condition characterized by peromelia (the congenital absence or malformation of the extremities caused by the abnormal formation and development of the limb bud from about the fourth to the eighth week after conception), absence or malformation of the external ear, fusion defects of the eye, and absence of the normal openings of the gastro-intestinal system of the body.

The United States escaped the thalidomide tragedy to a great extent because of the efforts of Frances O. Kelsey, M.D., of the U.S. Food and Drug Administration (FDA). She had serious doubts about the drug's safety and was instrumental in banning the approval of thalidomide for marketing in the United States. Other scientists such as Helen Brooke Taussig, a pioneer of pediatric cardiology and one of the physicians who outlined the surgery on babies with the Fallot (blue baby) syndrome, played a key role in preventing the approval of thalidomide by the FDA. It is estimated that about seven thousand births were affected by the ingestion of thalidomide.

The thalidomide incident made all scientists more skeptical about the final approval of any type of medicine, especially those likely to be used during pregnancy. The current trend has intensified the fight against any chemicals that may affect the fetus during the first trimester, when it is particularly vulnerable to teratogens. Alcohol and tobacco drew many headlines in the media in the 1990's. Both have been shown to create congenital problems in mental development and learning abilities. At the same time, regulation of new FDA-approved medicine has become much stricter, and efforts to study the long-term effects of various pharmaceuticals have increased. Surprisingly, thalidomide itself has been used successfully in leprosy cases and, in conjunction with cyclosporine, to treat cases of the immune reaction that appears in many bone-marrow transplant patients. There is also a movement to use thalidomide in the treatment of acquired immunodeficiency syndrome (AIDS).

Prevention of genetic defects is possible with the extensive development of genetics. Genetic counselors can provide parents with informa-

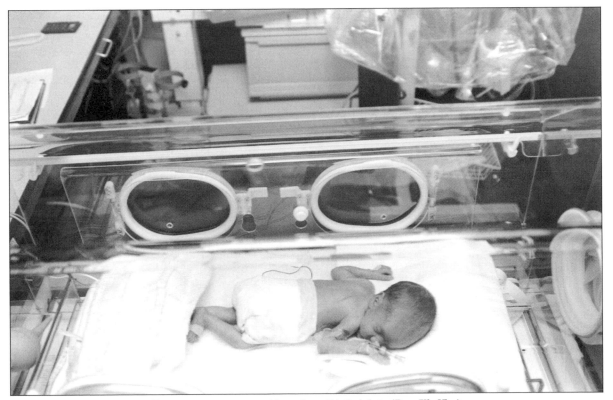

Teratogens can cause a wide variety of birth defects. (Ben Klaffke)

tion about possible genetic risks and can provide information about alternatives such as adoption if these genetic risks are too great. This, together with sonography and amniocentesis, has considerably reduced the number of malformed children.

—*Soraya Ghayourmanesh*

See Also: Congenital Defects; Developmental Genetics.

Further Reading: Mark Dowie, "Teratology: The Loneliest Science," *American Health* 9 (June, 1990), provides a thorough review of teratology. Max Sherman, "Infamous Drugs: A Quick History," *Drug and Cosmetic Industry* 142 (February, 1988), discusses dangerous drugs. Sasha Nemecek, "Transforming Hyde into Jekyll: Researchers Redesign Thalidomide," *Scientific American* 273 (November, 1995), discusses the pros and cons of thalidomide.

Transgenic Organisms

Field of study: Genetic engineering and biotechnology

Significance: *A transgenic organism is an organism that has had foreign DNA inserted into it. Such organisms are used in molecular genetics to study the regulation of gene expression, in medical research as models for diseases, and to produce better crops and make vaccines and biopharmaceuticals. Concerns have been raised about the release of transgenic organisms into the environment and the safety of eating transgenic (genetically engineered) food.*

Key terms

DEOXYRIBONUCLEIC ACID (DNA): the genetic material for most organisms; a long-chain macromolecule made of units called "nucleotides"

MARKER: a gene that encodes an easily detected product, used to indicate that foreign DNA is in an organism

GENETIC TRANSFORMATION: the uptake and maintenance of foreign DNA in a cell

EUKARYOTE: an organism made up of cells with true nuclei, which contain chromosomes

The Production of Transgenic Organisms

Transformation methods generate transgenic bacteria, fungi, animals, and plants for basic research and practical applications. To produce a transgenic organism, there must be an efficient way to introduce deoxyribonucleic acid (DNA) into a cell. The introduced DNA must integrate into the organism's DNA in order to be maintained. There are many different methods used to produce transgenic organisms. One method is electroporation, in which individual cells are suspended in a solution containing the DNA to be introduced into the cells. The cells and the DNA solution are placed in a chamber with two electrodes and are then subjected to a high-voltage electric discharge. The discharge causes small ruptures in the lipid membranes of the cells that result in pores that allow small molecules and macromolecules to enter or exit the cells. The pores of some of the cells that have taken up DNA reseal spontaneously; those cells will likely be transformed. Electroporation is used to transform bacteria, yeast, and cultures of eukaryotic cells in suspension.

Microinjection is used to introduce DNA into eukaryotic cells. To create transgenic mice by microinjection, for example, fertilized eggs are collected from a female mouse. The eggs are held with micromanipulators that allow small movements while DNA is injected into their nuclei through a very fine hollow needle (with an outer diameter of one micron, or one-millionth of a meter). Following microinjection, the eggs are implanted into other female mice. After the mice are born, some of their DNA is analyzed to determine if the added DNA sequences are present.

Biolistics (a term that comes from combining "biological" and "ballistic") instruments are used to introduce DNA into plants, animals, and bacteria. John Sanford performed the first biolistic transformation in the late 1980's. He propelled small (approximately one micron in diameter) tungsten beads coated with DNA into onion cells by placing the DNA-coated beads on a bullet and using a gun to fire the bullet into the onion cells. He placed a metal plate with small holes in it between the gun and the onions. The plate stopped the bullet, but the holes in the plate allowed the beads to continue forward to hit the target tissue to be transformed. Since Sanford's first biolistic transformation experiments, improvements have been made in the technique that makes it easier to regulate the speed at which the beads are propelled. Biolistic instruments that use an electric discharge or helium gas to propel the beads have also been used.

Some organisms can be transformed without the use of specialized instruments. For example, the single-celled alga *Chlamydomonas* (in a mutant lacking cell walls) can be transformed by shaking the cells with beads in a solution of DNA. For animal cells, transformation can occur if the cells are incubated in calcium phosphate and foreign DNA.

The bacterium *Agrobacterium tumefaciens* is an extremely useful vector for introducing DNA into many species of plants. When *Agrobacterium* infects a host plant, a part of the tumor-inducing (Ti) plasmid of the bacterium is transferred from the bacterium to the plant. The transferred DNA (called the T-DNA) is integrated into the plant's nuclear DNA. The normal T-DNA contains genes that encode plant growth hormones and cause the production of a plant tumor called a "crown gall." Scientists learned that any DNA within the T-DNA will be transferred to the plant and integrated into the plant's nuclear DNA. Using recombinant DNA methods, the tumor-causing genes are deleted from the T-DNA, and any DNA of interest can be inserted. The modified T-DNA is put back into *Agrobacterium* and transferred to the plant by the normal infection process. Intact plants can be produced.

Impact and Applications

In basic molecular genetics research, transgenic organisms are used to study the regulation of gene expression. DNA sequences needed for gene regulation are determined by deleting or mutating parts of the sequences, transforming the modified DNA into the organism, and analyzing the effect the changes have on the expression of the gene. Such methods have defined sequences important for tissue-specific and developmentally specific gene expression.

Transgenic plants with heightened insect and herbicide resistance have been made. The gene encoding *Bacillus thuringiensis* (*Bt*) toxin, an insecticidal protein produced by the bacterium, has been introduced into corn and cotton. These transgenic plants are resistant to feeding by certain insects, and their use should reduce the amount of insecticides that must be sprayed on these crops. However, there is a concern that the insects will eventually develop resistance to the *Bt* toxins in the crops.

Transgenic organisms can also be used to produce useful products. Charles Arntzen of the Boyce Thompson Institute has proposed the use of plant-based vaccines. In 1997, Kristian Dalsgaard and coworkers reported a successful plant-derived vaccine that protected target animals against a viral disease. Plant-based vaccines have the potential to be extremely useful. It would be easy to vaccinate a whole population if immunity could be achieved by simply eating a transgenic vaccine food.

Through recombinant DNA methods, genes for useful pharmaceutical proteins can be produced in the milk of transgenic livestock. To do this, a cloned gene of a useful pharmaceutical protein is fused to the transcriptional promoter of a gene that is expressed in milk, such as b-lactoglobin. Using microinjection, the gene is introduced into the livestock to produce the transgenic animals. Cows, sheep, and goats have been used. Cows produce large quantities of milk (10,000 liters per year), and it is relatively simple to isolate biopharmaceutical proteins from milk. An example of a transgenic biopharmaceutical is tpa (tissue plasminogen activator), which is administered to heart-attack victims to dissolve clots. The tpa gene has been cloned into goats, which then produce the tpa protein in their milk.

Transgenic animals are important models to use to study human diseases and the effects of different disease treatments. Transgenic animal models allow toxicological studies to be conducted in a shorter time and with fewer animals than with conventional animal bioassays. Transgenic mouse models have been used to help understand human genetic diseases such as Duchenne muscular dystrophy and cystic fibrosis (CF). For example, a transgenic

The transgenic bacterium Bacillus thuringiensis *is used to protect cotton and other crops.* (Linda K. Moore/ Rainbow)

mouse was made that lacked a gene involved in CF. This mouse was a null (lacking the gene) for the CF transmembrane conductance regulator. By introducing clones of pieces of human DNA into this null mouse and looking for the piece of DNA that made the CF model mouse normal, the piece of DNA that contained this gene was identified.

—Susan J. Karcher

See Also: Biopharmaceuticals; Biotechnology; Cloning; Genetic Engineering; Knockout Genetics and Knockout Mice.

Further Reading: William H. Velander et al. provide an overview of biopharmaceutical applications of transgenic animals in "Transgenic Livestock as Drug Factories," *Scientific American* 276 (January, 1997). Applications of transgenic animals are presented in Rudoff Jaenisch, "Transgenic Animals," *Science* 240 (June 10, 1988). Uses of transgenic crops are described in Charles S. Gasser and Robert T. Fraley, "Transgenic Crops," *Scientific American* 266 (June, 1992) and in Pamela C. Ronald, "Making Rice Disease-Resistant," *Scientific American* 277 (November, 1997). James D. Watson et al., *Recombinant DNA* (1992), provides an overview of methods for making transgenic organisms and applications of transgenic organisms.

Transposable Elements

Field of study: Bacterial genetics

Significance: *Transposable elements are discrete DNA sequences that have evolved the means to move (transpose) within the chromosomes. Transposition results in mutation and potentially large-scale genome rearrangements. Transposable elements contribute to the problem of multiple antibiotic resistance by mobilizing the genes for antibiotic resistance.*

Key terms

TRANSPOSASE: an enzyme encoded by a transposable element that initiates transposition by cutting specifically at the ends of the element and randomly at the site of insertion

COMPOSITE TRANSPOSON: a transposable element that contains genes other than those required for transposition

SELFISH DEOXYRIBONUCLEIC ACID (SELFISH DNA): a DNA sequence that spreads by forming additional copies of itself within the genome

RESISTANCE PLASMID (R PLASMID): a small, circular DNA molecule that replicates independently of the bacterial host chromosome and encodes a gene for antibiotic resistance

Jumping Genes

Transposable elements are deoxyribonucleic acid (DNA) sequences that are capable of moving from one chromosomal location to another in the same cell. In some senses, transposable elements have been likened to intracellular viruses. The first genetic evidence for transposable elements was described by Barbara McClintock in the 1940's. She was studying the genetics of the pigmentation of maize kernels and realized that the patterns of inheritance were not following Mendelian laws. Furthermore, she surmised that insertion and excision of DNA sequences was responsible for the genetic patterns she observed. McClintock was recognized for this pioneering work with a Nobel Prize in Physiology or Medicine in 1983. It was not until the 1960's that the jumping genes that McClintock postulated were isolated and characterized. The first transposable elements to be well characterized were found in the bacteria *Escherichia coli* but have subsequently been found in the cells of many bacteria, plants, and animals.

Transposable elements are discrete DNA sequences that encode a transposase, an enzyme that catalyzes transposition. Transposition refers to the movement within a genome. The borders of the transposable element are defined by specific DNA sequences; often the sequences at either end of the transposable element are inverted repeats of each other. The transposase enzyme cuts the DNA sequences at the ends of the transposable element to initiate transposition and cuts the DNA at the insertion site. The site for insertion of the transposable element is not specific. Therefore, transposition results in random insertion into chromosomes and often results in mutation and genome rearrangement. In many organisms, transposition accounts for a very significant fraction of all mutation. Although the details of the mechanism may vary, there are two basic mechanisms of transposition: conservative and replicative. In conservative transposition, the transposable element is excised from its original site and inserted at another. In replicative transposition, a copy of the transposable element is made and is inserted in a new location. The original transposable element remains at its initial site.

A subset of the replicative transposable elements includes the retrotransposons. These elements transpose through a ribonucleic acid (RNA) intermediate. Interestingly, their DNA sequence and organization is similar to retroviruses. It is likely that either retroviruses evolved from retrotransposons by gaining the genes to produce the proteins for a viral coat or retrotransposons evolved from retroviruses that lost the genes for a viral coat. This is one of the reasons that transposons are likened to viruses. Viruses can be thought of as transposons that gained the genes for a protein coat and thus the ability to leave one cell and infect others; conversely, transposons can be thought of as intracellular viruses.

Genetic Change and Selfish DNA

Transposition is a significant cause of mutation for many organisms. When McClintock studied the genetic patterns of maize kernel pigmentation, she saw the results of insertion and excision of transposable elements into and out of the pigment genes. Subsequently, it has been well established that mutations in many organisms are the result of insertion of transposable elements into and around genes. Transposition sometimes results in deletion mutations as well. Occasionally the transposase will cut at one end of the transposable element but skip the other end, cutting the DNA further downstream. This can result in a deletion of the DNA between the end of the transposable element and the cut site. In addition to these direct results, it is believed that transposable elements may be responsible for large-scale rearrangements of chromosomes. Genetic recombination, the exchange of genetic

1983 Nobel laureate Barbara McClintock, who first described transposable elements in the 1940's. (The Nobel Foundation)

information resulting in new combinations of DNA sequences, depends upon DNA sequence homology. Normally, recombination does not occur between nonhomologous chromosomes or between two parts of the same chromosome. However, transposition can create small regions of homology (the transposable element itself) spread throughout the chromosomes. Recombination occurring between homologous transposable elements can create deletions, inversions, and other large-scale rearrangements of chromosomes.

Scientists often take advantage of transposable elements to construct mutant organisms for study. The random nature of insertion ensures that many different genes can be mutated, the relatively large insertion makes it likely that there will be a complete loss of gene function, and the site of insertion is easy to locate to identify the mutated region.

Biologists often think of natural selection as working at the level of the organism. DNA sequences that confer a selective advantage to the organism are increased in number as a result of the increased reproductive success of the organisms that possess those sequences. It has been said that organisms are simply DNA's means of producing more DNA. In 1980, however, W. Ford Doolittle, Carmen Sapienza, Leslie Orgel, and Francis Crick elaborated on another kind of selection that occurs among DNA sequences within a cell. In this selection, DNA sequences are competing with each other to be replicated. DNA sequences that spread by forming additional copies of themselves will increase relative to other DNA sequences. There is selection for discrete DNA sequences to evolve the means to propagate themselves. One of the key points is that this selection does not work at the level of the organism's phenotype. There may be no advantage for the organism to have these DNA sequences. In fact, it may be that there is a slight disadvantage to having many of these DNA sequences. For this reason, DNA sequences that are selected for because of their tendency to make additional copies of themselves are referred to as "selfish" DNA. Transposable elements are often cited as examples of selfish DNA.

Impact and Applications

Some transposable elements have genes unrelated to the transposition process located between the inverted, repeat DNA sequences that define the ends of the element. These are referred to as composite transposons. Very frequently, bacterial composite transposons contain a gene that encodes resistance to antibiotics. The consequence is that the antibiotic resistance gene is mobilized: It will jump along with the rest of the transposable element to new DNA sites. Composite transposons may be generated when two of the same type of transposable elements end up near each other and flanking an antibiotic resistance gene. If mutations occurred to change the sequences at the "inside ends" of the transposable elements, the transposase would then only recognize and cut at the two "outside end" sequences to cause everything in between to be part of a new composite transposon.

Resistance to antibiotics is a growing public health problem that threatens to undo much of the progress that the antibiotic revolution made against infectious disease. Transposition of composite transposons is part of the problem. Transposition can occur between any two sites within the same cell, including between the chromosome and plasmid DNA. Plasmids are small, circular DNA molecules that replicate independently of the bacterial host chromosome. Resistance plasmids (R plasmids) are created when composite transposons carrying an antibiotic resistance gene insert into a plasmid. What makes this particularly serious is that some plasmids encode fertility factors (genes that promote the transfer of the plasmid from one bacteria to another). This provides a mechanism for rapid and widespread antibiotic resistance whenever antibiotics are used. The great selective pressure exerted by antibiotic use results in the spread of R plasmids throughout the bacterial population. This, in turn, increases the opportunities for composite transposon insertion into R plasmids to create multiple drug-resistant R plasmids. The first report of multiple antibiotic resistance caused by R plasmids was in Japan in 1957 when strains of *Shigella dysenteriae*, which causes dysentery, became resistant to four common antibiotics all at once. Some R plasmids encode resistance for up to eight different antibiotics. This often makes treatment of bacterial infection difficult. Furthermore, some plasmids are able to cause genetic transfer between bacterial species. This further limits the usefulness of many antibiotics.

—*Craig S. Laufer*

See Also: Bacterial Resistance and Super Bacteria; *Escherichia coli*; Genomics; Mutation and Mutagenesis.

Further Reading: L. E. Orgel and F. H. C. Crick, "Selfish DNA: The Ultimate Parasite," and W. Ford Doolittle and Carmen Sapienza, "Selfish Genes, the Phenotype Paradigm and Genome Evolution," both in *Nature* 284 (April 17, 1980), describe the fascinating hypothesis of selfish DNA. *The Antibiotic Resistance Paradox: How Miracle Drugs Are Destroying the Miracle* (1992), by S. B. Levy, examines the problem of antibiotic-resistant pathogens and how they came to be. For a more detailed overview of transposable elements, see S. N. Cohen and J. A. Shapiro, "Transposable Genetic Elements," *Scientific American* 40 (February, 1980).

Tumor-Suppressor Genes

Field of study: Molecular genetics
Significance: *Molecular analysis of tumor-suppressor genes has provided important information on mechanisms of cell cycle regulation and patterns of growth control in normal dividing cells and cancer cells. Tumor-suppressor genes represent cell cycle control genes that inhibit cell division and initiate cell death processes in abnormal cells. Mutations in these genes have been identified in many types of human cancer and play a critical role in the genetic destabilization and loss of growth control characteristic of malignancy.*

Key terms

CELL CYCLE: a highly regulated series of events critical to the initiation of cell division processes

MUTATION: any change in gene structure involving changes in nucleotide base sequence, including nucleotide substitutions, deletions, or genetic rearrangements

The Discovery of Tumor-Suppressor Genes

The existence of genes that play critical roles in cell cycle regulation by inhibiting cell division processes was predicted by several lines of evidence. In vitro studies involving the fusion of normal and cancer cell lines were often observed to result in suppression of the malignant phenotype, suggesting that normal cells contained inhibitors that could reprogram the abnormal growth behavior in the cancer cell lines. In addition, studies by Alfred Knudsen on inherited and noninherited forms of retinoblastoma, a childhood cancer associated with tumor formation in the eye, suggested that the inactivation of recessive genes as a consequence of mutation could result in the loss of function of inhibitory gene products critical to cell division control mechanisms. With the advent of molecular methods of genetic analysis, the gene whose inactivation is responsible for retinoblastoma was identified and designated *RB*.

Additional tumor-suppressor genes were identified by studies of deoxyribonucleic acid (DNA) tumor viruses whose cancer-causing properties were found to result, in part, from the ability of specific viral gene products to inactivate host cell inhibitory gene products involved in cell cycle regulation. By inactivating these host cell proteins, the tumor virus removes the constraints on viral and cellular proliferation. The most important cellular gene product to be identified in this way is the *p53* protein, named after its molecular weight. Genetic studies of human malignancies have implicated *p53* dysfunction in up to 75 percent of tumors of diverse tissue origin, including an inherited disorder called Li-Fraumeni syndrome, associated with many types of cancer. In addition, studies of other rare inherited malignancies have led to the identification of many other recessive genes whose inactivation contributes to oncogenic or cancer-causing mechanisms. Included in this list are the *BRCA1* and *BRCA2* genes in breast cancer, the *NF1* gene in neurofibromatosis, the *p16* gene in melanoma, and the *APC* gene in colorectal carcinoma. Each of these genes has also been implicated in nonhereditary cancers.

The Properties of Tumor-Suppressor Genes

Molecular analyses of the genetic and biochemical properties of tumor-suppressor genes have suggested that these gene products play critical but distinct roles in regulating processes involved in cellular proliferation. The *RB* gene represents a prototype tumor-suppressor gene that blocks cell cycle progression and cell division by binding to transcription factors in its active form. In order for cell division to occur in response to growth factor stimulation, elements of the signal cascade inactivate *RB*-mediated inhibition by a mechanism involving the addition of phosphate to the molecule, called phosphorylation. Loss of *RB* function as a consequence of mutation removes the brakes on this form of inhibitory control; the cell division machinery proceeds regardless of whether it has been appropriately initiated by growth factors or other mitogenic stimuli.

The *p53* tumor-suppressor gene product is a DNA-binding protein that regulates the expression of specific genes in response to genetic damage or other abnormal events that may occur during cell cycle progression. In response to *p53* activation, the cell may arrest the process of cell division (by indirectly blocking *RB* inactivation) to repair genetic damage before proceeding further along the cell cycle; alternatively, if the damage is too great, the *p53* gene product may initiate a process of cell death called apoptosis. The loss of *p53* activity in the cell as a consequence of mutation results in genetic destabilization and the failure of cell death mechanisms to eliminate damaged cells from the body; both events appear to be critical to late-stage oncogenic mechanisms.

Impact and Applications

The discovery of tumor-suppressor genes has revealed the existence of inhibitory mechanisms critical to the regulation of cellular proliferation. Mutations that destroy the functional activities of these gene products contribute to the cell cycle dysregulation that culminates in a loss of growth control characteristic of cancer cells. Taken together, research studies on the patterns of oncogene activation and the loss of tumor-suppressor gene function in many types of human malignancy suggest a general model of oncogenesis. Molecular analyses of many tumors show multiple genetic alterations involving both oncogenes and tumor-suppressor genes, suggesting that oncogenesis requires unregulated stimulation of cellular proliferation pathways and a concomitant loss of inhibitory activities that operate at cell cycle checkpoints.

With respect to clinical applications, restoration of *p53* tumor-suppressor gene function by gene therapy appears to result in tumor regression in some experimental systems; however, much more work needs to be done in this area to achieve clinical relevance. More important, research on the mechanism of action of standard chemotherapeutic drugs suggests that cytotoxicity may be mediated by *p53*-induced cell death; the absence of functional *p53* in many tumors may account for their resistance to chemotherapy. Promising research suggests that it may be possible to elicit cell death in tumor cells lacking functional *p53* gene product in response to chemotherapy. The clinical significance of activating these *p53*-independent cell death mechanisms may be extraordinary.

—*Sarah Crawford Martinelli*

See Also: Breast Cancer; Cancer; Oncogenes.

Further Reading: *Genes and the Biology of Cancer* (1993), by Harold Varmus and Robert Weinberg, provides a good general review of the topic. Alan Oliff et al., "New Molecular Targets for Cancer Therapy," *Scientific American* 275 (September, 1996), summarizes novel genetic approaches to cancer treatment. *Cancer Biology* (1995), by Raymond Ruddon, is a good general text.

Turner's Syndrome

Field of study: Human genetics

Significance: *Turner's syndrome is one of the most common genetic problems in women, affecting 1 out of every 2,000 to 2,500 women born. Short stature, infertility, and incomplete sexual development are the characteristics of this condition.*

Key terms

ESTROGENS: hormones or chemicals that stimulate the development of female sexual characteristics and control the reproductive cycle in women

GROWTH HORMONE: a chemical that plays a key role in promoting growth in body size

KARYOTYPE: a laboratory analysis or test that confirms the diagnosis of Turner's syndrome by documenting the absence or abnormality of one of the two X chromosomes normally found in women

SEX CHROMOSOMES: the chromosomes that control the sexual attributes of males and females; females have two X chromosomes and males have one X and one Y chromosome

SYNDROME: a set of features or symptoms often occurring together and believed to stem from the same cause

The Discovery of Turner's Syndrome

Henry H. Turner, an eminent clinical endocrinologist, is credited with first describing Turner's syndrome. In 1938, he published an article describing seven patients, ranging in age from fifteen to twenty-three years, who exhibited short stature, a lack of sexual development, arms that turned out slightly at the elbows, webbing of the neck, and low posterior hairline. He did not know what caused this condition. In 1959, C. E. Ford discovered that a chromosomal abnormality involving the sex chromosomes caused Turner's syndrome. He found that most girls with Turner's syndrome did not have all or part of one of their X chromosomes and argued that this missing genetic material accounted for the physical findings associated with the condition.

Turner's syndrome begins at conception. Scientists believe the disorder may result from an error that occurs during the division of the parents' sex cells, although the exact cause remains unknown. Girls suspected of having Turner's syndrome, usually because of their short stature, usually undergo chromosomal analysis. A simple blood test and laboratory analysis called a karyotype are done to document the existence of an abnormality.

Shortness is the most common characteristic of Turner's syndrome. The incidence of short stature among women with Turner's syndrome is virtually 100 percent. Women who have this condition are, on average, 4 feet 8 inches tall. The cause of the failure to grow is unclear. However, growth-promoting therapy with growth hormones has become standard therapy. Most women with the syndrome also experience ovarian failure. Since the ovaries normally produce estrogen, women with Turner's syndrome lack this essential hormone. This deficit results in infertility and incomplete sexual development. Cardiovascular disorders are the single source of increased mortality in women with this condition. High blood pressure is common.

Other physical features often associated with Turner's syndrome include puffy hands and feet at birth, a webbed neck, prominent ears, a small jaw, short fingers, a low hairline at the back of the neck, and soft fingernails that turn up at the end. Some women with Turner's syndrome have a tendency to become overweight. Many women will exhibit only a few of these distinctive features, and some may not show any of them. This condition does not affect general intelligence. Girls with Turner's syndrome follow a typical female developmental pattern with unambiguous female gender identification. However, another possible symptom is poor spatial perception abilities. For example, women with this condition may have difficulty driving, recognizing subtle social clues, and solving nonverbal mathematics problems; they may also suffer from clumsiness and attention-deficit disorder.

Treatments and Therapies

No treatment is available to correct the chromosome abnormality that causes this condition. However, injections of human growth hormone can restore most of the growth deficit. Unless they undergo hormone replacement therapy, girls with Turner's syndrome will not menstruate or develop breasts and pubic hair. In addition to estrogen replacement therapy, women with Turner's syndrome are often advised to take calcium and exercise regularly. Although infertility cannot be altered, pregnancy may be made possible through in vitro fertilization (fertilizing a woman's egg with sperm outside the body) and embryo transfer (moving the fertilized egg into a woman's uterus).

Individuals with Turner's syndrome can be healthy, happy, and productive members of society. Because of its relative rarity, a woman with Turner's syndrome may never meet another individual with this condition and may suffer from self-consciousness, embarrassment, and poor self-esteem. The attitudes of parents, siblings, and relatives are important in helping the girl develop a strong sense of identity and self-worth. The Turner's Syndrome Society of the United States is a key source of information and support groups. Advances in chromosomal analysis have proved helpful in the diagnosis and management of Turner's syndrome. In addition, new developments in hormonal therapy for short stature and ovarian failure, combined with advances in in vitro fertilization, have significantly improved the potential for growth, sexual development, and parenthood for afflicted individuals.

—*Fred Buchstein*

See Also: Congenital Defects; Genetic Testing; Heart Disease; In Vitro Fertilization and Embryo Transfer.

Further Reading: *Turner Syndrome: A Guide for Families* (1992), written by Patricia A. Rieser and Louis E. Underwood, and published by the Turner's Syndrome Society, describes Turner's syndrome, its causes, its features, and how families can help girls with the condition cope with the physical, social, and emotional concerns. The Turner's Syndrome Society also publishes other materials, including *Turner Syndrome: A Guide for Physicians* (1992), by Ron G. Rosenfeld, and "Recommendations for the Diagnosis, Treatment, and Management of Individuals with Turner's Syndrome," by Ron G. Rosenfeld et al., a reprint of a September, 1994, article appearing in *The Endocrinologist*. Additional information may be found in Paul Saenger, "Turner's Syndrome," *The New England Journal of Medicine* (December 5, 1996).

Twin Studies

Field of study: Human genetics
Significance: *Studies of twins are widely considered to be the best way to determine the extent to which human physical and psychological characteristics are caused by genetic factors rather than environmental factors.*

Key terms

DIZYGOTIC: developed from two separate zygotes; fraternal twins are dizygotic because they develop from two separate fertilized ova (eggs)

MONOZYGOTIC: developed from a single zygote; identical twins are monozygotic because they develop from a single fertilized ovum that splits in two

ZYGOSITY: the degree to which two individuals are genetically similar

ZYGOTE: a cell formed from the union of a sperm and an ovum

The Origin of Twin Studies

Sir Francis Galton, an early pioneer in the science of genetics and a founder of the theory of eugenics, conducted some of the earliest systematic studies of human twins in the 1870's. Galton recognized the difficulty of identifying the extent to which human traits are biologically inherited and the extent to which traits are produced by diet, upbringing, education, and other environmental influences. Borrowing a phrase from William Shakespeare, Galton called this the "nature versus nurture" problem. Galton reasoned that he could attempt to find an answer to this problem by comparing similarities among people who obviously shared a great deal of biological inheritance with similarities among people sharing less biological inheritance. Twins offered the clearest example of people who shared common biological backgrounds.

Galton contacted all of the twins he knew and asked them to supply him with the names of other twins. He obtained ninety-four sets of twins. Of these, thirty-five sets were very similar, people who would today be called identical twins. These thirty-five pairs reported that people often had difficulty telling them apart. Using questionnaires and interviews, Galton compared the thirty-five identical pairs with the other twins. He found that the identical twins were much more similar to one another in habits, interests, and personalities, as well as in appearance. They were even much more alike in physical health and susceptibility to illness. The one area in which all individuals seemed to differ markedly was in handwriting.

Modern Twin Studies

Since Galton's time, researchers have discovered how biological inheritance occurs, and this has made possible an understanding of why twins are similar. It has also enabled researchers to make more sophisticated use of twins in studies that address various aspects of the nature versus nurture problem. Parents pass their physical traits to their children by means of genes in chromosomes. Each chromosome carries two genes for every hereditary trait. One gene comes from the father and one comes from the mother. Any set of full brothers and sisters will have a good deal of heredity in common, since all of their genes come from the same parents. However, brothers and sisters usually also differ substantially. The cell that develops into a living creature, the zygote, is created when the ovum (egg) is fertilized by the sperm. Each zygote will combine genes from the father and the mother in a unique manner, so different zygotes will develop into quite different people. Even when two fertilized eggs are present at the same time, as in the case of dizygotic or fraternal twins, the two will have different combinations of genes from the mother and the father.

Identical twins are an exception to the rule of unique combinations of genes. Identical twins develop from a single zygote, a cell created by one union of egg and sperm. Therefore, monozygotic twins (from one zygote) will normally have the same genetic makeup. Differences between genetic twins, researchers argue, must therefore be produced by environmental factors following birth.

The ideal way to conduct twin studies is to compare monozygotic twins who have been reared apart from each other in vastly different types of families. This is rarely possible, however, because the numbers of twins separated at birth and adopted are quite limited. For this reason, researchers in most twin studies use fraternal twins as a comparison group, since the major difference between monozygotic and dizygotic twins is that the former are ge-

netically identical. Statistical similarities among monozygotic twins that are not found among dizygotic twins are therefore believed to be caused by genetic inheritance.

Researchers use several collections of information on twins to investigate the extent to which human characteristics appear to be a consequence of genetics. One of the main sources for twin studies is the Minnesota Twin Registry. In the 1990's, this registry consisted of about 10,500 twin individuals in Minnesota. They were found in Minnesota birth records of the years 1936 through 1955, and they were located and recruited by mail between 1985 and 1990. A second major source of twin studies is the Virginia Twin Registry. This is a register of twins constructed from a systematic review of public birth records in the Commonwealth of Virginia. A few other states also maintain records of twins. Some other organizations, such as the American Association of Retired People (AARP), keep records of twins who volunteer to participate and make these records available to researchers.

Zygosity, or degree of genetic similarity between twins, is usually measured by survey questions about physical similarity and by how often other people mistake one twin for the other. In some cases, zygosity may be determined more rigorously through deoxyribonucleic acid (DNA) samples.

Problems with Twin Studies

Although twin studies are one of the best available means for studying genetic influences on human beings, there are a number of problems with the twin approach. Although twin studies assume that monozygotic twins are biologically identical, some critics have claimed that there are reasons to question this assumption. Even though these twins tend to show greater uniformity than other people, developmental differences may emerge even in the womb after the splitting of the zygote.

Twins who show a great physical similarity may also be subject to environmental similarities so that traits believed to be caused by genetics may, in fact, be a result of upbringing. Some parents, for example, dress twins in matching clothing. Even when twins grow up in separate

Identical twins are often studied to measure the relative importance of genetic and environmental factors. (Ben Klaffke)

homes without being in contact with each other, their appearances and mannerisms may evoke the same kinds of responses from others. Physical attractiveness, height, and other characteristics often affect how individuals are treated by others so that the biologically based resemblances of twins can lead to common experiences.

Finally, critics of twin studies point out that twins constitute a special group of people and that it may be difficult to apply findings from twin studies to the population at large. Some studies have indicated that intelligence quotient (IQ) scores of twins, on average, are about five points below IQ scores in the general population, and twins may differ from the general population in other respects. It is conceivable that twins are more genetically determined than most other people.

Impact and Applications

Twin studies have provided evidence that a substantial amount of human character and behavior may be genetically determined. In 1976, psychologists John C. Loehlin and Robert C. Nichols published their analyses of the backgrounds and performances of 850 sets of twins who took the 1962 National Merit Scholarship test. Results showed that identical twins showed greater similarities than fraternal twins in abilities, personalities, opinions, and ambitions. A careful examination of backgrounds indicated that these similarities could not be explained by the similar treatment received by identical twins during upbringing.

Later twin studies continued to provide evidence that genes shape many areas of human life. Monozygotic twins tend to resemble each other in probabilities of developing mental illnesses, such as schizophrenia and depression, suggesting that these psychological problems are partly genetic in origin. A 1996 study published in the *Journal of Personality and Social Psychology* used a sample from the Minnesota Twin Registry to establish that identical twins are similar in probabilities of divorce. A 1997 study in the *American Journal of Psychiatry* indicated that there is even a great resemblance between twins in intensity of religious faith. Twin studies have offered evidence that homosexual or heterosexual orientation may be partly a genetic matter, although researcher Scott L. Hershberger has found that the genetic inheritance of sexual orientation may be greater among women than among men.

—*Carl L. Bankston III*

See Also: Behavior; Biological Determinism; Heredity and Environment; Intelligence.

Further Reading: One of the most influential books on modern twin studies is *Heredity, Environment, and Personality* (1976), by John C. Loehlin and Robert C. Nichols. Lawrence Wright's *Twins and What They Tell Us About Who We Are* (1997) is an overview of the use of twin studies in behavioral genetics, written for general readers. *Twins and Homosexuality: A Casebook* (1990), by Geoff Puterbaugh, describes ongoing twin studies that investigate genetic influences on sexual orientation. In *DNA and Destiny: Nurture and Nature in Human Behavior* (1996), medical researcher and popular science writer R. Grant Steen summarizes evidence regarding the relative contributions of genetics and environment to shaping the human personality.

Viral Genetics

Field of study: Bacterial genetics

Significance: *The genetic simplicity and diversity of viruses have provided unique tools for studying complex biological systems. Viruses have impacted and will continue to impact human health and disease.*

Key terms

DEOXYRIBONUCLEIC ACID (DNA): the genetic material for most organisms, a two-stranded chain of nucleotide subunits

ENZYME: a molecule (usually a protein) that accelerates chemical reactions in a cell

GENOME: all of the genetic information for an organism, consisting of one or more DNA molecules

PROTEIN: a complex molecule composed of a folded chain of amino acids

RIBONUCLEIC ACID (RNA): a molecule structurally related to DNA used to express genetic information

TRANSCRIPTION: the process of copying DNA into RNA during gene expression

TRANSLATION: the process of converting genetic information from RNA into protein

The Discovery of Viruses

The impact of viruses on human, plant, and animal health has been known for centuries. Efforts to control viral infections have their roots in the observations and experiments of English physician Edward Jenner in the late 1700's. He observed that women exposed to cowpox appeared to resist infection by smallpox. In 1796, Jenner experimentally exposed a boy to cowpox by scratching his skin with a needle containing fluid from a cowpox sore. The boy was able to resist a later exposure to smallpox. Even though Jenner did not know that related viruses are responsible for both cowpox and smallpox, his experiment is the basis for the understanding of vaccination for disease prevention. French chemist Louis Pasteur introduced systematic analysis of viral infections with his pioneering studies of the rabies virus in 1881. Toward the end of the nineteenth century, Adolf Mayer, Dimitri Ivanovsky, and Martinus Beijerinck each contributed observations that identified a new type of infectious agent as the cause of tobacco mosaic disease. The term "virus" would eventually be applied to these agents, and they were soon identified as the cause of disease in a wide variety of organisms, including humans.

Throughout the twentieth century, new techniques were developed that provided a more detailed picture of the structure and function of viruses. With understanding came new weapons to combat viral diseases, and some important victories were recorded. Through an unprecedented worldwide effort of vaccination, wild smallpox was eradicated from the planet in 1977. A similar effort was launched to eliminate poliovirus infections by the year 2000. Despite these high-profile successes, new viral diseases are still being uncovered, and well-known viruses continue to impact human health and world economy.

Virus Structure

Despite their complexity and diversity, two structural components are shared by all viruses: a nucleic acid molecule that carries its genetic information, and a protein coat that surrounds the genetic material. In addition, many viruses are enclosed by an envelope derived from a membrane of the infected cell. The fully assembled, infectious virus particle is called a "virion."

The genetic material of all living things consists of one or more molecules of deoxyribonucleic acid (DNA). The genome of an organism is the total content of its genetic material, including all of its genes. DNA is a long chain consisting of thousands to hundreds of millions of repeated nucleotide subunits. There are four different nucleotides, and their order along the DNA chain determines the genetic blueprint for each organism. The molecule resembles a ladder, with two chains each represented by a rail and with rungs composed of the interacting chemical bases. The chains can separate, like opening a zipper, a process that permits the molecule to be copied precisely. Many viruses also have a genome of double-stranded DNA, but viruses are unique in allowing variations in their genetic material.

A closely related molecule called ribonucleic acid (RNA) is an important molecule for many cell functions, particularly expression of genes. The genetic information stored in DNA is copied or transcribed into a message of RNA, which is then translated into protein. RNA does not normally store genetic information. Many viruses, however, use RNA instead of DNA as their genetic material. Viruses, then, can have genetic molecules of either RNA or DNA.

There are other features of the genome that show variability across different families of viruses. In addition to being either DNA or RNA, a virus genome can be single-stranded or double-stranded, linear or circular, and monopartite (one molecule) or multipartite (segmented). The genome may be only a few thousand nucleotides long and have as few as four genes, or it may be hundreds of thousands of units long and encode several hundred genes. This variability in genome structure means that viruses have evolved many unique and interesting strategies for expressing their genes.

Proteins are an extremely diverse class of molecules consisting of long chains of amino acids folded into precise, three-dimensional structures. They carry out many important functions in all cells. The protein capsid of a virus surrounds the nucleic acid and may participate in the infection process. Most capsids consist of only a few different proteins that are present in many copies. There are two common capsid shapes, helical and spherical, determined by the organization of proteins and nucleic acids in the capsid. Helical capsids are usually rod shaped. The genome is surrounded by proteins that form into a coil. Spherical capsids have their protein subunits arranged into an icosahedron, a twenty-sided polygon. Some of the simplest viruses have a single genome molecule encased in a capsid of a single type of protein, while many others have additional important components and exhibit considerable structural complexity.

The nucleocapsids of some viruses are surrounded by an envelope or membrane composed of fat-based lipid molecules. The membranes of cells and envelopes of viruses also have proteins attached, which protrude from the membrane. The membrane proteins of viruses are encoded in the virus genome and play an important role in virus infection.

Virus Infection

Viruses are unable to reproduce outside of a host cell. They use the raw materials and machinery of the infected cell to manufacture the components of new virus particles. The virus essentially converts the infected cell into a virus-making factory. Most plant, animal, and bacterial hosts have evolved elaborate mechanisms to fight off or prevent viral infections, so the virus must be quick or stealthy to succeed in its mission. There are as many different infection strategies as there are viruses, but most follow a general set of steps that have been well characterized.

In order to initiate an infection, a viral genome must get inside a suitable host cell. This entry step is preceded by attachment of the virus to the outside of the host cell. This attachment involves a very specific interaction between a molecule on the outside of the virus and a molecule on the outside of the cell. The virus component in this attachment is usually a protein, either part of the virus capsid or one of the envelope proteins. The cell component is called the receptor molecule and can be almost any type of molecule found on the outside of a cell. The receptor is very specific for each virus: The presence of the appropriate receptor is a major factor in determining which cells and organisms a particular virus can infect. Attachment can be thought of as a lock-and-key interaction: The attachment molecule on the virus is the "key" and will only work with a precise fit to the correct "lock" on the surface of the cell. Once attached, the cell often participates willingly in the entry of the virus. Many important molecules enter a cell via receptor-mediated attachment, so the virus uses a sort of "Trojan horse" strategy to gain entry to the cell. Once inside the cell, the viral genome is released from the capsid, or uncoated, permitting expression of the viral genes.

Once exposed, the genome of the virus is used for two different but related functions:

gene expression and replication. The viral genes are expressed, allowing viral proteins to be made by the cell. These proteins include capsid and envelope proteins, which become part of the new viruses, and other proteins required for copying the genome and assembling the new virus particles. Different viruses have different requirements and strategies for gene expression. The genome is also used as a template, or "mold," to make many new complete genomes that will become part of the progeny virions. This copying of the genetic material is called "genome replication." The timing and mechanics of gene expression and genome replication vary greatly for different families of viruses.

As newly synthesized genomes and capsid proteins accumulate, new virus particles are assembled in the infected cell. Capsid proteins can often self-assemble into the appropriate structure, but the assembly process can be quite different for different viruses. Incorporation of the viral genome into the capsid, or packaging, must be specific for the viral nucleic acid, and multipartite viruses must package the appropriate type and number of nucleic acid molecules.

Once assembled, new virions must escape the infected cell in order to continue the infection or find a new host organism. For many viruses, release is dependent on destruction of the host cell. As the cell disintegrates, the new virions are released to infect new cells. Most enveloped viruses complete their assembly and release in a process called "budding." Viral membrane proteins accumulate in the cell membrane and attract nucleocapsids that push out the membrane into a bud. The bud eventually pinches off, releasing the enveloped virion. The cell may survive this process or eventually be destroyed by the infection.

Virus Gene Expression: DNA Viruses

All of a cell's genes are found in its DNA genome, so viruses that have genomes of double-stranded DNA have the advantage of being able to use host-cell enzymes to express their genes. Often a cascade of viral gene expression is observed. As soon as the viral genome enters the cell nucleus, a subset of the viral genes referred to as "early" genes are expressed, using only host-cell enzymes for expression. One or more of the genes expressed at this early time make proteins that are required for the efficient expression of the next set of genes, the "intermediate" genes. These often include the genes for replication of the viral DNA. A third set, the "late" genes, is activated by other intermediate gene products. These are the proteins that are used in the assembly of new virus particles.

Most DNA viruses use host-cell transcription machinery and other enzymes and must replicate in the cell nucleus since that is where the host-cell DNA is found. The members of the poxvirus family are an exception to this rule. These viruses have very large genomes and encode all of their own enzymes for DNA and RNA synthesis and modification. Several families of viruses have smaller double-stranded DNA genomes, including the papovaviruses and polyomaviruses. Since their genomes are smaller, they have less capacity to encode enzymes and rely even more heavily on the enzymes of the host cell. They may express one or a few early proteins that recruit the cells' DNA replication machinery to the viral genome and trick the cells' DNA polymerase (synthesizing) enzymes into copying their own DNA.

Parvoviruses represent another class of DNA virus, those with single-stranded DNA genomes. The sequence and structure of the viral DNA permits the cellular DNA polymerases to use the viral genome as a template to make a double-stranded version. The double-stranded DNA can then be used for expression of genes and synthesis of viral proteins. New single-stranded genomes are made in a complex process that is directly coupled to the assembly of new virions.

Virus Gene Expression: RNA Viruses

Viruses with RNA genomes face numerous obstacles to gene expression and have evolved some unique and interesting strategies to overcome them. Since the host cell never uses RNA as a template to synthesize nucleic acids, RNA viruses must encode all the enzymes for replication of their genomes and expression of their genes. Which strategy is used for these impor-

Technicians isolating an insect virus that will be used to produce recombinant proteins. (Hank Morgan/ Rainbow)

tant functions depends on the nature of the RNA genome.

If the viral RNA is single stranded, it can be in one of two polarities. If the RNA corresponds to a genetic message, the viral genome is considered to be a plus-strand RNA virus; if the genome is complementary to a genetic message, the virus is a minus-strand RNA virus. As soon as a plus-strand RNA virus enters a cell, its genome can be recognized by the cell's protein synthesis machinery and translated into viral proteins. One of these proteins is a replicase enzyme, which can use the viral genome as a template to make more viral RNA. This replicase will usually convert the single-stranded viral genome to a double-stranded intermediate. This molecule can then be used as a template to make many copies of the plus-strand RNA genome for synthesizing more viral proteins and new virus particles. Picornaviruses, including poliovirus and rhinoviruses (common cold viruses), are examples of plus-strand RNA viruses, as are members of the togavirus family. Togaviruses include many of the mosquito-borne viruses such as those that cause dengue fever and Saint Louis encephalitis, as well as rubella.

Minus-strand viruses have even larger obstacles to gene expression. Their genomes cannot be modified by host-cell enzymes, nor can they be translated to protein. These viruses must carry with them, as part of the virus particle, a transcriptase enzyme to copy the minus-strand RNA genome into a plus strand. This plus-sense RNA can then be used by the cell to make viral proteins. Once viral proteins are synthesized, the viral genome is used to make a complete plus-strand copy, which is then copied back to many minus-strand genomes for assembling new virus particles. This gene-expression strategy is used by many virus families, including orthomyxoviruses (influenza), rhabdoviruses (rabies), filoviruses (ebola), paramyxoviruses (mumps and measles), bunyaviruses (hantavirus), and arenaviruses (Lassa fever). A similar type of strategy is employed by reoviruses, which have genomes of double-stranded RNA.

The reoviruses and many of the minus-strand RNA viruses have multipartite genomes.

How these viruses assemble the right number and type of genome segments is not well understood. If a cell becomes infected by more than one virion of a multipartite virus, the genome segments may be "shuffled" or reassorted during assembly. This allows for rapid and dramatic genetic evolution of these viruses, permitting them to elude the host's immune defenses. Furthermore, viral enzymes that replicate RNA genomes are notoriously error prone. As a result, mutations accumulate more rapidly than in DNA genomes, and new virus strains evolve quickly, often keeping ahead of the host immune response. Control of influenza virus infections is difficult for both of these reasons.

Retroviruses are another family of RNA genome viruses with an unusual lifestyle. These viruses have RNA genomes that are converted after infection into a DNA version of the genome. This viral DNA genome then inserts itself into the DNA of the host cell in a form called a "provirus" and becomes a permanent part of the host cell's genetic information. The provirus DNA is recognized and expressed by host-cell enzymes. This provirus expresses individual viral genes for protein synthesis and also expresses full-length, plus-sense RNA for packaging into new virions. The new virions are formed by budding, so the infected cell often remains intact. Human immunodeficiency virus (HIV) is the best-known example of the retrovirus family.

Viruses and Cancer

It has long been recognized that certain families of viruses have the capacity to cause cancer in their hosts. These tumor viruses fall into two broad categories: DNA tumor viruses and RNA tumor viruses. DNA tumor virus cancer mechanisms are best understood in the group of "small" DNA tumor viruses, which includes the papovavirus family and the adenovirus family. These viruses all contain oncogenes, which encode one or more viral oncoproteins ("onco-" means tumor). These proteins interact with and inactivate different members of the class of cellular proteins known as tumor suppressors, which function by helping to regulate cell growth and division. When these proteins are inactivated by viral oncoproteins or by mutation, the cell is released from growth restraints and may eventually progress to cancer. Tumor suppressor inactivation may aid the virus life cycle in some way. Cancer of the cervix is most likely caused by the oncoproteins of the human papillomaviruses.

Retroviruses are the only group of RNA viruses known to induce cancer. These cancer-causing or transforming retroviruses fall into two classes. Acute transforming retroviruses carry oncogenes, just as the DNA tumor viruses do. These genes are not normally part of the virus genome but are inadvertently incorporated into the genome during the virus life cycle. They are derived from host-cell genes (proto-oncogenes) that serve important functions in the normal host cell. The virus often carries a defective version of this gene. Expression of this defective oncogene results in aberrant protein function that leads to cancer. Nearly twenty retroviral oncogenes have been identified.

Nondefective retroviruses do not have oncogenes. They induce cancer by altering expression of normal cellular genes (proto-oncogenes). Recall that the retrovirus life cycle requires integration of the viral genome into the host-cell DNA. The insertion site is random and may affect expression of nearby cellular genes. If a nearby gene is expressed at abnormal levels because of signals from the viral genome, the protein may alter growth properties of the cell, leading to cancer. Alternatively, the viral genome may integrate in the middle of an important growth regulatory gene, shutting it off completely and resulting in abnormal growth. There are numerous examples of animal retroviruses that cause cancer in their hosts, and human T-cell leukemia viruses (HTLV-I and II) cause cancer of white blood cells in humans.

Impact and Applications

Viruses are and will continue to be a major threat to human health. New and threatening viruses are continually emerging, and rapid global travel could permit a serious outbreak to spread into a worldwide epidemic before it is even recognized. The impact of viruses goes far beyond their direct influence on human health, however. Infections such as the common cold and influenza have a tremendous impact on the global economy, costing billions of dollars per year in lost work time, doctor visits, vaccinations, and over-the-counter medications in the United States alone. Many people think of viruses only from the standpoint of their impact on humans, but most organisms on the planet are at some degree of risk. Viruses of agricultural animals and plants affect availability and cost of many common meats, fruits, vegetables, and grains at the supermarket. For example, yields of cultivated potatoes can be reduced by 10 to 20 percent through potexvirus infection. Cucumber mosaic virus infects over eight hundred plant species and can reduce infected crop yields by 30 to 60 percent. Periodic outbreaks of avian influenza viruses have decimated regional chicken farms in the United States and elsewhere.

Molecular techniques have revealed new strategies for interfering with viral infections. As viral genetic mechanisms and structure are better understood, rational approaches to new drug design have been developed. Promising "new generation" antiviral therapies are being tested for the viruses that cause acquired immunodeficiency syndrome (AIDS), influenza, and even the common cold. Molecular approaches are also paving the way for "high-tech" vaccines that promise to be more effective and safer, and to cause fewer side effects. The challenge for the future will be to make these new generation antiviral strategies available to developing areas of the world where viral infections are responsible for high mortality rates, particularly in infants.

Viruses are being used as tools in the fight against other human diseases with increasing frequency. Manipulation of viral genomes may one day permit a single viral vaccine to protect an individual from several different viruses and even protect against diseases caused by bacteria or other organisms. Engineered viruses may become a mechanism of choice in gene therapy (delivering human genes to patients in an effort to treat or correct important genetic diseases).

—Michael R. Lentz

See Also: DNA Replication; DNA Structure and Function; Gene Therapy; Molecular Genetics; Oncogenes; Protein Structure; RNA Structure and Function.

Further Reading: *The Coming Plague: Newly Emerging Diseases in a World out of Balance* (1994), by Laurie Garrett, and *The Hot Zone* (1994), by Richard M. Preston, provide a fascinating look at the world of viral epidemiology. In *The Science of Viruses: What They Are, Why They Make Us Sick,* *How They Will Change the Future* (1993), Ann Guidici explains in easy-to-understand language the basics of virology and viral disease. *A Field Guide to Germs* (1995), by Wayne Biddle, describes nearly one hundred common human pathogens. *Emerging Viruses* (1993), edited by Stephen Morse, is an excellent virology reference work including history, techniques, virus evolution, and other topics.

X Chromosome Inactivation

Field of study: Developmental genetics

Significance: *Normal females have two X chromosomes, and normal males have one X chromosome. In order to compensate for this apparent doubling of gene products in females, one X chromosome is randomly inactivated in each cell of females.*

Key terms

RIBONUCLEIC ACID (RNA): genetic material copied from deoxyribonucleic acid (DNA) by the process of transcription

BARR BODY: a highly condensed and inactivated X chromosome

CENTRAL DOGMA: the concept that genetic information is transferred from the code of DNA into RNA, which is then interpreted into a protein; there are now known to be exceptions to this generality

DOSAGE COMPENSATION: an equalization of gene products produced in males and females by the inactivation of one female X chromosome

MOSAIC: an individual possessing cells with more than one type of genetic constitution

SEX CHROMOSOMES: the X and Y chromosomes; females possess two X chromosomes, while males possess one X and one Y chromosome

The History of X Chromosome Inactivation

In 1961, Mary Lyon hypothesized that gene products were found in equal amounts in males and females because one of the X chromosomes in females became inactivated early in development. This hypothesis became known as the Lyon hypothesis, and the process became known as Lyonization, or X chromosome inactivation. Prior to this explanation, it was recognized that females had two X chromosomes and males had only one X chromosome, yet the proteins encoded by genes on the X chromosomes were found in equal amounts in females and males because of dosage compensation.

The principles of inheritance dictate that individuals receive half of their chromosomes from their fathers and the other half from their mothers at conception. Therefore, a female possesses two different X chromosomes (one

from each parent). In addition to hypothesizing the inactivation of one X chromosome in each cell, the Lyon hypothesis also implies that the event occurs randomly. In any individual, approximately one-half of the paternal X chromosomes and one-half of the maternal X chromosomes would be inactivated. It could be said that females represent a mosaic condition since half of their cells express the X chromosome genes inherited from the father and half of their cells express the X chromosome genes inherited from the mother. In fact, this situation can be seen in individuals who inherit a different form of a protein from each parent: Some cells express one parent's protein form, while other cells express the other parent's protein form.

Prior to Lyon's hypothesis, it was known that a densely staining material could be seen in cells from females that was absent in cells from males. This material was termed a "Barr body," after Murray Barr. Later, it was shown that Barr bodies were synonymous with the inactivated X chromosome. Other observations led scientists to understand that the number of Barr bodies in a cell always represented one less than the number of X chromosomes in the cell. For example, one Barr body indicated the presence of two X chromosomes, and two Barr bodies indicated the presence of three X chromosomes.

Clinical Significance

The significance of Barr bodies became apparent with the observation that females lacking one Barr body or possessing more than one Barr body developed an abnormal appearance. Particularly intriguing were females with Turner's syndrome. These females possess only one X chromosome per cell, a condition that is not analogous to normal females who possess only one functional X chromosome per cell as a result of inactivation. The difference in the development of a Turner's syndrome female and a normal female lies in the fact that both X chromosomes are active in normal females during the first few days of development. After this period, inactivation occurs randomly in each cell, as hypothesized by Lyon. In cases in which inactivation is not random, individuals may have a variety of developmental problems. Therefore, there is apparently a critical need for both X chromosomes to be active in females in early development for normal development to occur.

It is equally important that there not be more than two X chromosomes present during this early development. Females possessing three X chromosomes, and therefore two Barr bodies, are sometimes called "superfemales" because of a tendency to be taller than average. These females are also two to ten times more likely to suffer from mild to moderate mental retardation.

The same phenomenon has been observed in males who possess Barr bodies. Barr bodies are not normally present in males because they have only one X chromosome. The presence of Barr bodies indicates the existence of an extra X chromosome that has become inactive. Just as in females, extra X chromosomes are also expressed in early development, and abnormal amounts of gene products result in abnormal physical characteristics and mental retardation. Males with Klinefelter's syndrome have two X chromosomes and a Y chromosome. In cases in which males have more than two X chromosomes, the effects are even more remarkable.

Mechanism of X Inactivation

While it has been apparent since the 1960's that X inactivation is required for normal female development, the mechanism surrounding this process has been elusive. Only with the development of techniques to study the molecular events of the cell and its chromosomes has progress been made toward an understanding of the process of inactivation. One process involved in turning off a gene (thus "shutting down" the process of transcription) is the alteration of one of the molecules of deoxyribonucleic acid (DNA) known as cytosine. By adding a methyl group to the cytosine, the gene cannot produce the ribonucleic acid (RNA) necessary to make a protein. It is thought that this methyl group blocks the proteins that normally bind to the DNA so that

transcription cannot occur. When methyl groups are removed from cytosines, the block is removed and transcription begins. This is a common means of regulating transcription of genes. Methylation is significantly higher in the inactivated X chromosome than in the activated X chromosome. As the genes on the chromosome become inactive, the chromosome condenses into the tightly packed mass known as the Barr body. However, the process of methylation alone cannot account for all aspects of inactivation.

A region on the X chromosome called the X inactivation center (XIC) is considered the control center for X inactivation. Studies directed at this region revealed a gene called the X inactivation specific transcripts (XIST) gene. At the time of its discovery, this gene was the only gene does to be functional from an inactivated chromosome. Unexpectedly, this gene does not follow the tenet of the central dogma and produce a protein or enzyme; rather, it produces only an RNA that did not code for a protein but remained inside the nucleus.

Evidence in humans supports the hypothesis that the XIST gene is turned on and begins to make its RNA when the egg is fertilized. Studies with mice have shown that RNA is produced, at first, in low levels and from both X chromosomes. It has been shown in mice, but not humans, that prior to inactivation, Xist (lowercase when referring to mouse genes) RNA is localized at the XIC site only, thus suggesting a potential role prior to actual inactivation of the chromosome. At this point, one X chromosome will begin to increase its production of XIST RNA; shortly thereafter, XIST RNA from the other X chromosome ceases. It is not clear how XIST RNA might initiate the process of inactivation and condensing of the inactive chromosome, but XIST RNA can then be seen bound along the entire length of the inactive X chromosome in females. These results suggest that inactivation spreads from the XIC region toward the end of the chromosome and that XIST RNA is required to maintain an inactive state. If a mouse's Xist gene is mutated and cannot produce its RNA, inactivation of that X chromosome is blocked. Other studies have suggested that a product from a nonsex chromosome may interact with the XIC region on the X chromosome, causing it to remain active. This does not answer any questions about what prevents XIST RNA from binding to the active X chromosome or what has caused the XIST gene on this chromosome to suddenly be turned off. As expected, but again not explained, XIST RNA is repressed, or expressed at only very low levels, in males containing only one X chromosome.

No difference has been detected between maternally and paternally expressed XIST genes in humans. This has led scientists to suspect that XIST gene RNA may not be responsible for determining which X chromosome becomes inactivated. It is also not clear how the cell knows how many X chromosomes are present. The search for other candidates for these roles is underway. Finally, there are a few genes besides the XIST gene that are also active on the inactive X chromosome. How they escape the inactivation process and why this is necessary are also questions that must be resolved.

—*Linda R. Adkison*

See Also: Chromosome Theory of Heredity; Gender Identity; Klinefelter's Syndrome; Meiosis; One Gene-One Enzyme Hypothesis; Turner's Syndrome.

Further Reading: "X in a Cage," *Discover* 15 (March, 1994), summarizes a mechanism for X chromosome inactivation. Richard W. Erbe, "Single-Active-X Principle," *Scientific American Medicine* (volume 2, section 9:IV, 1995), reviews the significance of gene dosage compensation in humans. Keith E. Latham, "X Chromosome Imprinting and Inactivation in the Early Mammalian Embryo," *Trends in Genetics* (April, 1996), provides a more in-depth discussion of observations on embryos with sex chromosomes from only one parent.

XYY Syndrome

Field of study: Human genetics
Significance: *XYY syndrome occurs at fertilization and represents one of several human sex chromosome abnormalities. The resulting XYY male bears an extra Y chromosome that is associated with tall stature and possible intelligence and behavioral problems.*

Key terms
ANEUPLOIDY: an abnormal number of chromosomes
GENETIC SCREENING: a medical technique that uses either fetal or adult cells to directly view the chromosomes of an individual to detect abnormalities in number or structure
MEIOSIS: cell division that produces sperm and egg cells having half the original number of chromosomes
NONDISJUNCTION: abnormal separation of chromosome pairs or duplicates during cell division
SEX CHROMOSOMES: the X and Y chromosomes, which determine the gender of an individual

Causes and Effects of XYY Syndrome
All human cells contain forty-six chromosomes consisting of twenty-three pairs; one member of each pair is contributed by the female parent and one by the male. Of these forty-six chromosomes, two chromosomes, designated X and Y, are known as the sex chromosome pair. Individuals with an XX pair are female, while those with an XY pair are male. Unlike the other twenty-two chromosome pairs, the X and Y chromosomes are strikingly different from each other in both size and function. While the Y chromosome is primarily concerned with male biology, the X chromosome contains information important to both genders.

During formation of sperm and eggs in the testes and ovaries, respectively, a unique form of cell division known as meiosis (or reduction division) occurs that halves the chromosome number from forty-six to twenty-three. Sperm and eggs are thus carrying only one member of each pair of chromosomes, and the original number will be restored during fertilization. Because females only have the XX pair, their eggs can only have an X chromosome, while the male, having the XY pair, produces sperm bearing an X or a Y chromosome.

Many genetic errors occur during sperm and egg production, particularly because of incorrect pair separation or improper separation of duplicated chromosomes, producing abnormal chromosome numbers. These errors, known as "nondisjunction," result, at fertilization, in embryos without the normal forty-six chromosomes. XYY syndrome is one of several of these aneuploidy (incorrect number) conditions that involve the sex chromosome pair. While Klinefelter's syndrome (an XXY male) and Turner's syndrome (an X female) are more widely studied and recognized genetic diseases, the XYY male occurs with a frequency of 1 in 2,000 live births in the United States and 1 in 1,000 male births. Caused by a YY-bearing sperm fertilizing a normal X-bearing egg, the XYY embryo develops along a seemingly normal route and, unlike most other sex chromosome diseases, is not apparent at birth. In fact, these children are not known to be carriers of this disorder unless genetic testing or screening is administered for another purpose. The only physical clue is unusually tall stature; otherwise, an affected male will be normal in appearance. The XYY male also has the ability to reproduce, which is quite different from other sex chromosome aneuploidies, who are usually sterile.

Behavioral and Research Implications

Interest in the association between aggression and the Y chromosome began in the years following World War II. Both psychologists and geneticists began intensive scrutiny of the genes that were located on the male sex chromosome. Men with multiple copies of the Y chromosome thus became the subjects of much of this research. Genetic links to violent, aggressive, and even criminal behavior were found, although many argued that below-average intelligence played a greater role. Many males with XYY syndrome do perform lower than average on standard intelligence tests and have a greater incidence of behavioral problems. The majority, however, lead normal lives and are indistinguishable from XY males.

The controversy surrounding this research began with a study at Harvard University that began in the early 1960's and ended in 1973 because of pressure from both public and scientific communities. The researchers screened all boys born at a Boston hospital, identifying those with sex chromosomal abnormalities. Because the parents of XYY boys were told of their children's genetic makeup and the possibility of lower intelligence and bad behavior, critics claimed that the researchers had biased the parents against their sons causing the parents to treat the children differently. The environment would thus play a greater role than genetics in their behavior. Subsequent research has shown that the original hypothesis is at least partially accurate. There is a disproportionately large number of XYY males in prison populations, and they are usually of subaverage intelligence compared to other prisoners. It must be remembered, however, that the majority of XYY males show neither low intelligence nor criminal behavior.

Scientists, doctors, geneticists, and psychologists now agree that the extra Y chromosome does cause great height, reading and math difficulties, and, in some cases, severe acne, but the explanation of the high prevalence of XYY men in prison populations has changed its focus from genes to environment. Large body size during childhood, adolescence, and early adulthood will no doubt cause people to treat these individuals differently, and they may in turn have learned to use their size defensively. Aggressive behavior, coupled with academic difficulties, may lead to further problems. Clearly, however, the majority of XYY males do well. The issue would be much easier to resolve if a YY or Y male existed, but because the X chromosome has important information that involves genetic control of functions other than female sex factors, the X chromosome is necessary for life, and no YY or Y male embryo can survive.

—*Connie Rizzo*

See Also: Criminality; Klinefelter's Syndrome; Meiosis; Nondisjunction and Aneuploidy; Turner's Syndrome.

Further Reading: *Cecil's Textbook of Medicine* (1998), edited by C. Bennett, is a classic medical text that covers all aspects of sex chromosomal abnormalities. S. Mader, *Human Reproductive Biology* (1996), provides an excellent introduction to cell division, genetics, and sex from fertilization through birth. R. Tamarin, *Principles of Genetics* (1996), is a well-written reference text on genetics with complete discussions on aneuploidy, the sex chromosomes, genes, and abnormalities; it also includes a thorough reading list.

Time Line of Major Developments in Genetics

Date	Event	Significance
12,000 B.C.	Humans begin domesticating plants and animals.	Domestication involved selective breeding for certain traits. This genetic engineering allowed for transition from hunter-gatherer societies to agrarian civilizations.
1798	Edward Jenner develops vaccination.	Jenner used the cowpox virus as a vaccine to induce immunity against the genetically and structurally similar, but lethal, virus that causes smallpox in humans.
1838	G. J. Mulder precipitates a fibrous material from cells. He calls this material "protein."	Mulder believed protein was the most important of the known components of living matter.
1845-1847	Late blight, a fungal disease afflicting potatoes, ravages Ireland's potato crop. More than a million people die during the resulting famine. Many flee to America.	The famine was especially severe because all potatoes were derived from a single species and thus were all susceptible to the disease. Not enough genetic diversity existed to protect the crops.
1850	Theodore Schwann, Matthias Jakob Schleiden, and Rudolph Virchow recognize that tissues are made up of cells.	The cell theory disputed the prevailing view of "vitalism," which stated that no single part of an organism is alive (it was thought properties of living matter were somehow shared by the whole organism). The cell is considered to be the basic and most fundamental unit of life.
1857	Louis Pasteur begins research into fermentation. His "pasteurization" process is originally proposed as a means of preserving beer and wines.	Through his work, Pasteur made the important discovery that "life must be derived from life."
1859	Charles Darwin and Alfred Wallace propose theories of evolution.	The actual mechanism of evolution was not understood at the time. Once genetics was studied as a discipline, it became clear that genetics and evolution are intimately associated. Genetic theories helped to explain the process of evolution.
1862	The Organic Act establishes the U.S. Department of Agriculture (USDA). As one of its functions, the USDA is responsible for the collection of new and valuable seeds and plants and the distribution of them to agriculturists.	The preservation and dissemination of agriculturally important plants was a necessity for maintaining and increasing the world's food supply.

Date	Event	Significance
1865	Gregor Mendel, an Austrian monk, publishes a paper entitled "Experiments in Plant Hybridization," but his work lies unnoticed for more than thirty years.	Working with garden peas, Mendel used a systematic approach to study heredity. His mathematical analysis led to theories of inheritance that formed the basis of the discipline of genetics.
1869	Friedrich Miescher isolates "nuclein" from the nuclei of white blood cells.	This substance was later found to be the nucleic acids DNA and RNA.
1880	Walter Fleming first describes mitosis.	The processes of cell division, meiosis and mitosis, were key to the understanding of inheritance. Microscopic observations of dividing cells helped early researchers connect Mendelian genetics with cellular biology.
1883	E. van Beneden studies the processes of meiosis and fertilization in the parasitic worm *Ascaris*.	Van Beneden was the first to observe that the chromosome number in somatic, or body, cells is twice the number that exist in gametes, or sex cells. He also realized that when fertilization occurs (the combination of two gametes, the egg from the female and the sperm from the male), the chromosome number of somatic cells is established.
1883	The first absolutely pure yeast culture (yeast propagated from a single cell) is introduced at Denmark's Carlsberg Brewery.	The ability to propagate and maintain pure strains of organisms (genetically identical strains, or clones) would prove pivotal to future genetic research.
1897	Eduard Buchner shows that organic chemical transformations could be performed by cell extracts. He discovers that yeast extracts can convert glucose to ethyl alcohol.	Buchner's was one of the first in vitro experiments. Performing such experiments outside the body allowed researchers to control conditions and to observe the effects of individual variables.
1900	Hugo de Vries, Erich Von Tschermak, and Carl Correns rediscover and reproduce Mendel's work.	Mendel's theories provided a framework for other researchers. Studies in cytology, cellular biology, plant hybridization, and biochemistry supported Mendel's assertions.

Date	Event	Significance
1903	Walter Sutton is the first to relate genetics to the study of chromosome behavior, linking chromosomes to Mendelian heredity and forming the chromosomal theory of heredity. He also coins the term "gene" to describe heritable factors.	Mendel's notions of segregation and independent assortment coincided with Sutton's observations of how chromosomes segregated during cell division. This provided a cellular explanation for Mendel's observations.
1905	Nettie Stevens and Edmund Wilson independently describe the behavior of sex chromosomes.	The observations of Stevens and Wilson provided the first direct evidence to support the chromosomal theory of heredity.
1905-1933	The eugenics movement grows in popularity. It is influential in formulating social policies and immigration and sterilization laws in the United States and other countries.	The idea that human traits, most notably behavior, are governed by simple genetic rules was used to discriminate against the "mentally deficient," immigrants from specific countries, and even the poor and homeless. The U.S. eugenics movement effectively ended after the theory became associated with the policies of Nazi Germany.
1905	William Bateson, E. R. Saunders, and R. C. Punnett discover an exception to the rules of Mendelian inheritance. Specifically, they observe a violation of the principle of independent assortment.	These researchers also worked with pea plants and discovered two traits that did not assort independently. These genes were carried, or linked, on the same chromosome. Bateson also coined the term "genetics" to describe the science of heredity.
1908	Sir Archibald Garrod proposes that some human diseases are "inborn errors of metabolism."	By studying the inheritance of human disorders, Garrod provided the first evidence of a specific relationship between genes and enzymes.
1908	George Shull self-pollinates plants for many generations to produce pure-breeding lines. Donald Jones also performs similar experiments to increase productivity.	These two researchers developed the scientific basis of modern agricultural genetics.
1908	Godfrey Hardy and Wilhelm Weinberg discover mathematical relationships between genotypic and phenotypic frequencies in populations.	Known as the Hardy-Weinberg law, the rules governing these mathematical relationships helped researchers understand the dynamics of population genetics and the evolution of species.

Date	Event	Significance
1909	Hermann Nilsson-Ehle describes another violation of Mendelian inheritance. His studies with kernel color in wheat indicate this is a polygenic trait.	This was one of the first demonstrations that many genes could influence a single trait. Depending on the alleles, each gene contributes to the trait in an additive fashion. Other examples of polygenic inheritance include skin color and height in humans.
1909	Carl Correns discovers another class of exceptions to Mendelian inheritance. This is one of the first examples of extranuclear inheritance.	The notion that other cellular organelles besides the nucleus carry DNA was not recognized for decades. However, Correns's experiments in the plant *Mirabilis jalapa* showed inheritance of leaf color via the DNA in the chloroplasts.
1910	A white-eyed fruit fly is discovered in the lab of Thomas Hunt Morgan. His student, Calvin Bridges, saves the fly. It is found that this trait does not segregate exactly according to Mendelian principles, but rather is influenced by the sex of the fly.	This fly became the cornerstone upon which theories of Mendelian, chromosomal, and sexual inheritance were built into a cohesive whole. Morgan was awarded a Nobel Prize in 1933.
1911	Peyton Rous produces cell-free extracts from chicken tumors that, when injected, can induce tumors in other chickens.	The tumor-producing agent in the extract was later found to be a virus. Thus, Rous had discovered a link between cancer and viruses. He was awarded a Nobel Prize in 1966.
1913	Alfred Sturtevant, a student of Morgan, constructs the first gene maps of chromosomes. Maps indicate the order of genes as they exist physically on the chromosome.	Knowledge of gene locations on chromosomes provided insights into inheritance, genetic diseases, and the function and regulation of DNA. In addition, isolation of specific genes often required knowledge of their chromosomal location.
1913	Eleanor Carothers reports her discovery of the chromosomal basis of independent assortment.	By examining grasshopper chromosomes, Carothers observed the behavior of the X chromosome, responsible for sex determination, during cell division. These observations corresponded with Mendel's principle of independent assortment.

Date	Event	Significance
1917	Felix d'Herelle discovers bacteriophages, viruses that infect bacteria.	Bacteriophages played an important role in early genetics research, including confirmation that DNA is the hereditary material. Bacteriophages also became important in recombinant DNA applications.
1922-1932	Ronald Fisher, John B. S. Haldane, Sewall Wright, and S. S. Chetverikov independently publish papers on evolution, Mendelian inheritance, and natural selection.	These researchers combined Mendelian and Darwinian theories to explain the role of genetics in evolution.
1927	Hermann Müller, another student of Morgan, uses X rays to induce mutations in organisms.	The ability to mutate DNA was a powerful tool to determine the function of specific genes. Müller was awarded a Nobel Prize in Physiology or Medicine in 1945.
1928	Frederick Griffith uses the bacterium that causes pneumonia, *Diplococcus pneumoniae,* to initiate his investigations into the "transforming principle."	The hereditary material had not yet been identified. Griffith's experiments indicated that the transforming principle was DNA. Although not absolute proof, this experiment contributed significantly to the field and sparked ideas in other researchers.
1931	Barbara McClintock and Harriet Creighton discover physical exchange between chromosomes in corn, a process known as "crossing-over." Curt Stern also uses a similar approach in the study of the *Drosophila melanogaster* (fruit fly) X chromosome.	Crossing-over, or recombination, was vital to mapping genes on chromosomes and to understanding inheritance involving linkage.
1933	Theophilus Painter discovers polytene chromosomes in *Drosophila* salivary glands.	These special chromosomes, resulting from numerous rounds of DNA replications without separation, were large, with distinct banding patterns. These were used extensively in mapping genes to specific regions of the chromosome.

Date	Event	Significance
1933	Less than 1 percent of all the agricultural land in the Corn Belt has hybrid corn growing on it. However, by 1943, hybrids cover more than 78 percent of the same land.	Techniques used to produce crops with desired properties relied heavily on an understanding of genetics. Through the process of producing hybrids, researchers attempted to breed the best traits of several varieties into one. This time-consuming and inexact process was later superseded by the techniques of recombinant DNA technology.
1935	Ronald Aylmer publishes statistical analyses of Mendel's work. He finds errors in Mendel's interpretation of his data for a series of experiments.	Aylmer did not dispute Mendel's theories but instead implied that an assistant was ultimately responsible for the error.
1941	George Wells Beadle and Edward Tatum, working with a bread mold, *Neurospora*, publish results indicating that genes mediate cellular chemistry through the production of specific enzymes (the "one gene-one enzyme" experiment).	This established the use of "simple" organisms as model systems to study genetics. Beadle and Tatum were awarded a Nobel Prize in 1958.
1943	The Rockefeller Foundation, in collaboration with the Mexican government, initiates the Mexican Agricultural Program.	This was the first use of plant breeding as foreign aid.
1944	Oswald T. Avery, Colin MacLeod, and Maclyn McCarthy purify the "transforming principle" of Fred Griffith's work. This is DNA.	Although this experiment provided solid evidence that DNA is the hereditary material, most scientists still did not accept the notion.
1945	Max Delbruck, Salvador Luria, and Alfred Hershey work on bacteriophage as a model system to study the mechanism of heredity.	Delbruck organized a course at Cold Spring Harbor, New York, to introduce researchers to the methods of working with bacteriophage. Taught for twenty-six years, the course helped countless researchers to understand the use of model organisms in genetic investigations. Delbruck, Luria, and Hershey shared a 1969 Nobel Prize.

Date	Event	Significance
1946	Joshua Lederberg and Edward Tatum discover genetic recombination in bacteria, leading them to believe that bacteria, like eukaryotes, have a sexual reproductive cycle.	This discovery forced researchers to realize that bacteria are genetic organisms, similar to the eukaryotes studied at the time. Lederberg was awarded a Nobel Prize in 1958; Tatum also shared in this prize with George Beadle for their work with cellular chemistry, enzymes, and genetics.
1950	Barbara McClintock first describes the theory that DNA is mobile and that certain of its elements can insert into different regions on the chromosome. The technical name for this is transposition; however, it is often referred to as "jumping genes."	McClintock's ideas were far ahead of her time. While most scientists were still trying to determine just how DNA worked, McClintock was turning the field upside down. Her work was not well accepted until more evidence of transposons surfaced decades later. She was awarded a Nobel Prize in 1982.
1950	Erwin Chargaff discovers consistent one-to-one ratios of adenine to thymine and of guanine to cytosine in DNA.	These four chemicals are the basic building blocks of DNA. Chargaff's observations became an important clue in determining the exact structure of DNA.
1951	Maurice Wilkins and Rosalind Franklin obtain X-ray diffraction photographs of DNA.	Perhaps more important than Chargaff's contribution, these data indicated the exact shape of the DNA molecule.
1952	Alfred Hershey and Martha Chase use bacteriophage and a blender to identify the transforming principle as DNA.	Hershey and Chase were able to show that DNA, and not protein, is responsible for transforming organisms. This experiment was the conclusive piece of evidence confirming that DNA is the hereditary material.
1953	The three-dimensional structure of DNA is outlined by James Watson and Francis Crick in a 900-word manuscript.	This elegant and concise paper described the structure of DNA and also provided insight into its function. Watson, Crick, and Wilkins were awarded a Nobel Prize in 1962. Franklin did not share in the prize, as she had died several years earlier of cancer, almost certainly caused by her work with X rays.
1953	Jerome Lejeune discovers that Down syndrome is caused by the presence of an extra chromosome.	This was the first evidence that genetic disorders could be the result of changes in chromosome number, as opposed to changes in individual genes inherited in a Mendelian fashion.

Date	Event	Significance
1956	J. H. Tjio and A. Levan determine the chromosome number in humans to be forty-six.	Until that time, the chromosome number was thought to be forty-eight. The advances that Tjio and Levan pioneered were instrumental in obtaining good chromosome preparations, allowing for significant advances in the field of cytogenetics.
1957	Heinz Fraenkel-Conrat and B. Singer show that Tobacco Mosaic Virus contains RNA.	This was the first concrete evidence that RNA, in addition to DNA, could serve as the genetic material.
1958	Matthew Meselson and Frank Stahl determine how DNA replicates.	Meselson and Stahl showed that DNA replicates in a semiconservative manner. In other words, each strand of the molecule serves as a template for the synthesis of a new, complementary strand.
1958	Arthur Kornberg purifies the enzyme DNA polymerase I from *Escherichia coli*. This is the enzyme responsible for DNA replication.	The discovery of DNA polymerases provided insights into the enzymatic functions in the cell. In addition, the enzyme was used as a powerful tool in molecular genetics research. Kornberg shared a Nobel Prize with Severo Ochoa in 1959.
1961	Sol Spiegelman and Benjamin Hall discover that single-stranded DNA will hydrogen bond to its complementary RNA.	The discovery of the ability of DNA and RNA to form an association contributed greatly to the study of genes and their organization.
1961	François Jacob and Jacques Marod propose that certain genes regulate the activity of other genes.	In their experiments with the *E. coli* lac operon, Jacob and Marod for the first time elucidate how genes can be regulated based on environmental influences. The two, along with Andre Lwoff, receive a Nobel Prize in 1965.
1961	Sydney Brenner, François Jacob, and Matthew Meselson report that "messenger" RNA carries the information from DNA to create proteins.	This link between DNA and proteins was critical to the understanding of how proteins are made.

Date	Event	Significance
1962	Werner Arber finds bacteria that are resistant to infection by bacteriophage. It appears that some cellular enzymes destroy phage DNA, while others modify the bacterial DNA to prevent self-destruction. Several years later, Arber, Stuart Linn, Matthew Meselson, and Robert Yuan isolate the first restriction endonuclease and identify the modification of bacterial DNA as methylation.	By this time, genetic research took a bold new road. Scientists were starting to look at how DNA regulated, and was regulated by, cellular activities. This new aspect of genetics, often termed molecular genetics, provided insight into basic heredity, diseases (including cancer), and all biological processes.
1964	Robin Holliday proposes a model for the recombination of DNA.	Although recombination, or crossing-over, was not a new idea, the molecular mechanism behind the exchange of genetic information between DNA strands was not known. Holliday's model, which was widely accepted, explained the phenomenon.
1964	John Gurden transfers nuclei from adult toad cells into toad eggs. F. C. Steward grows single adult cells from a carrot into fully formed, normal plants.	These experiments produced viable organisms, thus ushering in the era of cloning.
1964	The International Rice Research Institute introduces new strains of rice that double the yield of previous strains.	This marked the beginning of the "Green Revolution," which sought to enable all nations to grow sufficient quantities of food to sustain their own populations.
1966	The genetic code is cracked by Marshal Nirenberg and H. Gobind Khorana.	These researchers discovered the code that elucidated the sequence of amino acids in a protein from the sequence of nucleic acids in a gene. Nirenberg and Khorana, along with Robert Holley, were awarded a Nobel Prize in 1968.
1970	M. Mandel and A. Higa discover a method to increase the efficiency of bacterial transformation.	They make the cells "competent" to take up DNA by treating bacteria with calcium chloride and then heat-shocking the cells. Introducing foreign DNA into cells was a key to the success of recombinant DNA methods.

Date	Event	Significance
1970	Hamilton Smith and Kent Wilcox isolate the first restriction endonuclease that cuts at a specific DNA sequence. Daniel Nathans uses this enzyme to create a restriction map of the virus SV40.	The use of restriction enzymes, those that cut DNA, allowed for the detailed mapping and analysis of genes. It also was pivotal for recombinant DNA techniques, including the production of transgenic organisms. Arber, Nathans, and Smith shared a Nobel Prize in 1978 for their work on restriction enzymes.
1972	Paul Berg is the first to create a recombinant DNA molecule.	Berg showed that restriction enzymes could be used to cut DNA in a predictable manner and that these DNA fragments could be joined together with fragments from different organisms. He was awarded a Nobel Prize in 1980.
1973	Joseph Sambrook and other researchers at Cold Spring Harbor improve the method of separating DNA fragments based on size, a technique called agarose gel electrophoresis.	These methods were central to accurate interpretation of information in DNA.
1973	Stanley Cohen and Annie Chang show that recombinant DNA molecules can be maintained and replicated in *E. coli*. Cohen and Herbert Boyer produce the first recombinant plasmid in bacteria.	Plasmids, or small, circular pieces of DNA, occur naturally in bacteria. Using the newly discovered tools of molecular biology, these researchers were able to insert a new piece of DNA into an existing plasmid and have it propagate in a bacterial cell.
1974	Critics call for a moratorium on recombinant DNA research.	Researchers in molecular biology realized that the manipulation of the genetic material of living organisms could have unforeseen consequences. They decided to tread carefully until more was known about the techniques and the results of their work.
1975-1976	The National Institutes of Health create guidelines for recombinant DNA research.	Guidelines were created to minimize potential hazards if genetically altered bacteria were released into the environment. The guidelines were relaxed by 1981.
1975	Edward Southern develops a method for transferring DNA from an agarose gel to a solid membrane.	This technique, known as Southern blotting, became one of the most important methods used to identify cloned genes.

Date	Event	Significance
1975	Renato Dubecco, David Baltimore, and Howard Temin receive the Nobel Prize.	The work of all three researchers centered on the interaction between tumor viruses and the genetic material of the cell. Dubecco applied phage genetic techniques to the study of animal viruses. Baltimore and Temin discovered reverse transcriptase, an enzyme used by viruses to convert their RNA into DNA. The reverse transcriptase enzyme became a useful tool in genetic research.
1976	Susumu Tonegawa discovers the genetic principles for generation of antibody diversity.	Tonegawa identified a novel mode of regulation of the genetic material. The genomic DNA of immune cells is actually cut and rejoined in different combinations. This explained how millions of different antibodies could be produced from a very small number of genes. Tonegawa was awarded a Nobel Prize in 1987.
1977	Allan Maxam and Walter Gilbert develop a method to determine the sequence of a piece of DNA. At the same time, Fred Sanger develops a different method. It is now possible, and relatively simple, to determine the exact sequence of A, G, T, and C in any DNA molecule.	Although both methods are effective, the Sanger method, often referred to as Sanger dideoxy sequencing, or the chain termination method, became the dominant technique. It did not involve toxic chemicals, as the Maxam and Gilbert method did. Gilbert and Sanger were awarded a Nobel Prize in 1980.
1977	The U.S. Court of Customs and Patent Appeals rules an inventor can patent new forms of microorganisms. The first patent granted for a recombinant organism, an oil-eating bacteria, is awarded in 1980.	The legality and ethics of patenting recombinant organisms and other biological systems was highly controversial.
1977	The human hormone somatostatin is produced in *E. coli.* by Herbert Boyer.	This was the first successful use of recombinant DNA to produce a substance from the gene of a "higher" organism. The first isolation of mammalian somatostatin required a half million sheep brains to produce 5 milligrams. Using recombinant DNA, only two gallons of bacterial culture were required to produce the same amount.

Date	Event	Significance
1977	Philip Sharp, Richard Roberts, and other researchers discover interruptions in the coding sequences of DNA. These regions are called introns, while the coding sequences are termed exons.	Introns were an intriguing mystery. Although they did not code for the final protein sequence, they were ubiquitous in eukaryotic DNA. Theories to explain their presence often focused on an evolutionary role.
1978	Several researchers report the discovery of splicing.	Splicing, the processing of messenger RNA after transcription to remove introns, revealed the mechanism behind the excision of these unused parts of genes.
1978	Genetech becomes the first company set up to commercialize the new emerging field of biotechnology.	A subsequent biotechnology boom in the 1980's saw many companies and products come and go.
1978	P. C. Steptoe and R. G. Edwards successfully use in vitro fertilization and artificial implantation in humans.	The process, commonly referred to as the production of "test-tube babies," gave hope to many childless couples who, prior to this development, had been unable to conceive.
1981	Several research teams determine that the origin of human cancer can be traced directly to the genetic material.	Each group isolated "oncogenes," genes that cause cancer. Michael Bishop and Harold Varmus were awarded a Nobel Prize in 1989 for their discovery of the origins of oncogenes.
1982	The first genetically engineered product, human insulin, is approved for sale by the U.S. government.	The production of pharmaceuticals through recombinant DNA technology is the driving force behind the biotechnology industry.
1983	Nancy Wexler, Michael Conneally, and James Gusella determine the chromosomal location of the Huntington's chorea gene.	Although close, they are unable to locate the gene itself.
1983	Thomas Cech and Sid Altman independently discover catalytic RNA.	The idea that RNA can have an enzymatic function changed researchers' views on the role of this molecule. This led to important new theories about the evolution of life. Cech and Altman were awarded a Nobel Prize in 1989.

Date	Event	Significance
1983	Bruce Cattanach provides evidence of genomic imprinting in mice.	The phenomenon of imprinting is the modification of genes in male and female gametes. This leads to differential expression of these genes in the embryo after fertilization. Imprinting represented another exception to the rules of Mendelian inheritance.
1983	John Sulston and others describe the cell lineage of the nematode, *Caenorhabditis elegans*.	The fixed developmental pattern of this small worm provided researchers with insights into how cells determine their own fates and how they influence the fates of neighboring cells.
1984	Alec Jeffreys is the first to use DNA in identifying individuals.	This technique, popularly known as "DNA fingerprinting," made identification of individuals and construction of genetic relationships virtually indisputable.
1984	A collaboration of more than twenty-five scientists isolates the gene that causes cystic fibrosis.	As a result of technological advances, the 1980's and 1990's saw the identification, isolation, and sequencing of genes as almost commonplace. Among the notable discoveries were genes implicated in cancer and in various genetic disorders.
1985	Kary Mullis invents the process of polymerase chain reaction (PCR).	This revolutionary method of copying DNA from extremely small amounts of material changed the way molecular research was done in only a few short years. It also became important in medical diagnostics and forensic analysis. Mullis was awarded a Nobel Prize in 1993.
1987	Frostban, a genetically engineered bacteria designed to prevent bacteria from freezing, is tested on strawberries in California. The bacteria is freely released outdoors.	It was hoped that the bacteria would grow on the strawberries and prevent the fruit from being destroyed by frost late in the growing season. The environmental release of recombinant organisms was an important and controversial step in the application of genetic engineering.

Date	Event	Significance
1987	Carol Greider and Elizabeth Blackburn, using the model organism *Tetrahymena* (a protozoan), report evidence that telomeres are regenerated through an enzyme with an RNA component.	Based on the action of DNA polymerase, telomeres (the very tips of chromosomes) should become shorter during each round of cell division. Another enzyme, called telomerase, was found to be necessary to maintain the telomeres. Research in this field sparked interest in the possibility that declining levels of telomerase may contribute to aging and that the inappropriate expression of this enzyme in cells may be a factor in cancer.
1988	The FDA approves the sale of recombinant TPA (tissue plasminogen activator) as a treatment for blood clots.	TPA showed promise in helping heart-attack and stroke victims.
1990	The Human Genome Project begins, headed by James Watson. The project is to be completed by the year 2005.	This ambitious project was designed to sequence the entire human genome in order to aid in identifying genes involved in diseases and biochemical processes. Also included as part of the Human Genome Project was the sequencing of many model organisms.
1990	The first human undergoes gene therapy.	The patient was a four-year-old girl who was born without a functioning immune system as a result of a faulty gene that makes an enzyme called ADA (adenosine deaminase).
1993	The mutation that causes Huntington's chorea is found.	Fifty-eight scientists collaborated on the project. The mutation was discovered ten years after its chromosomal location was first identified.
1994	The FDA approves the bovine hormone known as BST or BGH.	The hormone was made from recombinant bacteria containing the bovine gene for BST. When injected into cows, the hormone increased milk production by up to 20 percent. Many supermarkets and manufacturers of dairy products refused to carry or use milk from BST-injected cows, uncertain of what the long-term effects of this recombinant drug might be.

Date	Event	Significance
1994	The FDA gives approval for the marketing of the Flavr Savr tomato.	This genetically altered tomato could be ripened on the vine before being picked and transported. Because the ripening process took longer, the tomatoes would not rot on their way to the market.
1995	A mutation in the gene *BRCA1* is implemented in breast cancer by Mark Skolnick and others.	More than any other gene previously identified, this discovery had wide potential for assessing cancer risk.
1995	The first complete DNA sequence of a bacterium (*Haemophilus influenzae*) is reported.	As part of the Human Genome Project, the sequencing of many model organisms, including *H. influenzae*, helped researchers to study the functions of genes found in humans and in other living things.
1995	Edward Lewis, Christiane Nusslein-Volhard, and Eric Wiechaus win the Nobel Prize for their work on the genetic control of early development in *Drosophila*.	These researchers took the fruit fly, a research tool from the age of classical genetics, into the age of molecular biology and discovered how genetics controls development. The same developmental mechanisms appeared to be at work in other organisms, including humans.
1995	Nusslein-Volhard completes a genetic mutation project involving zebra fish.	Repeating her earlier work with *Drosophila*, Nusslein-Volhard used similar techniques to begin an intensive study of development in a vertebrate system. This involved screening thousand of mutants to determine if any had developmental defects.
1996	A group of more than six hundred researchers sequences the DNA of the yeast *Saccharomyces cereviseae*.	This was the first eukaryotic organism to be sequenced.
1997	Ian Wilmut and others clone the first mammal, a sheep named Dolly.	It was hoped that successful cloning of a mammal would allow for easier and cheaper development and propagation of transgenic animals.
1997	The genomic sequence of the bacterium *Escherichia coli* is reported by Frederick Blattner and colleagues.	Although *E. coli* was not the first complete bacterial sequence reported, because of the importance of *E. coli* in genetics research, the event represented a critical step.

—*Nancy Morvillo*

Biographical Dictionary of Important Geneticists

Arber, Werner (1929-   ): First to isolate enzymes that modify DNA and enzymes that cut DNA at specific sites. Such restriction enzymes were critical in the developing field of molecular biology. Arber was awarded the 1978 Nobel Prize in Physiology or Medicine.

Aristotle (c. 384-322 B.C.): Greek philosopher and scientist. Aristotle's *De Generatione* was devoted in part to his theories on heredity. Aristotle believed the semen of the male contributes a form-giving principle (*eidos*), while the menstrual blood of the female is shaped by the *eidos*. The philosophy implied it was the father only who supplied form to the offspring.

Avery, Oswald Theodore (1877-1955): Immunologist and biologist who determined DNA to be the genetic material of cells. Avery's early work involved classification of the pneumococci, the common cause of pneumonia in the elderly. In 1944, he reported that the genetic information in these bacteria is DNA.

Baltimore, David (1938-   ): Along with Howard Temin, Baltimore isolated the enzyme RNA-directed DNA polymerase (reverse transcriptase), demonstrating the mechanism by which RNA tumor viruses can integrate their genetic material into the cell chromosome. Baltimore was awarded the 1975 Nobel Prize in Physiology or Medicine.

Bateson, William (1861-1926): Plant and animal geneticist who popularized the earlier work of Gregor Mendel. In his classic *Mendel's Principles of Heredity* (1909), Bateson introduced much of the modern terminology used in the field of genetics. Bateson suggested the term "genetics" (from the Greek word meaning "descent") to apply to the field of the study of heredity.

Beadle, George Wells (1903-1989): Beadle's studies of the bread mold *Neurospora* demonstrated that the function of a gene is to encode an enzyme. Beadle and Edward Tatum were awarded the 1958 Nobel Prize in Physiology or Medicine for their one gene-one enzyme hypothesis.

Berg, Paul (1926-   ): Developed DNA recombination techniques for insertion of genes in chromosomes. The techniques became an important procedure in understanding gene function and for the field of genetic engineering. Berg was awarded the 1980 Nobel Prize in Chemistry.

Bishop, John Michael (1936-   ): Determined that oncogenes, genetic information initially isolated from RNA tumor viruses, actually originate in normal host cells. Bishop was awarded the 1989 Nobel Prize in Physiology or Medicine for his discovery.

Brenner, Sydney (1927-   ): Molecular geneticist whose observations of mutations in nematodes—long, unsegmented worms—helped in understanding the design of the nervous system. Brenner was among the first to clone specific genes.

Burnet, Frank Macfarlane (1899-1985): Proposed a theory of clonal selection to explain regulation of the immune response. Burnet was awarded the 1960 Nobel Prize in Physiology or Medicine.

Cairns, Hugh John (1922-   ): British virologist whose investigations of rates and mechanisms of DNA replication helped to lay the groundwork in studying the replication process.

Chargaff, Erwin (1905-   ): Determined that the DNA composition in a cell is characteristic of that particular organism. His discovery of base ratios, in which the concentration of adenine is equal to that of thymine, and guanine to that of cytosine, provided an important clue to the structure of DNA.

Collins, Francis Sellers (1950-   ): In 1989, Collins identified the gene that, when mutated, results in the genetic disease cystic fibrosis. Collins was instrumental in the identification of a number of genes associated with genetic diseases.

Correns, Carl Erich (1864-1935): German botanist who confirmed Gregor Mendel's laws through his own work on the garden pea. Correns was one of several geneticists who rediscovered Mendel's work in the early 1900's.

Crick, Francis Harry Compton (1916-   ): Along with James Watson, Crick determined the double-helix structure of DNA. Crick was awarded the Nobel Prize in Physiology or Medicine in 1962.

Darwin, Charles Robert (1809-1882): Naturalist whose theory of evolution established natural selection as the basis for descent with modification, more commonly referred to as evolution. His classic work on the subject, *On the Origin of Species* (1859), based on his five-year voyage during the 1830's on the British ship HMS *Beagle*, summarized the studies and observations that initially led to the theory. Darwin's pangenesis theory, first noted in *The Variation of Animals and Plants Under Domestication* (1868), later became the basis for the concept of the gene.

Delbruck, Max (1906-1981): A leading figure in the application of genetics to bacteriophage research, and later, with *Phycomyces*, a fungal organism. His bacteriophage course, taught for decades at Cold Spring Harbor, New York, provided training for a generation of biologists. He was awarded the 1969 Nobel Prize in Physiology or Medicine.

Demerec, Milislav (1895-1966): Croatian-born geneticist who was among the scientists who brought the United States to the forefront of genetics research. Demerec's experiments based upon the genetics of corn addressed the question of what a gene represents. Demerec was director of the biological laboratories in Cold Spring Harbor, New York, for many years among the most important sites of genetic research.

De Vries, Hugo (1848-1935): Dutch botanist whose hypothesis of intracellular pangenesis postulated the existence of pangenes, factors which determined characteristics of a species. De Vries established the concept of mutation as a basis for variation in plants. In 1900, de Vries was one of several scientists who rediscovered Mendel's work.

Dobzhansky, Theodosius (1900-1975): Russian-born American geneticist who established evolutionary genetics as a viable discipline. His book *Genetics and the Origin of Species* (1937) represented the first application of Mendelian theory to Darwinian evolution.

Dulbecco, Renato (1914-   ): Among the first to study the genetics of tumor viruses. Dulbecco was awarded the 1975 Nobel Prize in Physiology or Medicine.

Fisher, Ronald Aylmer (1890-1962): British biologist whose application of statistics provided a means by which use of small sampling size could be applied to larger interpretations. Fisher's breeding of small animals led to an understanding of genetic dominance. He later applied his work to the study of inheritance of blood types in humans.

Franklin, Rosalind Elsie (1920-1958): British crystallographer whose X-ray diffraction studies helped confirm the double-helix nature of DNA. Franklin's work, along with that of Maurice Wilkins, was instrumental in confirming the structure of DNA as proposed by James Watson and Francis Crick. Franklin's early death precluded her receiving a Nobel Prize for the research.

Galton, Francis (1822-1911): British scientist who was an advocate of eugenics, the belief that human populations could be improved through "breeding" of desired traits. Galton was also the first to observe that fingerprints were unique to the individual.

Garrod, Archibald Edward (1857-   ): Applying his work on alkaptonuria, Garrod proposed that some human diseases result from a lack of specific enzymes. His theory of inborn errors of metabolism, published in 1908, established the genetic basis for certain hereditary diseases.

Gartner, Carl Friedrich von (1772-1850): German plant biologist and geneticist. Though Gartner did not generalize as to the significance of his work, his results provided the experimental basis for questions later developed by Gregor Mendel and Charles Darwin.

Griffith, Frederick (1877-1941): British microbiologist who in 1928 reported the existence of a "Transforming Principle," an unknown substance that could change the genetic properties of bacteria. In 1944, Oswald Avery determined the substance to be DNA.

Haeckel, Ernst Heinrich (1834-1919): German zoologist whose writings were instrumental in the dissemination of Charles Darwin's theories. Haeckel's "biogenetic law," since discarded, stated that "ontogeny repeats phylogeny," suggesting that embryonic development mirrors the evolutionary relationship of organisms.

Haldane, John Burdon (1892-1964): British physiologist and geneticist who proposed that natural selection, and not mutation per se, was the driving force of evolution. Haldane was the first to determine an accurate rate of mutation for human genes, and he later demonstrated the genetic linkage of hemophilia and color blindness.

Hardy, Godfrey Harold (1877-1947): British mathematician who, along with Wilhelm Weinberg, developed the Hardy-Weinberg law of population genetics. In a 1908 letter to the journal *Science*, Hardy used algebraic principles to confirm Mendel's theories as applied to populations, an issue then currently in dispute.

Hershey, Alfred Day (1908-1997): Molecular biologist who played a key role in understanding the replication and genetic structure of viruses. His experiments with Martha Chase confirmed that DNA carried the genetic information in some viruses. Hershey was awarded the 1969 Nobel Prize in Physiology or Medicine.

Hippocrates (c. 460-377 B.C.): Greek physician who proposed the earliest theory of inheritance. Hippocrates believed that "seed material" was carried by body humors to the reproductive organs.

Holley, Robert William (1922-1993): Determined the sequence of nucleotide bases in transfer RNA, the molecule that carries amino acids to ribosomes for protein synthesis. Holley's work provided a means for demonstrating the reading of the genetic code. He was awarded the Nobel Prize in Physiology or Medicine in 1968.

Jacob, François (1920-   ): French geneticist and molecular biologist who, along with Jacques Monod, elucidated a mechanism of gene and enzyme regulation in bacteria. The Jacob-Monod theory of gene regulation became the basis for understanding a wide range of genetic processes; they were awarded the 1965 Nobel Prize in Physiology or Medicine.

Johannsen, Wilhelm L. (1857-1927): Danish botanist who introduced the term "genes," derived from "pangenes," factors suggested by Hugo de Vries to determine hereditary characteristics in plants. Johannsen also introduced the concepts of phenotype and genotype to distinguish between physical and hereditary traits.

Khorana, Har Gobind (1922-   ): Developed methods for investigating the structure of DNA and deciphering the genetic code. Khorana synthesized the first artificial gene in the 1960's. He was awarded the Nobel Prize in Physiology or Medicine in 1968.

Knight, Thomas Andrew (1759-1853): Plant biologist who first recognized the usefulness of the garden pea for genetic studies because of its distinctive traits. Knight was the first to characterize dominant and recessive traits in the pea, though, unlike Gregor Mendel, he never determined the mathematical relationships among his crosses.

Kolreuter, Joseph Gottlieb (1733-1806): A precursor of Gregor Mendel, Kolreuter demonstrated the sexual nature of plant fertilization, in which characteristics were derived from each member of the parental generation in equivalent amounts.

Kornberg, Arthur (1918-   ): Carried out the first purification of DNA polymerase, the enzyme that replicates DNA. His work on the synthesis of biologically active DNA in a test tube culminated with his being awarded the 1959 Nobel Prize in Physiology or Medicine.

Lamarck, Jean-Baptiste Pierre Antoine de Monet (1744-1829): French botanist and evolutionist who introduced many of the earliest concepts of inheritance. Lamarck proposed that hereditary changes occur as a result of an organism's needs; his theory of inherited characteristics, since discredited, postulated that organisms transmit acquired characteristics to their offspring.

Lederberg, Joshua (1925-   ): Established the occurrence of sexual reproduction in bacteria. Lederberg demonstrated that genetic manipulation of the DNA during bacterial conjugation could be used to map bacterial genes. He was awarded the 1958 Nobel Prize in Physiology or Medicine.

Levene, Phoebus Aaron (1869-1940): American biochemist who determined the components found in DNA and RNA. Levene described the presence of ribose sugar in RNA and of 2′-deoxyribose in DNA, thereby differentiating the two molecules. He also identified the nitrogen bases found in nucleic acid, though he was never able to determine the acid's molecular structure.

Linnaeus, Carolus (1707-1778): Swedish naturalist and botanist most noted for establishing the modern method for classification of plants and animals. In his *Philosophia Botanica* (1751), Linnaeus proposed that variations in plants or animals are induced by environments such as soil.

Luria, Salvador Edward (1912-1991): Pioneer in understanding replication and genetic structure in viruses. The Luria-Delbruck fluctuation test, pioneered by Luria and Max Delbruck, demonstrated that genetic mutations precede environmental selection. Luria was awarded the 1969 Nobel Prize in Physiology or Medicine.

Lwoff, Andre (1902-1994): French biochemist and protozoologist. Lwoff's early work demonstrated that vitamins function as components of living organisms. He is best known for demonstrating that the genetic material of bacteriophage can become part of the host bacterium's DNA, a process known as lysogeny. Lwoff was awarded the 1965 Nobel Prize in Physiology or Medicine.

McClintock, Barbara (1902-1992): Demonstrated the existence in plants of transposons, genes that "jump" from one place on a chromosome to another. The process was discovered to be widespread in nature. McClintock was awarded the 1983 Nobel Prize in Physiology or Medicine.

Mendel, Johann Gregor (1822-1884): The "father of genetics," Mendel was an Austrian monk whose studies on the transmission of traits in the garden pea established the mathematical basis of inheritance. Mendel's pioneering theories, including such fundamental genetics principles as the law of segregation and the law of independent assortment, were published in 1866 but received scant attention until the beginning of the twentieth century.

Meselson, Matthew Stanley (1930-): Demonstrated the nature of DNA replication, in which the two parental DNA strands are separated, each passing into one of the two daughter molecules.

Miescher, Johann Friedrich (1844-1895): In 1869, Miescher discovered and purified DNA from cell-free nuclei obtained from white blood cells and gave the name "nuclein" to the extract.

Monod, Jacques Lucien (1910-1976): French geneticist and molecular biologist who with François Jacob demonstrated a method of gene regulation in bacteria that came to be known as the Jacob-Monod model. Jacob and Monod were jointly awarded the 1965 Nobel Prize in Physiology or Medicine.

Morgan, Thomas Hunt (1866-1945): Embryologist whose studies of fruit flies established the existence of genes on chromosomes. Through his selective breeding of flies, Morgan also established concepts such as gene linkage, sex-linked characteristics, and genetic recombination.

Müller, Hermann Joseph (1890-1967): Geneticist and colleague of Thomas Hunt Morgan. Müller's experimental work with fruit flies established the gene as the site of mutation. His work with X rays demonstrated a means of artificially introducing mutations into an organism.

Mullis, Kary Banks (1944-): Devised the polymerase chain reaction (PCR), a method for duplicating small quantities of DNA. The PCR procedure became a major tool in research in the fields of genetics and molecular biology. Mullis was awarded the 1993 Nobel Prize in Chemistry.

Nathans, Daniel (1928-): Applied the use of restriction enzymes to the study of genetics. Nathans developed the first genetic map of SV40, among the first DNA viruses shown to transform normal cells into cancer. Nathans was awarded the 1978 Nobel Prize in Physiology or Medicine.

Nirenberg, Marshall Warren (1927-   ): Molecular biologist who was among the first to decipher the genetic code. He later demonstrated the process of ribosome binding in protein synthesis and carried out the first cell-free synthesis of protein. Nirenberg was awarded the 1968 Nobel Prize in Physiology or Medicine.

Pauling, Linus (1901-1994): American chemist who received the Nobel Prize in Chemistry in 1954 for his work on the nature of the chemical bond and the 1962 Nobel Peace Prize for his antinuclear activism. His 1950's investigations of protein structure contributed to the determination of the structure of DNA.

Punnett, Reginald C. (1875-1967): English biologist who collaborated with William Bateson in a series of important breeding experiments that confirmed the principles of Mendelian inheritance. Punnett also introduced the Punnett square, the standard graphical method of depicting hybrid crosses.

Sageret, Augustin (1763-1851): French botanist who discovered the ability of different traits to segregate independently in plants.

Simpson, George Gaylord (1902-1984): American paleontologist who applied population genetics to the study of the evolution of animals. Simpson was instrumental in establishing a Neo-Darwinian Theory of Evolution (the rejection of Lamarck's inheritance of acquired characteristics) during the early twentieth century.

Smith, Hamilton Othanel (1931-   ): American researcher who pioneered the purification of restriction enzymes. The isolation and use of restriction enzymes were critical to application of molecular biology to genetics and genetic engineering. Smith was awarded the 1978 Nobel Prize in Physiology or Medicine.

Sonneborn, Tracy Morton (1905-1981): Discovered crossbreeding and mating types in paramecia, integrating the genetic principles as applied to multicellular organisms with single-celled organisms such as protozoa.

Spencer, Herbert (1820-1903): English philosopher influenced by the work of Charles Darwin. Spencer proposed the first general theory of inheritance, postulating the existence of self-replicating units within the individual which determine the traits. Spencer is more popularly known as the source of "Survival of the Fittest" as applied to natural selection.

Stanley, Wendell Meredith (1904-1971): American biochemist who was the first to crystallize a virus (tobacco mosaic virus), demonstrating their protein nature. Stanley was later a member of the team which determined the amino acid sequence of the TMV protein. Stanley spent the last years of his long career studying the relationship of viruses and cancer.

Sturtevant, Alfred Henry (1891-1970): Colleague of Thomas Hunt Morgan, and among the pioneers in application of the fruit fly in the study of genetics. In 1913, Sturtevant constructed the first genetic map of a fruit fly chromosome. His work became a major factor in chromosome theory. In the 1930's, his work with George Beadle led to important observations and understanding of meiosis.

Sutton, Walter Stanborough (1877-1916): Biologist and geneticist who demonstrated the role of chromosomes during meiosis in gametes, and demonstrated their relationship to Mendel's laws. Sutton observed that chromosomes form homologous pairs during meiosis, with one member of each pair appearing in gametes. The particular member of each pair was subject to Mendel's law of independent assortment.

Tatum, Edward Lawrie (1909-1975): Along with George Beadle, demonstrated that the function of a gene is to encode an enzyme. Beadle and Tatum were awarded the 1958 Nobel Prize in Physiology or Medicine for their one gene-one enzyme hypothesis.

Temin, Howard Martin (1934-1994): Proposed that RNA tumor viruses replicate by means of a DNA intermediate. Temin's theory, initially discounted, became instrumental in understanding the process of infection and replication by such viruses. He later isolated the replicating enzyme, the RNA-directed DNA polymerase (reverse transcriptase). He was awarded the 1975 Nobel Prize in Physiology or Medicine, along with David Baltimore, for the work.

Varmus, Harold Elliot (1939-   ): Elucidated the molecular mechanisms by which retroviruses (RNA tumor viruses) transform cells. Varmus was awarded the 1989 Nobel Prize in Physiology or Medicine.

Watson, James Dewey (1928-   ): Along with Francis Crick, determined the double helix structure of DNA. Together with Crick and Maurice Wilkins, Watson was awarded the 1962 Nobel Prize in Physiology or Medicine for their work in determining the structure of DNA.

Weinberg, Robert Allan (1942-   ): Molecular biologist who isolated the first human oncogene, the *ras* gene, associated with a variety of cancers including those of the colon and brain. Weinberg later isolated the first tumor suppressor gene, the retinoblastoma gene.

Weinberg, Wilhelm (1862-1937): German obstetrician who demonstrated that hereditary characteristics of humans such as multiple births and genetic diseases were subject to Mendel's laws of heredity. The mathematical application of such characteristics, published simultaneously (and independently) by Godfrey Hardy, became known as the Hardy-Weinberg Equilibrium. The equation demonstrates that dominant genes do not replace recessive genes in a population; gene frequencies would not change from one generation to the next if certain criteria such as random mating and lack of natural selection were met.

Weismann, August (1834-1914): German zoologist noted for his chromosome theory of heredity. Weismann proposed that the source of heredity is in the nucleus only, and that inheritance is based upon transmission of a chemical or molecular substance from one generation to the next. Weismann's theory, which rejected the inheritance of acquired characteristics, came to be called neo-Darwinism. Though portions of Weismann's theory were later disproved, the nature of the chromosome was subsequently demonstrated by Thomas Hunt Morgan and his colleagues.

Wilkins, Maurice Hugh Frederick (1916-   ): Studies on the X-ray diffraction patterns exhibited by DNA confirmed the double-helix structure of the molecule. Wilkins was a colleague of Rosalind Franklin, and it was their work which confirmed the nature of DNA as proposed by Watson and Crick. Wilkins was awarded the Nobel Prize for Physiology and Medicine in 1962, along with Watson and Crick.

Wilmut, Ian (1944-   ): Scottish embryologist and leader of a research team at the Roslyn Institute near Edinburgh. In 1996, Wilmut and his colleagues succeeded in cloning an adult sheep, the first adult mammal to be successfully produced by cloning.

—Richard Adler

Glossary

acentric chromosome: a chromosome that does not have a centromere and that is unable to participate properly in cell division; often the result of a chromosomal mutation during recombination

acrocentric chromosome: a chromosome with its centromere near one end

active site: the region of an enzyme that interacts with a substrate molecule; any alteration in the three-dimensional shape of the active site usually has an adverse effect on the enzyme's activity

adenine (A): a purine nitrogenous base found in the structure of both DNA and RNA

adenosine triphosphate (ATP): the major energy molecule of cells, produced either through the process of cellular respiration or fermentation; used to drive most of the cell's activities, including genetic processes such as DNA replication

agarose: a chemical substance used to create gels for the electrophoresis of nucleic acids

albino: a genetic condition in which an individual does not produce the pigment melanin in the skin; other manifestations of the trait may be seen in the pigmentation of the hair or eyes

alkaptonuria: a genetic disorder, first characterized by geneticist Archibald Garrod, in which a compound called homogentisic acid accumulates in the cartilage and is excreted in the urine of affected individuals, turning both of these black (the name of the disorder literally means "black urine"); the specific genetic defect involves an inability to process by-products of phenylalanine and tyrosine metabolism

allele: a form of a gene; most genes have at least two naturally occurring alleles

allergy: an abnormal immune response to a substance that does not normally provoke an immune response or that is not inherently dangerous to the body (such as plant pollens or animal hair)

Alu sequence: a repetitive DNA sequence of unknown function, approximately three hundred nucleotides long, scattered throughout the genome of primates; the name comes from the presence of recognition sites for the restriction endonuclease Alu I in these sequences

amber codon: a stop codon (UAG) found in messenger RNA molecules that signals termination of translation

Ames test: a test devised by molecular biologist Bruce Ames for determining the mutagenic or carcinogenic properties of various compounds based on their ability to affect the nutritional characteristics of the bacterium *Salmonella typhimurium*

amino acid: a nitrogen-containing compound used as the building block of proteins; in nature, there are twenty amino acids that can be used to build proteins

aminoacyl tRNA: a transfer RNA (tRNA) molecule with an appropriate amino acid molecule attached; in this form, the tRNA molecule is ready to participate in translation

amniocentesis: a procedure in which a small amount of amniotic fluid containing fetal cells is withdrawn from the amniotic sac surrounding a fetus; the fetal cells are then tested for the presence of genetic abnormalities that may have an impact on the development of the fetus

anabolism: the part of the cell's metabolism concerned with synthesis of complex molecules and cell structures

anaphase: the third phase in the process of mitosis; in anaphase, sister chromatids separate at the centromere and migrate toward the poles of the cell

aneuploid: having an abnormal number of chromosomes in a cell or organism

angstrom: a unit of measurement equal to one ten-millionth of a millimeter; a DNA molecule is 20 angstroms wide

annealing: the process by which two single-stranded nucleic acid molecules are converted into a double-stranded molecule through hydrogen bonding between complementary base pairs

antibiotic: any substance produced naturally by a microorganism that inhibits the growth of other microorganisms; antibiotics are important in the treatment of bacterial infections

antibody: an immune protein (immunoglobulin) that specifically recognizes an antigen; produced by B cells of the immune system

anticodon: the portion of a transfer RNA molecule that is complementary in sequence to a codon in a messenger RNA molecule; because of this complementarity, the transfer RNA molecule can bind briefly to messenger RNA during translation and direct the placement of amino acids in a polypeptide chain

antigen: any molecule that is capable of being recognized by an antibody molecule or of provoking an immune response

antiparallel: a characteristic of the Watson-Crick double-helix model of DNA, in which the two strands of the molecule can be visualized as oriented in opposite directions; this characteristic is based on the orientation of the deoxyribose molecules in the sugar-phosphate backbone of the double helix

apoptosis: cell "suicide" occurring after a cell is too old to function properly or as a response to irreparable genetic damage; apoptosis prevents such cells from developing into a cancerous state

ascospore: a haploid spore produced by meiosis in some species of fungi

ascus: a reproductive structure, found in some species of fungi, that contains ascospores

asexual reproduction: reproduction of cells or organisms without the transfer or reassortment of genetic material; results in offspring that are genetically identical to the parent

ATP. *See* adenosine triphosphate

autosomes: non-sex chromosomes; humans have forty-four autosomes

B-DNA: the predominant form of DNA in solution and in the cell; a right-handed double helix most similar to the Watson-Crick model

back-cross: a cross involving an offspring individual crossed with one of its parents

Barr body: a darkly staining structure primarily present in female cells, believed to be an inactive X chromosome; used as a demonstration of the Lyon hypothesis

base: a chemical subunit of DNA or RNA that encodes gentic information; in DNA, the bases are adenine (A), cytosine (C), guanine (G), and thymine (T); in RNA, thymine is replaced by uracil (U)

base-pairing: the process by which bases link up to form molecules of DNA or RNA; in DNA, adenine (A) always pairs with thymine (T) and cytosine (C) pairs with guanine (G); in RNA, uracil (U) replaces thymine

bidirectional replication: a characteristic of DNA replication involving synthesis of DNA in both directions away from an origin of replication

biotechnology: the use of biological molecules or organisms in industrial or commercial products and techniques

C terminus: the end of a polypeptide with an amino acid that has a free carboxyl group

cAMP. *See* cyclic adenosine monophosphate

carcinogen: any physical or chemical cancer-causing agent

catabolism: the part of the cell's metabolism concerned with the breakdown of complex molecules, usually as an energy-generating mechanism

catabolite repression: a mechanism of operon regulation involving an enzyme reaction's product used as a regulatory molecule for the operon that encodes the enzyme; a kind of feedback inhibition

cDNA. *See* complementary DNA

cell cycle: the various growth phases of a cell, which include (in order) G1 (gap phase 1), S (DNA synthesis), G2 (gap phase 2), and M (mitosis)

centimorgan: a unit of genetic distance between genes on the same chromosome, equal to a recombination frequency of 1 percent; also called a map unit, since these distances can be used to construct genetic maps of chromosomes

central dogma: a foundational concept in modern genetics stating that genetic information present in the form of DNA can be converted to the form of messenger RNA (or other types of RNA) through transcription and that the information in the form of mRNA can be converted into the form of a protein through translation

centriole: a eukaryotic cell organelle involved in cell division, possibly with the assembly or disassembly of the spindle apparatus during mitosis and meiosis; another name for this organelle is the microtubule organizing center (MTOC)

centromere: a central region where a pair of chromatids are joined before being separated during anaphase of mitosis or meiosis; also, the region of the chromatids where the microtubules of the spindle apparatus attach

chi-square analysis: a statistical analysis of data from an experiment to determine how well the observed data correlate with the expected data

chiasmata: the point at which two homologous chromosomes exchange genetic material during the process of recombination; the word literally means "crosses," which refers to the appearance of these structures when viewed with a microscope

chorionic villus sampling: a procedure in which fetal cells are obtained from an embryonic structure called the chorion and analyzed for the presence of genetic abnormalities in the fetus

chromatid: one half of a chromosome that has been duplicated in preparation for mitosis or meiosis; each chromatid is connected to its sister chromatid by a centromere

chromosome: the form in which genetic material is found in the nucleus of a cell; composed of a single DNA molecule that is extremely tightly coiled, usually visible only during the processes of mitosis and meiosis

chromosome map: a diagram showing the locations of genes on a particular chromosome; generated through analysis of linkage experiments involving those genes

chromosome puff: an extremely unwound or uncoiled region of a chromosome indicative of a transcriptionally active region of the chromosome

chromosome theory of inheritance: a concept, first proposed by geneticists Walter Sutton and Theodore Boveri, that genes are located on chromosomes and that the inheritance and movement of chromosomes during meiosis explain Mendelian principles on the cellular level

chromosome walking: a molecular genetics technique used for analysis of long DNA fragments; the name comes from the technique of using previously cloned and characterized fragments of DNA to "walk" into uncharacterized regions of the chromosome that overlap with these fragments

cistron: a region of a DNA molecule that codes for a protein; synonymous with the term "gene"

clone: a molecule, cell, or organism that is a perfect genetic copy of another

cloning: the technique of making a perfect genetic copy of an item such as a DNA molecule, a cell, or an entire organism

cloning vector: a DNA molecule that can be used to transport genes of interest into cells, where these genes can then be copied

codominance: a genetic condition involving two alleles in a heterozygous organism; each of these alleles is fully expressed in the phenotype of the organism

codon: a group of three nucleotides in messenger RNA that represent a single amino acid in the genetic code; this is mediated through binding of a transfer RNA anticodon to the codon during translation

complementary DNA (cDNA): a molecule of DNA that has been copied from a messenger RNA molecule; important in construction of cDNA libraries, which are used to locate particular genes being expressed in a cell

congenital: a genetic condition or defect present at birth

conjugation: a form of genetic transfer among bacterial cells involving the F pilus

consanguine: of the same blood or origin; in genetics, the term implies the sharing of genetic traits or characteristics via a close relation (as cousins, for example)

consensus sequence: a sequence commonly found in DNA molecules from various sources, implying that the sequence has been actively conserved and plays an important role in some genetic process

cosmid: a cloning vector partially derived from genetic sequences of lambda, a bacteriophage; cosmids are useful in cloning relatively large fragments of DNA

crossing over: the exchange of genetic material

between two homologous chromosomes during prophase I of meiosis, providing an important source of genetic variation; also called "recombination"

cyclic adenosine monophosphate (cAMP): an important cellular molecule involved in cell signaling and regulation pathways

cyclins: a group of eukaryotic proteins with characteristic patterns of synthesis and degradation during the cell cycle; part of an elaborate mechanism of cell cycle regulation, and a key to the understanding of cancer

cytogenetics: the study of inheritance focusing on cellular mechanisms using techniques of cellular and molecular biology

cytokinesis: the division of the cytoplasm occurring as a last step in cell division

cytoplasmic inheritance. *See* extranuclear inheritance

cytosine (C): a pyrimidine nitrogenous base found in the structure of both DNA and RNA

cytoskeleton: the structure, composed of microtubules and microfilaments, that gives shape to a eukaryotic cell, enables some cells to move, and assists in such processes as cell division

dalton: a unit of molecular weight equal to the mass of a hydrogen atom; cellular molecules such as proteins are often measured in terms of a kilodalton, equal to 1,000 daltons

degenerate: refers to a property of the genetic code via which two or more codons can code for the same amino acid

deletion: a type of chromosomal mutation in which a genetic sequence is lost from a chromosome, usually through an error in recombination

denaturation: changes in the physical shape of a molecule caused by changes in the immediate environment such as temperature or pH level; denaturation usually involves the alteration or breaking of various bonds within the molecule and is important in DNA and protein molecules

deoxyribonucleic acid (DNA): the genetic material found in all cells; DNA consists of nitrogenous bases (adenine, guanine, cyto-

sine, and thymine), sugar (deoxyribose), and phosphate

deoxyribose: a five-carbon sugar used in the structure of DNA

diakinesis: a subphase of prophase I in meiosis in which chromosomes are completely condensed and position themselves in preparation for metaphase

dicentric chromosome: a chromosome with two centromeres, usually resulting from an error of recombination

differentiation: the series of changes necessary to convert an embryonic cell into its final adult form, usually with highly specialized structures and functions

dihybrid: an organism that is hybrid for each of two genes, for example "*AaBb*"; when two dihybrid organisms are mated, the offspring will appear in a 9:3:3:1 ratio with respect to the traits controlled by the two genes

diploid: a cell or organism with two complete sets of chromosomes, usually represented as "$2N$" where N stands for one set of chromosomes

diplotene: a subphase of prophase I in meiosis in which synapsed chromosomes begin to move apart and the chiasmata are clearly visible

discontinuous replication: replication on the lagging strand of a DNA molecule, resulting in the formation of Okazaki fragments

discontinuous variation: refers to a set of related phenotypes that are distinct from one another, with no overlapping

disjunction: the normal division of chromosomes that occurs during meiosis or mitosis; the related term "nondisjunction" refers to problems with this process

DNA. *See* deoxyribonucleic acid

DNA footprinting: a molecular biology technique involving DNA-binding proteins that are allowed to bind to DNA; the DNA is then degraded by DNases, and the binding sites of the proteins are revealed by the nucleotide sequences protected from degradation

DNA gyrase: a bacterial enzyme that reduces tension in DNA molecules that are being unwound during replication; a type of cellular enzyme called a topoisomerase

DNA ligase: a cellular enzyme used to connect

pieces of DNA together using a phosphodiester bond; important in genetic engineering procedures

DNA polymerase: the cellular enzyme responsible for making new copies of DNA molecules through replication of single-stranded DNA template molecules

DNase: refers to a class of enzymes, deoxyribonucleases, which specifically degrade DNA molecules

dominant: an allele or a trait that will mask the presence of a recessive allele or trait

double helix: a model of DNA structure proposed by molecular biologists James Watson and Francis Crick; the major features of this model are two strands of DNA wound around each other and connected by hydrogen bonds between complementary base pairs

duplication: a type of chromosomal mutation in which a chromosome region is duplicated because of an error in recombination during prophase I of meiosis; thought to play an important role in gene evolution

E. coli. See *Escherichia coli*

electrophoresis. *See* gel electrophoresis

embryo: the earliest period of development for complex organisms; in humans, the stage of development that begins at fertilization and ends with the eighth week of development, after which the embryo is called a fetus

endonuclease: an enzyme that degrades a nucleic acid molecule by breaking phosphodiester bonds within the molecule

endosymbiotic hypothesis: a hypothesis stating that mitochondria and chloroplasts were once free-living bacteria that entered into a symbiotic relationship with early preeukaryotic cells; structural and genetic similarities between these organelles and bacteria provide support for this hypothesis

enhancer: a region of a DNA molecule that facilitates the transcription of a gene, usually by stimulating the interaction of RNA polymerase with the gene's promoter

enzyme: a protein that speeds up or facilitates a specific biochemical reaction in a cell

epistasis: a genetic phenomenon in which one gene influences the expression of a second gene, usually by masking the effect of the second gene; however, only one trait is being controlled by these two genes, so epistasis is characterized by modified dihybrid ratios

equational division: refers to meiosis II, in which the basic number of chromosome types remains the same although sister chromatids are separated from one another; after equational division occurs, functional haploid gametes are present

Escherichia coli: a bacterium widely studied in genetics research and extensively used in biotechnological applications

ethidium bromide: a chemical substance that inserts itself into the DNA double helix; when exposed to ultraviolet light, ethidium bromide fluoresces, making it useful for the visualization of DNA molecules in molecular biology techniques

euchromatin: chromatin that is loosely coiled during interphase; thought to contain transcriptionally active genes

eugenics: a largely discredited field of genetics that seeks to improve humankind by selective breeding; can be positive eugenics, in which individuals with desirable traits are encouraged or forced to breed, or negative eugenics, in which individuals with undesirable traits are discouraged or prevented from breeding

eukaryote: a cell with a nuclear membrane surrounding its genetic material, a characteristic of a true nucleus; eukaryotic organisms include all known organisms except bacteria, which are prokaryotic

euploid: the normal number of chromosomes for a cell or organism

exon: protein coding sequences in eukaryotic genes, usually flanked by introns

exonuclease: an enzyme that degrades a nucleic acid molecule by breaking phosphodiester bonds at either end of the molecule

expression vector: a DNA cloning vector designed to allow genetic expression of inserted genes via promoters engineered into the vector sequence

extranuclear inheritance: inheritance involving genetic material located in the mitochondria or chloroplasts of a eukaryotic cell; also known as maternal inheritance (because these organelles are always inherited from the mother) and cytoplasmic inheritance

F pilus: also called the fertility pilus; a reproductive structure found on the surface of some bacterial cells that allows the cells to exchange a DNA molecule called an F plasmid during the process of conjugation

F1 generation: first filial generation; offspring produced from a mating of P generation individuals

F2 generation: second filial generation; offspring produced from a mating of F1 generation individuals

formylmethionine (fMet): the amino acid used to start all bacterial proteins; attached to the initiator transfer RNA molecule

frameshift mutation: a DNA mutation involving the insertion or deletion of one or two nucleotides, resulting in a shift of the codon reading frame; usually produces nonfunctional proteins

fraternal twins: twins that develop and are born simultaneously but are genetically unique, being produced from the fertilization of two separate eggs; a synonymous term is "dizygotic twins"

G0: a point in the cell cycle at which a cell is no longer progressing toward cell division; can be considered a "resting" stage

G1 checkpoint: a point in the cell cycle at which a cell commits either to progressing toward cell division (by replicating its DNA and eventually engaging in mitosis) or to entering the G0 phase, thereby withdrawing from the cell cycle either temporarily or permanently

gamete: a sex cell, either sperm or egg, containing half the genetic material of a normal cell

gel electrophoresis: a molecular biology technique in which nucleic acids are placed into a gel-like matrix (such as agarose) and then subjected to an electric current; using this technique, researchers can separate nucleic acid fragments of varying sizes

gene: a portion of a DNA molecule containing the genetic information necessary to pro-

duce a molecule of messenger RNA (via the process of transcription) that can then be used to produce a protein (via the process of translation)

gene frequency: the occurrence of a particular allele present in a population, expressed as a percentage of the total number of alleles present

gene pool: the complete assortment of genes possessed by the members of a population that are eligible to reproduce

genetic code: the correspondence between the sequence of nucleotides in DNA or mRNA molecules and the amino acids used to construct a protein

genetic counseling: a discipline concerned with analyzing the inheritance patterns of a particular genetic defect within a given family, including the determination of the risk associated with the presence of the genetic defect in future generations

genetic engineering: a term encompassing a wide variety of molecular biology techniques, all concerned with the modification of genetic characteristics of cells or organisms to accomplish a desired effect

genetics: an area of biology involving the scientific study of heredity

genome: the complete amount of DNA found in the nucleus of a normal cell, expressed as a particular number of chromosomes; for example, a human cell has a genome of forty-six chromosomes

genomic imprinting: a genetic phenomenon in which the phenotype associated with a particular allele depends on which parent donated the allele

genotype: the genetic characteristics of a cell or organism, expressed as a set of symbols representing the alleles present

guanine (G): a purine nitrogenous base found in the structure of both DNA and RNA

H substance: a carbohydrate molecule on the surface of red blood cells; when modified by certain monosaccharides, this molecule provides the basis of the ABO blood groups

haploid: refers to a cell or an organism with one set of chromosomes; usually represented as the "N" number of chromosomes, with "$2N$"

standing for the diploid number of chromosomes

Hardy-Weinberg law: a concept in population genetics stating that, given an infinitely large population that experiences random mating without mutation or any other such affecting factor, the frequency of particular alleles will reach a state of equilibrium, after which their frequency will not change from one generation to the next

helicase: a cellular enzyme that breaks hydrogen bonds between the strands of the DNA double helix, thus unwinding the helix and facilitating DNA replication

hemizygous: a gene present in a single copy, such as any gene on the X chromosome in a human male

hemophilia: an X-linked recessive disorder in which an individual's blood does not clot properly because of a lack of blood clotting factors; as in all X-linked recessive traits, the disease is most common in males, the allele for the disease being passed from mother to son

heredity: the overall mechanism by which characteristics or traits are passed from one generation of organisms to the next; genetics is the scientific study of heredity

heterochromatin: a highly condensed form of chromatin, usually transcriptionally inactive

heteroduplex: a double-stranded molecule of nucleic acid with each strand from a different source, formed either through natural means such as recombination or through artificial means in the laboratory

heterogametic sex: the particular sex of an organism that produces gametes containing two types of sex chromosome; in humans, males are the heterogametic sex, producing sperm that can carry either an X chromosome or a Y chromosome

heterogeneous nuclear RNA (hnRNA): an assortment of RNA molecules of various types found in the nucleus of the cell and in various stages of processing prior to their export to the cytoplasm

heterozygous: a genotype composed of two alleles that are different, for example *Aa*; synonymous with "hybrid"

histones: specialized proteins in eukaryotic cells that bind to DNA molecules and cause them to become more compact; thought to be involved in regulation of gene expression as well

HLA. *See* human leukocyte antigens

hnRNA. *See* heterogeneous nuclear RNA

holandric: refers to a trait passed from father to son via a sex chromosome such as the Y chromosome in human males

homeobox: a DNA sequence encoding a highly basic protein known as a homeodomain; a homeodomain functions as a transcription factor and is thought to help regulate major events in the embryonic development of higher organisms

homogametic sex: the particular sex of an organism that produces gametes containing only one type of sex chromosome; in humans, females are the homogametic sex, producing eggs with X chromosomes

homologous: refers to chromosomes that are identical in terms of types of genes present and the location of the centromere; because of their high degree of similarity, homologous chromosomes can synapse and recombine during prophase I of meiosis

homozygous: a genotype composed of two alleles that are the same, for example *AA* or *aa*; synonymous with "pure-bred"

human leukocyte antigens (HLA): molecules found on the surface of cells that allow the immune system to differentiate between foreign, invading cells and the body's own cells

hybrid: any cell or organism with genetic material from two different sources, either through natural processes such as sexual reproduction or more artificial processes such as genetic engineering

hybridization: a process of base-pairing involving two single-stranded nucleic acid molecules with complementary sequences; the extent to which two unrelated nucleic acid molecules will hybridize is often used as a way to determine the amount of similarity between the sequences of the two molecules

hybridoma: a type of hybrid cancer cell created by artificially joining a cancer cell with an antibody-producing cell; useful to research techniques in immunology

hydrogen bond: a bond formed between molecules containing hydrogen atoms with positive charges and molecules containing atoms such as nitrogen or oxygen that can possess a negative charge; a relatively weak but important bond in nature that, among other things, connects water molecules, allows DNA strands to base-pair, and contributes to the three-dimensional shape of proteins

identical twins: a pair of genetically identical offspring that develop from a single fertilized egg; also known as monozygotic twins

in vitro: an event occurring in an artificial setting such as in a test tube, as opposed to inside a living organism; literally, "in glass"

in vivo: an event occurring in a living organism, as opposed to an artificial setting; literally, "in the living"

inborn error of metabolism: a genetic defect in a one of a cell's metabolic pathways, usually at the level of an enzyme, that causes the pathway to malfunction; results in phenotypic alterations at the cellular or organism level

incomplete dominance: a phenomenon involving two alleles, neither of which masks the expression of the other; instead, the combination of the alleles in the heterozygous state produces a new phenotype that is usually intermediate to the phenotypes produced by either allele alone in the homozygous state

independent assortment: a characteristic of standard Mendelian genetics referring to the random assortment or shuffling of alleles and chromosomes that occurs during meiosis I; independent assortment is responsible for the offspring ratios observed in Mendelian genetics

inducer: a molecule that activates some bacterial operons, usually by interacting with regulatory proteins bound to the operator region

initiation codon: also called the "start codon," this is a codon, composed of the nucleotides AUG, signaling the beginning of a protein coding sequence in a messenger RNA molecule; in the genetic code, AUG always represents the amino acid methionine

intercalary deletion: a type of chromosome deletion in which DNA has been lost from within the chromosome (as opposed to a terminal deletion involving a region of DNA lost from the end of the chromosome)

interference: in genetic linkage, a mathematical expression that represents the difference between the expected and the observed number of double recombinant offspring; this can be a clue to the physical location of linked genes on the chromosome

interphase: the period of the cell cycle in which the cell is preparing to divide, consisting of two distinct growth phases (G1 and G2) separated by a period of DNA replication (S phase)

intron: an intervening sequence within eukaryotic DNA, transcribed as part of an mRNA precursor but then removed by splicing before the mRNA molecule is translated; introns are thought to play an important role in the evolution of genes

inversion: a chromosomal abnormality resulting in a region of the chromosome where the normal order of genes is reversed

isotope: an alternative form of an element with a variant number of neutrons in its atomic nucleus; isotopes are frequently radioactive and are important tools for numerous molecular biology techniques

karyokinesis: division of the cell's nuclear contents, as opposed to cytokinesis (division of the cytoplasm)

karyotype: the complete set of chromosomes possessed by an individual, usually isolated during metaphase and arranged by size and type as a method of detecting chromosomal abnormalities

kilobase: a unit of measurement for nucleic acid molecules, equal to one thousand bases or nucleotides; abbreviated kb

kinetochore: a chromosome structure found in the region of the centromere and used as an attachment point for the microtubules of the spindle apparatus during cell division

Klinefelter's syndrome: a human genetic disorder in males who possess an extra X chromosome; Klinefelter's males have forty-seven chromosomes instead of the normal forty-six and suffer from abnormalities such as sterility, body feminization, and mental retardation

lactose: a disaccharide that is an important part of the metabolism of many bacterial species; lactose metabolism in these species is genetically regulated via the lac operon

lagging strand: in DNA replication, the strand of DNA being synthesized in a direction opposite to that of replication fork movement; this strand is synthesized in a discontinuous fashion as a series of Okazaki fragments later joined together

leading strand: in DNA replication, the strand of DNA being synthesized in the same direction as the movement of the replication fork; this strand is synthesized in a continuous fashion

leptotene: a subphase of prophase I of meiosis in which chromosomes begin to condense and become visible

lethal allele: an allele capable of causing the death of an organism; a lethal allele can be recessive (two copies of the allele are required before death results) or dominant (one copy of the allele produces death)

leucine zipper: an amino acid sequence, found in some DNA-binding proteins, characterized by leucine residues separated by sets of seven amino acids; two molecules of this amino acid sequence can combine via the leucine residues and "zip" together, creating a structure that can then bind to a specific DNA sequence

linkage: a genetic phenomenon involving two or more genes inherited together because they are physically located on the same chromosome; Mendel's principle of independent assortment does not apply to linked genes, but genotypic and phenotypic variation is possible through crossing over

locus: the specific location of a particular gene on a chromosome

Lyon hypothesis: a hypothesis stating that one X chromosome of the pair found in all female cells must be inactivated in order for those cells to be normal; the inactivated X chromosome is visible by light microscopy and stains as a Barr body

lysogeny: a viral process involving repression and integration of the viral genome into the genome of the host bacterial cell

major histocompatibility complex (MHC): a group of molecules found on the surface of cells, allowing the immune system to differentiate between foreign, invading cells and the body's own cells; in humans, this group of molecules is called HLA (human leukocyte antigens)

map unit. *See* centimorgan

maternal inheritance. *See* extranuclear inheritance

meiosis: a process of cell division in which the cell's genetic material is reduced by half and sex cells called gametes are produced; important as the basis of sexual reproduction

melting: a term used to describe the denaturation of a DNA molecule as it is heated in solution; as the temperature rises, hydrogen bonds between the DNA strands are broken until the double-strand molecule has been completely converted into two single-strand molecules

messenger RNA (mRNA): a type of RNA molecule containing the genetic information necessary to produce a protein through the process of translation; produced from the DNA sequence of a gene in the process of transcription

metacentric: a chromosome with the centromere located at or near the middle of the chromosome

metafemale: a term used to describe *Drosophila* (fruit fly) females that have more X chromosomes than sets of autosomes (for example, a female that has two sets of autosomes and three X chromosomes)

metaphase: the second phase in the process of mitosis, involving chromosomes lined up in the middle of the cell on a line known as the equator

microtubule: a cell structure involved in the movement and division of chromosomes during mitosis and meiosis; part of the cell's cytoskeleton, microtubules can be rapidly assembled and disassembled

miscegenation: sexual activity or marriage between members of two different human races

mismatch repair: a cellular DNA repair process in which improperly base-paired nucleotides are enzymatically removed and replaced with the proper sequence

missense mutation: a DNA mutation that changes an existing amino acid codon in a gene to some other amino acid codon; depending on the nature of the change, this can be a harmless or a serious mutation (for example, sickle-cell anemia in humans is the result of a missense mutation)

mitochondrion: the organelle responsible for production of ATP through the process of cellular respiration in a eukaryotic cell; sometimes referred to as the "powerhouse of the cell"

mitosis: the process of cell division in which a cell's duplicated genetic material is evenly divided between two daughter cells, so that each daughter cell is genetically identical to the original parent cell

monoclonal antibodies: identical antibodies (having specificity for the same antigen) produced by a single type of antibody-producing cell, either a B cell or a hybridoma cell line; important in various types of immunology research techniques

monohybrid: an organism that is hybrid with respect to a single gene, for example *Aa*; when two monohybrid organisms are mated, the offspring will generally appear in a 3:1 ratio involving the trait controlled by the gene in question

monosomy: a genetic condition in which one chromosome from a homologous chromosome pair is missing, producing a $2n$-1 genotype; usually causes significant problems in the phenotype of the organism

mRNA. *See* messenger RNA

multiple alleles: a genetic phenomenon in which a particular gene exists in more than two alleles; the greater the number of alleles, the greater the genetic diversity

mutagen: any chemical or physical substance capable of increasing mutations in a DNA sequence

mutant: a trait or organism different from the normal, or wild-type, trait or organism seen commonly in nature; mutants can arise either through expression of particular al-

leles in the organism or through spontaneous or intentional mutations in the genome

mutation: a change in the genetic sequence of an organism, usually leading to an altered phenotype

N terminus: the end of a polypeptide with an amino acid that has a free amino group

natural selection: a process involving genetic variation on the genotypic and phenotypic levels that contributes to the success or failure of various species in reproduction; thought to be the primary engine behind evolution

neutral mutation: a mutation with no observable effect on the phenotype of the cell or organism in which it occurs

nondisjunction: refers to the improper division of chromosomes during anaphase of mitosis or meiosis, resulting in cells with abnormal numbers of chromosomes and sometimes seriously altered phenotypes

nonsense codon: another term for a termination or stop codon (UAA, UAG, or UGA)

nonsense mutation: a DNA mutation that changes an existing amino acid codon in a message to one of the three termination, or stop, codons; this results in an abnormally short protein that is usually nonfunctional

Northern blot: a molecular biology procedure in which a labeled single-stranded DNA probe is exposed to cellular RNA immobilized on a filter; under the proper conditions, the DNA probe will seek out and bind to its complementary sequence in the RNA molecules if such a sequence is present

nuclease: an enzyme that degrades nucleic acids by breaking the phosphodiester bond that connects nucleosides

nucleic acid: the genetic material of cells, found in two forms: deoxyribonucleic acid (DNA) and ribonucleic acid (RNA); composed of repeating subunits called nucleotides

nucleocapsid: a viral structure including the capsid, or outer protein coat, and the nucleic acid of the virus

nucleoid: a region of a prokaryotic cell containing the cell's genetic material

nucleolus: a eukaryotic organelle located in the

nucleus of the cell; site of ribosomal RNA synthesis

nucleoside: a building block of nucleic acids, composed of a sugar (deoxyribose or ribose) and one of the nitrogenous bases A, C, G, T, or U

nucleosomes: the basic unit molecule of chromatin, composed of a segment of a DNA molecule that is bound to and wound around histone molecules; appears as beads on a string when viewed by electron microscopy

nucleotide: a building block of nucleic acids, composed of a sugar (deoxyribose or ribose), one of the nitrogenous bases (A, C, G, T, or U), and one or more phosphate groups

nucleus: the "control center" of eukaryotic cells, where the genetic material is separated from the rest of the cell by a membrane; site of DNA replication and transcription

nullisomy: a genetic condition in which both members of a homologous chromosome pair are absent; usually, embryos with this type of genetic defect are not viable

ochre codon: a stop codon (UAA) found in mRNA molecules; signals termination of translation

Okazaki fragments: short DNA fragments, approximately two thousand bases long or less, produced during discontinuous replication of a DNA molecule

oligonucleotide: a short molecule of DNA, generally fewer than twenty bases long and usually synthesized artificially; an important tool for numerous molecular biology procedures, including site-directed mutagenesis

oncogene: any gene capable of stimulating cell division, thereby being a potential cause of cancer if unregulated; found in all cells and in many cancer-causing viruses

oogenesis: the process of producing eggs in a sexually mature female organism; another term for meiosis in females

opal codon: a stop codon (UGA) found in mRNA molecules; signals termination of translation

open reading frame (ORF): a putative protein-coding DNA sequence, marked by a start codon at one end and a stop codon at the other end

operator: a region of a bacterial operon serving as a control point for transcription of the operon; a regulatory protein of some type usually binds to the operator

operon: a genetic structure found only in bacteria, whereby a set of genes are controlled together by the same control elements; usually these genes have a common function, such as the genes of the lactose operon in *E. coli* for the metabolism of lactose

P generation: parental generation; the original individuals mated in a genetic cross

pachytene: a subphase of prophase I in meiosis in which tetrads become visible

palindrome: in genetics, a DNA sequence that reads the same on each strand of the DNA molecule, although in opposite directions because of the antiparallel nature of the double helix; most DNA palindromes serve as recognition sites for restriction endonucleases

paracentric inversion: an inversion of a chromosome's sequence that does not involve the centromere, taking place on a single arm of the chromosome

parthenogenesis: production of an organism from an unfertilized egg

pedigree: a diagram of a particular family, showing the relationships between all members of the family and the inheritance pattern of a particular trait or genetic defect; especially useful for research into human traits that may otherwise be difficult to study

penetrance: a quantitative term referring to the percentage of individuals with a certain genotype that also exhibit the associated phenotype

peptide bond: a bond found in proteins; occurs between the carboxyl group of one amino acid and the amino group of the next, linking them together

pericentric inversion: an inversion of a chromosome's sequence involving the centromere

phenotype: the physical characteristics of an organism, such as "tall" versus "short"

phenylketonuria (PKU): a genetic defect causing an absence of the enzyme phenylalanine hydroxylase, which enables individuals to metabolize the amino acid phenylalanine; this defect results in mental retardation if the individual's diet is not carefully controlled with regard to phenylalanine levels

phosphodiester bond: in DNA, the phosphate group connecting one nucleoside to the next in the polynucleotide chain

photoreactivation repair: a cellular enzyme system responsible for repairing DNA damage caused by ultraviolet light; the system is activated by light

pilus: a hairlike reproductive structure possessed by some species of bacterial cells that allows them to engage in a transfer of genetic material known as conjugation

plasmid: a small, circular DNA molecule commonly found in bacteria and responsible for carrying various genes such as antibiotic resistance genes; important as a cloning vector for genetic engineering

pleiotropy: a genetic phenomenon in which a single gene has an effect on two or more traits

ploidy: refers to a chromosome set; the normal chromosome set number for a cell or organisms is called euploidy, whereas any departure from this number is called aneuploidy

point mutation: a DNA mutation involving a single nucleotide

polar body: a by-product of oogenesis used to dispose of extra, unnecessary chromosomes while preserving the cytoplasm of the developing ovum

polycistronic: usually refers to mRNA molecules that contain coding sequences for more than one protein; common in prokaryotic cells

polygenic inheritance: expression of a trait depending on the cumulative effect of multiple genes; human traits such as skin color, obesity, and intelligence are thought to be examples of polygenic inheritance

polymerase: a cellular enzyme capable of creating a phosphodiester bond between two nucleotides, producing a polynucleotide chain complementary to a single-stranded nucleic acid template; the enzyme DNA polymerase is important for DNA replication, and the enzyme RNA polymerase is involved in transcription

polymerase chain reaction (PCR): a molecular biology technique in which millions of copies of a single DNA molecule can be artificially produced in a relatively short period of time; important for a wide variety of applications when the source of DNA to be copied is either scarce or impure

polypeptide: a single chain of amino acids connected to one another by peptide bonds; can be used synonymously with the term "protein"

polyploid: a cell or organism that possesses multiple sets of chromosomes, usually more than two

polysome: a group of ribosomes attached to the same mRNA molecule and producing the same protein product in varying stages of completion

primer: a short nucleic acid molecule used as a beginning point for the enzyme DNA polymerase as it attempts to replicate a single-stranded template

prion: an infectious agent composed solely of protein; thought to be the cause of various human and animal diseases characterized by neurological degeneration, including scrapie in sheep, mad cow disease in cattle, and Creutzfeldt-Jakob disease in humans

probe: in genetics research, typically a single-stranded nucleic acid molecule that has been labeled in some way, either with radioactive isotopes or fluorescent dyes; this molecule is then used to seek out its complementary nucleic acid molecule in a variety of molecular biology techniques such as Southern or Northern blotting

product rule: a rule of probability stating that the probability associated with two simultaneous yet independent events is the product of the events' individual probabilities

prokaryote: a cell that lacks a nuclear membrane and therefore has no true nucleus; the bacteria are the only known prokaryotic organisms

promoter: a region of a gene that controls transcription of that gene; a physical binding site for RNA polymerase

prophase: the first phase in the process of mitosis or meiosis, in which the nuclear membrane disappears, the spindle apparatus begins to form, and chromatin takes on the form of chromosomes by becoming shorter and thicker

propositus: the individual in a human pedigree who is the focus of the pedigree, usually by being the first person who came to the attention of the geneticist

protein: a biological molecule composed of amino acids linked together by peptide bonds; used as structural components of the cell or as enzymes; the term "protein" can refer to a single chain of amino acids or to multiple chains of amino acids functioning in a concerted way, as in the molecule hemoglobin

proto-oncogene: a gene, found in eukaryotic cells, that stimulates cell division; ordinarily, expression of this type of gene is tightly controlled by the cell, but in cancer cells, proto-oncogenes have been converted into oncogenes through alteration or elimination of controlled gene expression

pseudodominance: a genetic phenomenon involving a recessive allele on one chromosome that is automatically expressed because of the deletion of its corresponding dominant allele on the other chromosome of the homologous pair

purine: either of the nitrogenous bases adenine or guanine; used in the structure of nucleic acids

pyrimidine: any of the nitrogenous bases cytosine, thymine, or uracil; used in the structure of nucleic acids

reading frame: refers to the manner in which an mRNA sequence is interpreted as a series of amino acid codons by the ribosome; because of the triplet nature of the genetic code, a typical mRNA molecule has three possible reading frames, although usually only one of these will actually code for a functional protein

recessive: a term referring to an allele or trait that will only be expressed if another, dominant, trait or allele is not also present

reciprocal cross: a mating that is the reverse of another with respect to the sex of the organisms that possess certain traits; for example, if a particular cross were tall male X short

female, then the reciprocal cross would be short male X tall female

reciprocal translocation: a two-way exchange of genetic material between two non-homologous chromosomes, resulting in a wide variety of genetic problems depending on which chromosomes are involved in the translocation

recombinant DNA: DNA molecules that are the products of artificial recombination between DNA molecules from two different sources; important as a foundation of genetic engineering

recombination: an exchange of genetic material, usually between two homologous chromosomes; provides one of the foundations for the genetic reassortment observed during sexual reproduction

reductional division: refers to meiosis I, in which the amount of genetic material in the cell is reduced by half through nuclear division; it is at this stage that the diploid cell is converted to an essentially haploid state

replication: the process by which a DNA molecule is enzymatically copied

replicon: a region of a chromosome under control of a single origin of replication

replisome: a multiprotein complex that functions at the replication fork during DNA replication; contains all the enzymes and other proteins necessary for replication, including DNA polymerase

repressor: a protein molecule capable of preventing transcription of a gene, usually by binding to a regulatory region close to the gene

restriction endonuclease: a bacterial enzyme that cuts DNA molecules at specific sites; part of a bacterial cell's built-in protection against infection by viruses; an important tool of genetic engineering

restriction fragment length polymorphism (RFLP): a genetic marker, consisting of variations in the length of restriction fragments in DNA from individuals being tested, allowing researchers to compare genetic sequences from various sources; used in a variety of fields, including forensics and the Human Genome Project

retinoblastoma: a cancer of the retina occurring primarily in young children; the study of the retinoblastoma gene, classified as a tumor-suppressor gene, has yielded important insights into the nature and causes of cancer

reverse transcriptase: a form of DNA polymerase, discovered in retroviruses, that uses an RNA template to produce a DNA molecule; the name indicates that this process is the reverse of the transcription process occurring naturally in the cell

Rh factor: a human red-blood-cell antigen, first characterized in rhesus monkeys, that contributes to blood typing; individuals can be either Rh positive (possessing the antigen on their red blood cells) or Rh negative (lacking the antigen)

ribonucleic acid (RNA): a form of nucleic acid in the cell used primarily for genetic expression through transcription and translation; in structure, it is virtually identical to DNA, except that ribose is used as the sugar in each nucleotide, and the nitrogenous base thymine is replaced by uracil; present in three major forms in the cell: messenger RNA, transfer RNA, and ribosomal RNA

ribose: a five-carbon sugar used in the structure of ribonucleic acid (RNA)

ribosomal RNA (rRNA): a type of ribonucleic acid in the cell that constitutes some of the structure of the ribosome and participates in the process of translation

ribosome: a cellular organelle, composed of ribosomal RNA and proteins, that is the site of translation

RNA polymerase: the cellular enzyme required for making an RNA copy of genetic information contained in a gene; an integral part of transcription

RNase: refers to a group of enzymes, ribonucleases, capable of specifically degrading RNA molecules

rRNA. *See* ribosomal RNA

segregation: a characteristic of Mendelian genetics, resulting in the division of homologous chromosomes into separate gametes during the process of meiosis

semiconservative replication: a characteristic of DNA replication, in which every new DNA molecule is actually a hybrid molecule, being composed of a parental, pre-existing strand and a newly synthesized strand

sex chromosome: a chromosome carrying genes responsible for determination of an organism's sex; in humans, the sex chromosomes are designated *X* and *Y*

sex-influenced inheritance: refers to traits the expression of which is influenced or altered relative to the sex of the individual possessing the trait; pattern baldness is an example of this type of inheritance in humans

sex-limited inheritance: refers to traits expressed in only one sex, although these traits are usually produced by non-sex-linked genes (that is, they are genes located on autosomes instead of sex chromosomes)

sexual reproduction: reproduction of cells or organisms involving the transfer and reassortment of genetic information, resulting in offspring that can be phenotypically and genotypically distinct from either of the parents; mediated by the fusion of gametes produced during meiosis

Shine-Dalgarno sequence: a short sequence in prokaryotic mRNA molecules complementary to a sequence in the prokaryotic ribosome; important for proper positioning of the start codon of the mRNA relative to the P site of the ribosome

shotgun cloning: a technique by which random DNA fragments from an organism's genome are inserted into a collection of vectors to produce a library of clones, which can then be used in a variety of molecular biology procedures

sigma factor: a molecule that is part of RNA polymerase molecules in bacterial cells; allows RNA polymerase to select the genes that will be transcribed

site-directed mutagenesis: a molecular genetics procedure in which synthetic oligonucleotide molecules are used to induce carefully planned mutations in a cloned DNA molecule

small nuclear RNA (snRNA): small, numerous RNA molecules found in the nuclei of eukaryotic cells; involved in splicing of mRNA precursors to prepare them for translation

sociobiology: the study of social structures, organizations, and actions in terms of underlying biological principles

solenoid: a complex, highly compacted DNA structure consisting of many nucleosomes packed together in a bundle

somatic mutation: a mutation occurring in a somatic, or nonsex, cell; because of this, somatic mutations cannot be passed to the next generation

Southern blot: a molecular biology technique in which a labeled single-stranded DNA probe is exposed to denatured cellular DNA immobilized on a filter; under the proper conditions, the DNA probe will seek out and bind to its complementary sequence among the cellular DNA molecules, if such a sequence is present

species: a group of organisms that can interbreed with one another but not with organisms outside the group; generally, members of a particular species share the same gene pool

spermatogenesis: the process of producing sperm in a sexually mature male organism; another term for meiosis in males

spindle apparatus: a structure, composed of microtubules and microfilaments, important for the proper orientation and movement of chromosomes during mitosis and meiosis; appears during prophase and begins to disappear during anaphase

spliceosome: a complex of nuclear RNA and protein molecules responsible for the excision of introns from mRNA precursors before they are translated

SRY: the sex-determining region of the Y chromosome; a gene encoding a protein product called testis determining factor (TDF), responsible for conversion of a female embryo to a male embryo through the development of the testes

sum rule: a rule of probability theory stating that the probability of either of two mutually exclusive events occurring is the sum of the events' individual probabilities

supercoil: a complex DNA structure in which the DNA double helix is itself coiled into a helix; usually observed in circular DNA molecules such as bacterial plasmids

synapsis: refers to the close association of homologous chromosomes occurring during early prophase I of meiosis; during synapsis, recombination between these chromosomes can occur

Taq polymerase: DNA polymerase from the bacterium *Thermus aquaticus*; an integral component of polymerase chain reaction

telocentric chromosome: a chromosome with a centromere at the end

telomere: the end of a eukaryotic chromosome, protected and replaced by the cellular enzyme telomerase

telophase: the final phase in the process of mitosis or meiosis, in which division of the cell's nuclear contents has been completed and division of the cell itself occurs

template: a single-stranded DNA molecule used to create a complementary strand of nucleic acid through the activity of a polymerase

teratogen: any chemical or physical substance that creates birth defects in offspring

testcross: a mating involving an organism with a recessive genotype for desired traits crossed with an organism that has an incompletely determined genotype; the types and ratio of offspring produced allow geneticists to determine the genotype of the second organism

tetrad: a group of four chromosomes formed as a result of the synapsis of homologous chromosomes that takes place early in meiosis

tetranucleotide hypothesis: a disproven hypothesis, formulated by geneticist P. A. Levene, stating that DNA is a structurally simple molecule composed of a repeating unit known as a tetranucleotide (composed, in turn, of equal amounts of the bases A, C, G, and T)

theta structure: an intermediate structure in the bidirectional replication of a circular DNA molecule; the name comes from the resemblance of this structure to the Greek letter theta

thymine (T): a pyrimidine nitrogenous base found in the structure of DNA; in RNA, thymine is replaced by uracil

thymine dimer: a pair of thymine bases in a DNA molecule connected by an abnormal chemical bond induced by ultraviolet light; prevents DNA replication in the cell unless it is removed by specialized enzymes

topoisomerases: cellular enzymes that relieve tension in replicating DNA molecules by introducing single- or double-stranded breaks into the DNA molecule; without these enzymes, replicating DNA becomes progressively more supercoiled until it can no longer unwind, and DNA replication is halted

totipotent: the ability of a cell to produce an entire adult organism through successive cell divisions and development; as cells become progressively differentiated, they lose this characteristic

trait: a phenotypic characteristic that is heritable

transcription: the cellular process by which genetic information in the form of a gene in a DNA molecule is converted into the form of an mRNA molecule; dependent on the enzyme RNA polymerase

transduction: DNA transfer between cells, with a virus serving as the genetic vector

transfer RNA (tRNA): a type of RNA molecule necessary for translation to occur properly; provides the basis of the genetic code, in which codons in a messenger RNA molecule are used to direct the sequence of amino acids in a polypeptide; contains a binding site for a particular amino acid and a region complementary to a messenger RNA codon (an anticodon)

transformation: the process by which a normal cell is converted into a cancer cell; also refers to the change in phenotype accompanying entry of foreign DNA into a cell, such as in bacterial cells being used in recombinant DNA procedures

transgenic organism: an organism possessing one or more genes from another organism, such as mice that possess human genes; important for the study of genes in a living organism, especially in the study of mutations within these genes

transition mutation: a DNA mutation in which one pyrimidine (C or T) takes the place of another, or a purine (A or G) takes the place of another

translation: the cellular process by which genetic information in the form of an mRNA molecule is converted into the amino acid sequence of a protein, using ribosomes and RNA molecules as accessory molecules

translocation: the movement of a chromosome segment to a nonhomologous chromosome as a result of an error in recombination; also refers to the movement of an mRNA codon from the A site of the ribosome to the P site during translation

transposon: a DNA sequence capable of moving to various places in a chromosome, discovered by geneticist Barbara McClintock; transposons are thought to be important as mediators of genetic variability in both prokaryotes and eukaryotes

transversion: a DNA mutation in which a pyrimidine (C or T) takes the place of a purine (A or G), or vice versa

triploid: possessing three complete sets of chromosomes, or $3N$; important in the development of desirable characteristics in the flowers or fruit of some plants

trisomy: a genetic condition involving one chromosome of a homologous chromosome pair that has been duplicated in some way, giving rise to a $2N + 1$ genotype and causing serious phenotypic abnormalities; a well-known example is trisomy 21, or Down syndrome, in which the individual possesses three copies of chromosome 21 instead of the normal two copies

tumor-suppressor genes: any of a number of genes that limit or halt cell division under certain circumstances, thereby preventing the formation of tumors in an organism; two well-studied examples are the retinoblastoma gene and the *p53* gene

Turner syndrome: a human genetic defect in which an individual has only forty-five chromosomes, lacking one sex chromosome; the sex chromosome present is an X chromosome, making these individuals phenotypically female, although with serious abnormalities such as sterility and anatomical defects

uracil (U): a pyrimidine nitrogenous base found in the structure of RNA; in DNA, uracil is replaced by thymine

variable number tandem repeats (VNTR): repetitive DNA sequences of approximately fifty to one hundred nucleotides; important in the process of forensic identification known as DNA fingerprinting

vector: a DNA molecule, such as a bacterial plasmid, into which foreign DNA can be inserted and then transported into a cell for further manipulation; important in a wide variety of recombinant DNA techniques

virus: a microscopic infectious particle composed primarily of protein and nucleic acid; bacterial viruses, or bacteriophages, have been important tools of study in the history of molecular genetics

Western blot: a molecular biology technique involving labeled antibodies exposed to cellular proteins immobilized on a filter; under the proper conditions, the antibodies will seek out and bind to the proteins for which they are specific, if such proteins are present

wild-type: a trait common in nature; usually contrasted with variants of the trait, which are known as mutants

wobble hypothesis: a concept stating that the anticodon of a transfer RNA molecule is capable of interacting with more than one mRNA codon by virtue of the inherent flexibility present in the third base of the anticodon; first proposed by molecular biologist Francis Crick

X linkage: a genetic phenomenon involving a gene located on the X chromosome; the typical pattern of X linkage involves recessive alleles, such as that for hemophilia, which exert their effects when passed from mother to son and are more likely to be exhibited by males than females

Y linkage: a genetic phenomenon involving a gene located on the Y chromosome; as a result, such a condition can be passed only from father to son

yeast artificial chromosome (YAC): a cloning vector that has been engineered with all of the major genetic characteristics of a eukaryotic chromosome so that it will behave as such during cell division; YACs are used to clone extremely large DNA fragments from eukaryotic cells and are an integral part of the Human Genome Project

Z-DNA: a zigzag form of DNA in which the strands form a left-handed helix instead of the normal right-handed helix of B-DNA; Z-DNA is known to be present in cells and is thought to be involved in genetic regulation

zinc finger: an amino acid sequence, found in some DNA-binding proteins, that complexes with zinc ions to create polypeptide "fingers" that can then wrap around a specific portion of a DNA molecule

zygote: a diploid cell produced by the union of a male gamete (sperm) with a female gamete (egg); through successive cell divisions, the zygote will eventually give rise to the adult form of the organism

zygotene: a subphase of prophase I of meiosis involving synapsis between homologous chromosomes

—*Randall K. Harris*

Bibliography

Bacterial Genetics

Adelberg, Edward A., ed. *Papers on Bacterial Genetics*. Boston: Little, Brown, 1966.

Birge, Edward A. *Bacterial and Bacteriophage Genetics*. New York: Springer-Verlag, 1994.

Brock, Thomas D. *The Emergence of Bacterial Genetics*. Cold Spring Harbor, N.Y.: Cold Spring Harbor Laboratory Press, 1990.

Day, Martin J. *Plasmids*. London: Edward Arnold, 1982.

Dean, Alastair Campbell Ross, and Sir Cyril Hinshelwood. *Growth, Function, and Regulation in Bacterial Cells*. Oxford: Clarendon Press, 1966.

De Bruijn, Frans J., et al., eds. *Bacterial Genomes: Physical Structure and Analysis*. New York: Chapman & Hall, 1998.

Dorman, Charles J. *Genetics of Bacterial Virulence*. Oxford: Blackwell Scientific, 1994.

Drlica, Karl, and Monica Riley, eds. *The Bacterial Chromosome*. Washington, D.C.: American Society for Microbiology, 1990.

Fry, John C., and Martin J. Day, eds. *Bacterial Genetics in Natural Environments*. London: Chapman & Hall, 1990.

Garrett, Laurie. *The Coming Plague: Newly Emerging Diseases in a World out of Balance*. New York: Penguin Books, 1994.

Jacob, François, and Elie L. Wollman. *Sexuality and the Genetics of Bacteria*. New York: Academic Press, 1961.

Joset, Françoise, et al. *Prokaryotic Genetics: Genome Organization, Transfer, and Plasticity*. Boston: Blackwell Scientific Publications, 1993.

Lappe, Marc. *Breakout: The Evolving Threat of Drug-Resistant Disease*. San Francisco: Sierra Club Books, 1996.

Office of Technology Assessment. *New Developments in Biotechnology—Field-Testing Engineered Organisms: Genetic and Ecological Issues*. Washington, D.C.: National Technical Information Service, 1988.

Snyder, Larry, and Wendy Champness. *Molecular Genetics of Bacteria*. Washington, D.C.: ASM Press, 1997.

Zubay, Geoffrey L., ed. *Papers in Biochemical Genetics*. New York: Holt, Rinehart and Winston, 1968.

Classical Transmission Genetics

Berg, Paul, and Maxine Singer. *Dealing with Genes: The Language of Heredity*. Mill Valley, Calif.: University Science Books, 1992.

Cittadino, E. *Nature as the Laboratory*. New York: Columbia University Press, 1990.

Dunn, L. C. *A Short History of Genetics: The Development of Some of the Main Lines of Thought, 1864-1939*. New York: McGraw-Hill, 1965.

Fast, Julius. *Blueprint for Life: The Story of Modern Genetics*. New York: St. Martin's Press, 1965.

Ford, E. B. *Understanding Genetics*. London: Faber & Faber, 1979.

Gardner, Eldon J. *Principles of Genetics*. 5th ed. New York: John Wiley & Sons, 1975.

Goodenough, Ursula. *Genetics*. 2d ed. New York: Holt, Rinehart and Winston, 1978.

Iltis, Hugo. *Life of Mendel*. Translated by Eden Paul and Cedar Paul. London: Allen & Unwin, 1932.

Jacob, François. *The Logic of Life: A History of Heredity*. New York: Pantheon, 1973.

Klug, William, and Michael Cummings. *Concepts of Genetics*. 4th ed. New York: Macmillan College, 1994.

Lewin, Benjamin. *Genes IV*. Cambridge, Mass.: Oxford University Press and Cell Press, 1990.

Mendel, Gregor. *Experiments in Plant-Hybridization*. Foreword by Paul C. Mangelsdorf. Cambridge, Mass.: Harvard University Press, 1965.

Olby, Robert C. *Origins of Mendelism*. New York: Schocken Books, 1966.

Rothwell, Norman V. *Understanding Genetics*. 4th ed. New York: Oxford University Press, 1988.

Russell, Peter. *Genetics*. 2d ed. Boston: Scott, Foresman, 1990.

Singer, Sam. *Human Genetics: An Introduction to the Principles of Heredity*. 2d ed. New York: W. H. Freeman, 1985.

Stubbe, H. *A History of Genetics*. Cambridge, Mass.: MIT Press, 1968.

Sturtevant, Alfred H. *A History of Genetics*. New York: Harper & Row, 1965.

Suzuki, David T., et al., eds. *An Introduction to Genetic Analysis.* 2d ed. San Francisco: W. H. Freeman, 1981.

Weaver, Robert F., and Philip W. Hendrick. 3d ed. *Genetics.* Dubuque, Iowa: Wm. C. Brown, 1997.

Wilson, Edward O. *The Diversity of Life.* New York: W. W. Norton, 1992.

Developmental Genetics

Davidson, Eric H. *Gene Activity in Early Development.* Orlando, Fla.: Academic Press, 1986.

DePomerai, David. *From Gene to Animal: An Introduction to the Molecular Biology of Animal Development.* New York: Cambridge University Press, 1985.

Dyban, A. P., and V. S. Baranov. *Cytogenetics of Mammalian Embryonic Development.* Translated by V. S. Baranov. New York: Oxford University Press, 1987.

Gottlieb, Frederick J. *Developmental Genetics.* New York: Reinhold, 1966.

Gurdon, John B. *The Control of Gene Expression in Animal Development.* Cambridge, Mass.: Harvard University Press, 1974.

Hahn, Martin E., et al., eds. *Developmental Behavior Genetics: Neural, Biometrical, and Evolutionary Approaches.* New York: Oxford University Press, 1990.

Hennig, W., ed. *Early Embryonic Development of Animals.* New York: Springer-Verlag, 1992.

Hsia, David Yi-Yung. *Human Developmental Genetics.* Chicago: Year Book Medical Publishers, 1968.

Leighton, Terrance, and William F. Loomis, Jr. *The Molecular Genetics of Development.* New York: Academic Press, 1980.

Pritchard, Dorian J. *Foundations of Developmental Genetics.* Philadelphia: Taylor & Francis, 1986.

Malacinski, George M., ed. *Developmental Genetics of Higher Organisms: A Primer in Developmental Biology.* New York: Macmillan, 1988.

Raff, Rudolf A., and Thomas C. Kaufman. *Embryos, Genes, and Evolution : The Developmental-Genetic Basis of Evolutionary Change.* New York: Macmillan, 1983.

Sang, James H. *Genetics and Development.* New York: Longman, 1984.

Saunders, John Warren, Jr. *Patterns and Principles of Animal Development.* New York: Macmillan, 1970.

Stewart, Alistair D., and David M. Hunt. *The Genetic Basis of Development.* New York: John Wiley & Sons, 1982.

Wilkins, Adam S. *Genetic Analysis of Animal Development.* New York: Wiley-Liss, 1993.

Genetic Engineering

Aldridge, Susan. *The Thread of Life: The Story of Genes and Genetic Engineering.* Cambridge, England: Cambridge University Press, 1996.

Baskin, Yvonne. *The Gene Doctors: Medical Genetics at the Frontier.* New York: William Morrow, 1984.

Breitenbach, M., et al., eds. *The Sixteenth International Conference on Yeast Genetics and Molecular Biology, Vienna, Austria, August 15-21, 1992: Book of Abstracts.* New York: John Wiley & Sons, 1992.

Carsiotis, Michael, and Sunil Khanna. *Genetic Engineering of Enhanced Microbial Nitrification.* Cincinnati, Ohio: U.S. Environmental Protection Agency, Risk Reduction Engineering Laboratory, 1989.

Chakrabarty, Ananda M., ed. *Genetic Engineering.* West Palm Beach, Fla.: CRC Press, 1978.

Cherfas, Jeremy. *Man-Made Life: An Overview of the Science, Technology, and Commerce of Genetic Engineering.* New York: Pantheon Books, 1982.

Crispeels, M. J., and D. E. Sadava. *Plants, Genes, and Agriculture.* Boston: Jones & Bartlett, 1994.

Crocomo, O. J., ed. *Biotechnology of Plants and Microorganisms.* Columbus: Ohio State University Press, 1986.

Dale, Jeremy. *Molecular Genetics of Bacteria.* New York: John Wiley & Sons, 1994.

Davis, Bernard D., ed. *The Genetic Revolution: Scientific Prospects and Public Perceptions.* Baltimore: The Johns Hopkins University Press, 1991.

Drlica, Karl. *Understanding DNA and Gene Cloning: A Guide for the Curious.* New York: John Wiley & Sons, 1997.

Fincham, J. R. S. *Genetically Engineered Organisms: Benefits and Risks.* Toronto: University of Toronto Press, 1991.

Gaillardin, Claude, and Henri Heslot. *Molecu-*

lar Biology and Genetic Engineering of Yeasts. Boca Raton, Fla.: CRC Press, 1992.

Goodman, David, et al. *From Farming to Biotechnology: A Theory of Agro-Industrial Development.* New York: Basil Blackwell, 1987.

Grange, J. M., et al., eds. *Genetic Manipulation: Techniques and Applications.* Boston: Blackwell Scientific, 1991.

Hatch, Randolph T., ed. *Expression Systems and Processes for rDNA Products.* Washington, D.C.: American Chemical Society, 1991.

Impacts of Applied Genetics: Microorganisms, Plants, and Animals. Washington, D.C.: Office of Technology Assessment, 1981.

Jacobson, G. K., and S. O. Jolly. *Gene Technology.* New York: VCH Verlagsgesellschaft, 1989.

Joyner, Alexandra L., ed. *Gene Targeting: A Practical Approach.* New York: Oxford University Press, 1993.

Kevles, Daniel J. *In the Name of Eugenics: Genetics and the Uses of Human Heredity.* New York: Alfred A. Knopf, 1985.

Kirby, L. T. *DNA Fingerprinting: An Introduction.* New York: Stockton Press, 1990.

Lappe, Marc. *Broken Code: The Exploitation of DNA.* San Francisco: Sierra Club Books, 1984.

McGee, Glenn. *The Perfect Baby: A Pragmatic Approach to Genetics.* Lanham, Md.: Rowman & Littlefield, 1997.

McKelvey, Maureen D. *Evolutionary Innovations: The Business of Biotechnology.* New York : Oxford University Press, 1996.

Mantell, S. H., et al., eds. *Principles of Plant Biotechnology: An Introduction to Genetic Engineering in Plants.* Boston: Blackwell Scientific, 1985.

Mantell, S. H., and H. Smith, eds. *Plant Biotechnology.* New York: Cambridge University Press, 1983.

Miklos, D., and G. Freyer. *DNA Science: A First Course in Recombinant DNA Technology.* Cold Spring Harbor, N.Y.: Cold Spring Harbor Laboratory Press, 1990.

National Academy of Engineering. *Engineering and the Advancement of Human Welfare: Ten Outstanding Achievements, 1964-1989.* Washington, D.C.: Author, 1989.

Nicholl, Desmond S. T. *An Introduction to Genetic Engineering.* New York: Cambridge University Press, 1994.

Nossal, Gustav J. V., and Ross L. Coppel. *Reshaping Life: Key Issues in Genetic Engineering.* New York: Cambridge University Press, 1985.

Old, R. W., and S. B. Primrose. *Principles of Gene Manipulation: An Introduction to Genetic Engineering.* Boston: Blackwell Scientific, 1994.

Olson, Steve. *Biotechnology: An Industry Comes of Age.* Washington, D.C.: National Academy Press, 1986.

Oxender, Dale L., and C. Fred Fox, eds. *Protein Engineering.* New York: Liss, 1987.

Prokop, Ales, and Rakesh K. Bajpai. *Recombinant DNA Technology I.* New York: New York Academy of Sciences, 1991.

Rifkin, Jeremy. *Declaration of a Heretic.* Boston: Routledge & Kegan Paul, 1985.

Rodrigues-Pousada, C., et al., eds. *The Seventeenth International Conference on Yeast Genetics and Molecular Biology, Lisbon, Portugal, June 10-16, 1995: Book of Abstracts.* New York: John Wiley & Sons, 1995.

Russo, V. E. A. and David Cove. *Genetic Engineering: Dreams and Nightmares.* New York: W. H. Freeman, 1995.

Santos, Miguel A. *Genetics and Man's Future: Legal, Social, and Moral Implications of Genetic Engineering.* Springfield, Ill.: Thomas, 1981.

Singer, Maxine, and Paul Berg, eds. *Exploring Genetic Mechanisms.* Sausalito, Calif.: University Science Books, 1997.

Singer, Peter, and Deane Wells. *Making Babies: The New Science and Ethics of Conception.* New York: Charles Scribner's Sons, 1985.

Spallone, Patricia. *Generation Games: Genetic Engineering and the Future for Our Lives.* Philadelphia: Temple University Press, 1992.

Sylvester, Edward J., and Lynn C. Klotz. *The Gene Age: Genetic Engineering and the Next Industrial Revolution.* New York: Scribner, 1983.

Technology Assessment Office, U.S. Congress. *Genetic Technology: A New Frontier.* Boulder, Colo.: Westview Press, 1982.

U.S. Congress. House. Committee on Science. Subcommittee on Technology. *Review of the President's Commission's Recommendations on Cloning: Hearing Before the Committee on Science, Subcommittee on Technology, U.S. House of Representatives, One Hundred Fifth Congress, First Session, June 12, 1997.* Washington, D.C.: Government Printing Office, 1997.

Vega, Manuel A., ed. *Gene Targeting*. Boca Raton, Fla.: CRC Press, 1995.

Wade, Nicholas. *The Ultimate Experiment: Man-Made Evolution*. New York: Walker, 1977.

Walker, Matthew R., with Ralph Rapley. *Route Maps in Gene Technology*. Oxford, England: Blackwell Scientific, 1997.

Warr, J. Roger. *Genetic Engineering in Higher Organisms*. Baltimore: E. Arnold, 1984.

Watson, James D., et al. *Recombinant DNA*. 2d ed. New York: W. H. Freeman, 1992.

Williams, Jeff G., and R. K. Patient. *Genetic Engineering*. Washington, D.C.: IRL Press, 1988.

Human Genetics

Adolph, Kenneth W., ed. *Human Molecular Genetics*. New York: Academic Press, 1996.

Boyer, Samuel, ed. *Papers on Human Genetics*. Englewood Cliffs, N.J.: Prentice-Hall, 1963.

Brierley, John Keith. *The Thinking Machine: Genes, Brain, Endocrines, and Human Nature*. Rutherford, N.J.: Fairleigh Dickinson University Press, 1973.

Burnet, Sir Frank Macfarlane. *Endurance of Life: The Implications of Genetics for Human Life*. New York: Cambridge University Press, 1978.

Carter, Cedric O. *Human Heredity*. Baltimore: Penguin Books, 1962.

Cavalli-Sforza, L. L., and Francesco Cavalli-Sforza. *The Great Human Diasporas: The History of Diversity and Evolution*. Translated by Sarah Thorne. Reading, Mass.: Addison-Wesley, 1995.

Clegg, Edward J. *The Study of Man: An Introduction to Human Biology*. London: English Universities Press, 1968.

Cooper, Necia Grant, ed. *The Human Genome Project: Deciphering the Blueprint of Heredity*. Mill Valley, Calif.: University Science Books, 1994.

Curran, Charles E. *Politics, Medicine, and Christian Ethics: A Dialogue with Paul Ramsey*. Philadelphia: Fortress Press, 1973.

Dobzhansky, Theodosius G. *Mankind Evolving: The Evolution of the Human Species*. New Haven: Yale University Press, 1967.

Edlin, Gordon. *Human Genetics: A Modern Synthesis*. Boston: Jones & Bartlett, 1990.

Fooden, Myra, et al., eds. *The Second X and Women's Health*. New York: Gordian Press, 1983.

Haldane, J. B. S. *Selected Genetic Papers of J. B. S. Haldane*. Edited with an introduction by Krishna R. Dronamraju; foreword by James F. Crow. New York: Garland, 1990.

Harris, Harry. *The Principles of Human Biochemical Genetics*. New York: Elsevier/North-Holland Biomedical Press, 1980.

Hsia, David Yi-Yung. *Human Developmental Genetics*. Chicago: Year Book Medical Publishers, 1968.

Jacquard, Albert. *In Praise of Difference: Genetics and Human Affairs*. Translated by Margaret M. Moriarty. New York: Columbia University Press, 1984.

Jensen, Arthur Robert. *Genetics and Education*. New York: Harper & Row, 1972.

Karlsson, Jon L. *Inheritance of Creative Intelligence*. Chicago: Nelson-Hall, 1978.

Korn, Noel, and Harry Reece Smith, eds. *Human Evolution: Readings in Physical Anthropology*. New York: Holt, 1959.

Lasker, Gabriel W., ed. *The Processes of Ongoing Human Evolution*. Detroit: Wayne State University Press, 1960.

Levitan, Max. *Textbook of Human Genetics*. New York: Oxford University Press, 1988.

Lewontin, Richard C. *Human Diversity*. New York: Scientific American Library, 1995.

Ludmerer, Kenneth M. *Genetics and American Society: A Historical Appraisal*. Baltimore: The Johns Hopkins University Press, 1972.

Lynn, Richard. *Dysgenics: Genetic Deterioration in Modern Populations*. Westport, Conn.: Praeger, 1996.

Mange, Arthur P., and Elaine Johansen Mange. *Genetics: Human Aspects*. Sunderland, Mass.: Sinauer Associates, 1990.

Mielke, James H., and Michael H. Crawford, eds. *Current Developments in Anthropological Genetics*. New York: Plenum Press, 1980.

Moody, Paul Amos. *Genetics of Man*. New York: W. W. Norton, 1967.

Nelkin, Dorothy, and Laurence Tancredi. *Dangerous Diagnostics: The Social Power of Biological Information*. New York: Basic Books, 1989.

Office of Technology Assessment. *Mapping Our Genes: Genome Projects—How Big, How Fast?*

Baltimore: The Johns Hopkins University Press, 1988.

Ostrer, Harry. *Non-Mendelian Genetics in Humans*. New York: Oxford University Press, 1998.

Ott, Jurg. *Analysis of Human Genetic Linkage*. Baltimore: The Johns Hopkins University Press, 1991.

Pearson, Roger. *Eugenics and Race*. Los Angeles: Noontide Press, 1966.

Puterbaugh, Geoff. *Twins and Homosexuality: A Casebook*. New York: Garland, 1990.

Roderick, Gordon Wynne. *Man and Heredity*. New York: St. Martin's Press, 1968.

Rosenberg, Charles, ed. *The History of Hereditarian Thought: A Thirty-two Volume Reprint Series Presenting Some of the Classic Books in This Intellectual Tradition*. New York: Garland, 1984.

Santos, Miguel A. *Genetics and Man's Future: Legal, Social, and Moral Implications of Genetic Engineering*. Springfield, Ill.: Thomas, 1981.

Singh, Jai Rup, ed. *Current Concepts in Human Genetics*. Amritsar, India: Guru Nanak Dev University, 1996.

Suzuki, D., and P. Knudtson. *Genethics: The Clash Between the New Genetics and Human Value*. Cambridge, Mass.: Harvard University Press, 1989.

Underwood, Jane H. *Human Variation and Human Microevolution*. Englewood Cliffs, N.J.: Prentice-Hall, 1979.

Vandenberg, Steven G., ed. *Methods and Goals in Human Behavior Genetics*. New York: Academic Press, 1965.

Varmus, Harold, and Robert Weinberg. *Genes and the Biology of Cancer*. New York: W. H. Freeman, 1993.

Vogel, Friedrich, and A. G. Motulsky. *Human Genetics: Problems and Approaches*. New York: Springer-Verlag, 1997.

Weiss, Kenneth M. *Genetic Variation and Human Disease: Principles and Evolutionary Approaches*. New York: Cambridge University Press, 1993.

Whittinghill, Maurice. *Human Genetics and Its Foundations*. New York: Reinhold, 1965.

Immunogenetics

Bell, John I., et al., eds. *T Cell Receptors*. New York: Oxford University Press, 1995.

Bibel, Debra Jan. *Milestones in Immunology*. New York: Springer-Verlag, 1988.

Clark, William R. *At War Within: The Double-Edged Sword of Immunity*. New York: Oxford University Press, 1995.

_____. *The Experimental Foundations of Modern Immunology*. 3d ed. New York: John Wiley & Sons, 1986.

Coleman, Robert M., et al. *Fundamental Immunology*. Dubuque, Iowa: Wm. C. Brown, 1989.

Dwyer, John M. *The Body at War: The Miracle of the Immune System*. New York: New American Library, 1988.

Fudenberg, H. Hugh, et al. *Basic Immunogenetics*. New York: Oxford University Press, 1984.

Gallo, Robert C., and Flossie Wong-Staal, eds. *Retrovirus Biology and Human Disease*. New York: Marcel Dekker, 1990.

Kimball, John W. *Introduction to Immunology*. 3d ed. New York: Macmillan, 1990.

Kreier, Julius P., and Richard F. Mortensen. *Infection, Resistance, and Immunity*. New York: Harper & Row, 1990.

Mizel, Steven B., and Peter Jaret. *In Self-Defense*. San Diego: Harcourt Brace Jovanovich, 1985.

Roitt, Ivan. *Immunology*. St. Louis: C. V. Mosby, 1996.

Samter, Max. *Immunological Diseases*. 4th ed. Boston: Little, Brown, 1988.

Silverstein, Arthur M. *A History of Immunology*. San Diego: Academic Press, 1989.

Smith, George P. *The Variation and Adaptive Expression of Antibodies*. Cambridge, Mass.: Harvard University Press, 1973.

Stewart, John. *The Primordial VRM System and the Evolution of Vertebrate Immunity*. Austin, Tex.: R. G. Landes, 1994.

Tizard, Ian R. *Immunology: An Introduction*. 2d ed. New York: W. B. Saunders, 1988.

Molecular Genetics

Adolph, Kenneth W., ed. *Gene and Chromosome Analysis*. San Diego: Academic Press, 1993.

_____, ed. *Human Molecular Genetics*. San Diego: Academic Press, 1996.

Alberts, B., ed. *Mapping and Sequencing the Human Genome*. Washington, D.C.: National Academy of Sciences, 1988.

Alberts, Bruce, et al. *Molecular Biology of the Cell*. 2d ed. New York: Garland, 1989.

Baltimore, David, ed. Nobel Lectures in Molecular Biology, 1933-1975. New York: Elsevier North-Holland, 1977.

Barry, John Michael, and E. M. Barry. *Molecular Biology: An Introduction to Chemical Genetics.* Englewood Cliffs, N.J.: Prentice-Hall, 1973.

Brandon, L. *Introduction to Protein Structure.* New York: Garland Press, 1991.

Bryant, J. A., ed. *Molecular Aspects of Gene Expression in Plants.* New York: Academic Press, 1976.

Crick, Francis. *Life Itself: Its Origin and Nature.* New York: Simon & Schuster, 1981.

————. *What Mad Pursuit.* New York: Basic Books, 1988.

De Pomerai, David. *From Gene to Animal: An Introduction to the Molecular Biology of Animal Development.* New York: Cambridge University Press, 1985.

Dillon, Lawrence S. *The Gene: Its Structure, Function, and Evolution.* New York: Plenum Press, 1987.

Gesteland, Raymond, and John Atkins, eds. *The RNA World.* Cold Spring Harbor, N.Y.: Cold Spring Harbor Laboratory Press, 1993.

Gros, François. *The Gene Civilization.* New York: McGraw Hill, 1992.

————. *Uniqueness and Universality in a Biological World: Report of a Symposium, Held on 10-12 January 1995 at the UNESCO Headquarters, Paris, France.* Paris: International Union of Biological Sciences, 1995.

Gwatkin, Ralph B. L., ed. *Genes in Mammalian Reproduction.* New York: Wiley-Liss, 1993.

Hawkins, John D. *Gene Structure and Expression.* New York: Cambridge University Press, 1996.

Hoelzel, A. R., ed. *Molecular Genetic Analysis of Populations: A Practical Approach.* New York: IRL Press at Oxford University Press, 1992.

Kimura, Motoo, and Naoyuki Takahata, eds. *New Aspects of the Genetics of Molecular Evolution.* New York: Springer-Verlag, 1991.

Langridge, John. *Molecular Genetics and Comparative Evolution.* New York: John Wiley & Sons, 1991.

Lewin, Benjamin M. *Gene Expression.* New York: John Wiley & Sons, 1980.

MacIntyre, Ross J., ed. *Molecular Evolutionary Genetics.* New York: Plenum Press, 1985.

Marks, Jonathan M. *Human Biodiversity: Genes, Race, and History.* New York: Aldine de Gruyter, 1995.

National Research Council. *DNA Technology in Forensic Science.* National Academy of Sciences Press, 1992.

Nei, Masatoshi. *Molecular Evolutionary Genetics.* New York: Columbia University Press, 1987.

Pollack, Robert. *Signs of Life: The Language and Meanings of DNA.* New York: Houghton Mifflin, 1994.

Sarma, Ramaswamy H., and M. H. Sarma. *DNA Double Helix and the Chemistry of Cancer.* Schenectady, N.Y.: Adenine Press, 1988.

Schleif, Robert F. *Genetics and Molecular Biology.* Reading, Mass.: Addison-Wesley, 1986.

Selander, Robert K., et al., eds. *Evolution at the Molecular Level.* Sunderland, Mass.: Sinauer Associates, 1991.

Smith, Thomas B., and Robert K. Wayne. *Molecular Genetic Approaches in Conservation.* New York: Oxford University Press, 1996.

Snyder, Larry, and Wendy Champness. *Molecular Genetics of Bacteria.* Washington, D.C.: ASM Press, 1997.

Stone, Edwin M., and Robert J. Schwartz. *Intervening Sequences in Evolution and Development.* New York: Oxford University Press, 1990.

Strachan, Tom, and Andrew P. Read. *Human Molecular Genetics.* New York: Wiley-Liss, 1996.

Watson, J. D. *The Double Helix.* New York: Atheneum, 1968.

Population Genetics

Ayala, Francisco J. *Population and Evolutionary Genetics: A Primer.* Menlo Park, Calif.: Benjamin/Cummings, 1982.

Barrett, P. H., et al. *Charles Darwin's Notebooks, 1836-1844.* Ithaca, N.Y.: Cornell University Press, 1987.

Bateson, P. *Mate Choice.* Cambridge, England: Cambridge University Press, 1983.

Boorman, Scott A., and Paul R. Levitt. *The Genetics of Altruism.* New York: Academic Press, 1980.

Charlesworth, Brian. *Evolution in Age-Structured Populations.* New York: Cambridge University Press, 1980.

Costantino, Robert F., and Robert A. Desharnais. *Population Dynamics and the Tribolium Model: Genetics and Demography.* New York: Springer-Verlag, 1991.

Crow, James F. *Basic Concepts in Population, Quantitative, and Evolutionary Genetics.* New York: W. H. Freeman, 1986.

Crow, James F., and Motoo Kimura. *An Introduction to Population Genetics Theory.* New York: Harper & Row, 1970.

Darwin, Charles. *The Correspondence of Charles Darwin.* Cambridge, England: Cambridge University Press, 1994.

_____. *The Descent of Man and Selection in Relation to Sex.* London: John Murray, 1871. Reprint. Princeton, N.J.: Princeton University Press, 1981.

_____. *On the Origin of Species by Means of Natural Selection or the Preservation of Favored Races in the Struggle for Life.* London: John Murray, 1859. Reprint. New York: Penguin Classics, 1984.

Dawkins, Richard. *The Selfish Gene.* Rev. ed. Oxford, England: Oxford University Press, 1989.

Dawson, Peter S., and Charles E. King, eds. *Readings in Population Biology.* Englewood Cliffs, N.J.: Prentice-Hall, 1971.

De Waal, Franz. *Good Natured: The Origins of Right and Wrong in Humans and Other Animals.* Cambridge, Mass.: Harvard University Press, 1996.

Dobzhansky, Theodosius G. *Genetics and the Origin of Species.* New York: Columbia University Press, 1937.

_____. *Genetics of the Evolutionary Process.* New York: Columbia University Press, 1970.

Fisher, R. A. *The Genetical Theory of Natural Selection.* Rev. ed. New York: Dover, 1958.

Gale, J. S. *Population Genetics.* New York: John Wiley & Sons, 1980.

Gould, Steven J. *Ontogeny and Phylogeny.* Cambridge, Mass.: Belknap Press, 1977.

_____. *The Panda's Thumb.* New York: W. W. Norton, 1980.

Harper, J. L. *Population Biology of Plants.* New York: Academic Press, 1977.

Hartl, Daniel. *A Primer of Population Genetics.* Sunderland, Mass.: Sinauer Associates, 1981.

Hartl, Daniel, and Andrew Clark. *Principles of Population Genetics.* 2d ed. Sunderland, Mass.: Sinauer Associates, 1989.

Kingsland, S. E. *Modeling Nature: Episodes in the History of Population Ecology.* Chicago: University of Chicago Press, 1985.

Lack, D. *The Natural Regulation of Animal Numbers.* Oxford, England: Clarendon Press, 1954.

Lewontin, R. *The Genetic Basis of Evolutionary Change.* New York: Columbia University Press, 1974.

Lloyd, E. *The Structure of Evolutionary Theory.* Westport, Conn.: Greenwood Press, 1987.

McKinney, M. L., and K. J. McNamara. *Heterochrony: The Evolution of Ontogeny.* New York: Plenum Press, 1991.

Mayr, E. *One Long Argument: Charles Darwin and the Nature of Modern Evolutionary Thought.* Cambridge, Mass.: Harvard University Press, 1991.

_____. *Populations, Species, and Evolution.* Cambridge, Mass.: Belknap Press, 1970.

Provine, William B. *Sewall Wright and Evolutionary Biology.* Chicago: University of Chicago Press, 1986.

Real, Leslie A., ed. *Ecological Genetics.* Princeton, N.J.: Princeton University Press, 1994.

Ruse, M. *Sociobiology: Sense or Nonsense?* Boston: D. Riedel, 1979.

Schonewald-Cox, Christine M., et al., eds. *Genetics and Conservation: A Reference for Managing Wild Animal and Plant Populations.* Menlo Park, Calif.: Benjamin/Cummings, 1983.

Wilson, E. O. *Sociobiology: The New Synthesis.* Cambridge, Mass.: Belknap Press, 1975.

Encyclopedia of
Genetics

Category List

Articles are listed under the following categories:

Bacterial Genetics
Classical Transmission Genetics
Cloning
Developmental Genetics
Diseases
Gender and Sexuality
Genetic Engineering and Biotechnology
Health and Medicine
Heredity and Inheritance
History of Genetics
Human Genetics
Immunogenetics
Molecular Genetics
Population Genetics
Racial and Ethnic Issues
Social, Ethical, and Legal Issues
Techniques, Methods, and Applications

For ease of reference, articles may be listed under more than one category.

Bacterial Genetics
Archaebacteria
Bacterial Genetics and Bacteria Structure
Bacterial Resistance and Super Bacteria
Cholera
Diphtheria
Escherichia coli
Gene Regulation: Bacteria
Gene Regulation: Lac Operon
Methane-Producing Bacteria
Transposable Elements
Viral Genetics

Classical Transmission Genetics
Chloroplast Genes
Chromatin Packaging
Chromosome Structure
Chromosome Theory of Heredity
Classical Transmission Genetics
Dihybrid Inheritance
Drosophila melanogaster
Epistasis
Extrachromosomal Inheritance
Genomics
Heredity and Environment
Hybridization and Introgression

Incomplete Dominance
Lamarckianism
Lethal Alleles
Linkage Maps
Meiosis
Mendel, Gregor, and Mendelism
Mitochondrial Genes
Mitosis
Monohybrid Inheritance
Multiple Alleles
Nondisjunction and Aneuploidy
Parthenogenesis
Polyploidy
Telomeres

Cloning
Cloning
Cloning: Ethical Issues
Cloning Vectors
Sheep Cloning
Shotgun Cloning

Developmental Genetics
Developmental Genetics
Homeotic Genes
Knockout Genetics and Knockout Mice
X-Chromosome Inactivation

Huntington's Chorea
Hybridomas and Monoclonal Antibodies
Immune Deficiency Disorders
Immunogenetics
In Vitro Fertilization and Embryo Transfer
Inborn Errors of Metabolism
Infertility
Isolates and Genetic Disease
Klinefelter's Syndrome
Lactose Intolerance
Mitochondrial Diseases
Neural Tube Defects
Oncogenes
Organ Transplants and HLA Genes
Paternity Tests
Prenatal Diagnosis
Prion Diseases: Kuru and Creutzfeldt-Jacob
    Syndrome
Sickle-Cell Anemia
Synthetic Antibodies
Tay-Sachs Disease
Testicular Feminization Syndrome
Thalidomide and Other Teratogens
Tumor-Suppressor Genes
Turner's Syndrome
Viral Genetics
XYY Syndrome

Heredity and Inheritance
Biological Determinism
Chromosome Theory of Heredity
Classical Transmission Genetics
Dihybrid Inheritance
Extrachromosomal Inheritance
Hereditary Diseases
Heredity and Environment
Monohybrid Inheritance
Quantitative Inheritance

History of Genetics
Biotechnology
Chromosome Theory of Heredity
Classical Transmission Genetics
Cloning
Diamond v. Chakrabarty
Eugenics
Eugenics: Nazi Germany
Evolutionary Biology
Genetic Code, Cracking of
Genetic Engineering: Historical Development
Genetics, Historical Development of

Human Genome Project
Lamarckianism
Mendel, Gregor, and Mendelism
Miscegenation and Antimiscegenation Laws
Natural Selection
Race
Sheep Cloning
Sociobiology

Human Genetics
Aggression
Aging
Albinism
Alcoholism
Altruism
Alzheimer's Disease
Amniocentesis and Chorionic Villus Sampling
Behavior
Biological Determinism
Breast Cancer
Burkitt's Lymphoma
Cancer
Congenital Defects
Criminality
Cystic Fibrosis
Diabetes
Down Syndrome
Eugenics
Eugenics: Nazi Germany
Forensic Genetics
Gender Identity
Gene Therapy
Gene Therapy: Ethical and Economic Issues
Genetic Clocks
Genetic Counseling
Genetic Medicine and "Magic Bullets"
Genetic Screening
Genetic Testing
Genetic Testing: Ethical and Economic Issues
Heart Disease
Hemophilia
Hereditary Diseases
Hermaphrodites
Human Genetics
Human Genome Project
Human Growth Hormone
Huntington's Chorea
In Vitro Fertilization and Embryo Transfer
Inborn Errors of Metabolism
Infertility

Genetic Screening
Genetic Testing
Gender Identity
Heredity and Environment
Human Genome Project
Infertility
Insurance
Intelligence
Lamarckianism
Miscegenation and Antimiscegenation Laws
Natural Selection
Paternity Tests
Population Genetics
Race
Sociobiology
Sterilization Laws
Twin Studies

Techniques, Methods, and Applications

Amniocentesis and Chorionic Villus Sampling
Biofertilizers
Biopesticides
Biopharmaceuticals
Biotechnology
Cloning
Cloning Vectors
DNA Fingerprinting
DNA Isolation
Forensic Genetics
Gel Electrophoresis

Gene Therapy
Genetic Engineering
Genetic Engineering: Agricultural
    Applications
Genetic Engineering: Industrial Applications
Genetic Engineering: Medical Applications
Genetic Medicine and "Magic Bullets"
Genetic Screening
Genetic Testing
Genetically Engineered Foods
Genomic Libraries
High-Yield Crops
Human Genome Project
Hybridomas and Monoclonal Antibodies
In Vitro Fertilization and Embryo Transfer
Knockout Genetics and Knockout Mice
Linkage Maps
Organ Transplants and HLA Genes
Paternity Tests
Polymerase Chain Reaction
Prenatal Diagnosis
Sheep Cloning
Shotgun Cloning
Synthetic Antibodies
Synthetic Genes
Transgenic Organisms
Twin Studies
X-Chromosome Inactivation

Index

A page range in **boldface** type indicates a full article devoted to that topic